65.00

THE STATISTICAL HANDBOOK ON TECHNOLOGY

THE STATISTICAL HANDBOOK ON TECHNOLOGY

by
Paula Berinstein

Oryx Press
1999

The rare Arabian Oryx is believed to have inspired the myth of the unicorn. This desert antelope became virtually extinct in the early 1960s. At that time, several groups of international conservationists arranged to have nine animals sent to the Phoenix Zoo to be the nucleus of a captive breeding herd. Today, the Oryx population is over 1,000, and over 500 have been returned to the Middle East.

© 1999 by Paula Berinstein
Published by The Oryx Press
4041 North Central at Indian School Road
Phoenix, Arizona 85012-3397

Published simultaneously in Canada
Printed and bound in the United States of America

∞ The paper used in this publication meets the minimum requirements of American National Standard for Information Science—Permanence of Paper for Printed Library Materials, ANSI Z39.48, 1984.

Library of Congress Cataloging-in-Publication Data

Berinstein, Paula.
    The statistical handbook on technology / by Paula Berinstein.
    ISBN 1-57356-208-4 (alk. paper)
    1. Technology—United States Statistics Handbooks, manuals, etc.
I. Title.
T21.B47   1999
609.73'021—dc21   99-41232
                                                                        CIP

For the kids: Marina, Karen, Cate, Kayleigh, Kelsey, and Kyra

Contents

List of Tables and Figures ix

Preface xix

Acknowledgments xxiii

Introduction xxv

Chapter 1. Basic Technology Indicators 1

    Research and Development 11

    Employment in Science and Technology 19

    Patents, High-Tech Trade, and High-Tech Company Formation 23

Chapter 2. Agriculture 33

    Agricultural Methods 35

    Farm Equipment 39

    Genetics and Farming 43

Chapter 3. Business, Manufacturing, and Materials 45

    Business Technology 47

    Manufacturing Technology 48

    Materials Technology 51

Chapter 4. Communications 66

    Communication Cables 67

    Communication Equipment 75

    Communication Services, Costs, and Usage 79

    Broadcasting 93

Chapter 5. Computers 98

    Computer Equipment 100

    Networks and the Internet 112

    Databases, Programming Languages, and Security 115

Chapter 6. Construction and Infrastructure 119

    Buildings and Components 120

    Infrastructure 125

Chapter 7. Consumer Products and Entertainment 130

    Household Appliances and Other Consumer Products 131

    Entertainment Products 136

Chapter 8. Education and Libraries 146

    Libraries and Technology 147

    Technology in Schools 151

Chapter 9. Energy and Environment 158

    Energy 160

    Environment 174

Chapter 10. The Military and Law Enforcement 182

    Military 184

    Law Enforcement 197

Chapter 11. Medicine and Biotechnology 203

    Medical Services and Procedures 205

    Drugs and Nutrition 211

    Medical Equipment 225

    Biotechnology 229

Chapter 12. Space 230

    Global Positioning Systems and Space Commercialization 231

    Space Debris 234

    Spacecraft and Space Programs 235

Chapter 13. Transportation 241

    Transportaton System 243

    Air Transportation 245

    Rail and Urban Transport 256

    Cars and Trucks 257

    Alternative Fuels, Fuel Efficiency, and Vehicle Emissions 260

Glossary 266

Index 269

List of Tables and Figures

CHAPTER 1. BASIC TECHNOLOGY INDICATORS

1.1.   United States National Critical Technologies.

Research and Development

1.2.   Trends in Industrial R&D Performance, by Source of Funds, in Current and Constant Dollars, United States, 1953–97.

1.3.   Total (Company, Federal, and Other) Funds for Industrial R&D Performance, by Industry and Size of Company, United States, 1986–97.

1.4.   Federal Funds for Industrial R&D Performance, by Industry and Size of Company, United States, 1986–97.

1.5.   Total (Company, Federal, and Other) Funds for Performance of Basic Research, Applied Research, and Development, in Current and in Constant Dollars, United States, 1953–97.

1.6.   Number of Full-Time-Equivalent (FTE) R&D Scientists and Engineers in R&D-Performing Companies, by Industry and Size of Company, United States, 1986–98.

Employment in Science and Technology

1.7.   Employment Status of Scientists and Engineers, by Broad Occupation and Highest Degree Received, United States, 1995.

1.8.   Number of Employed Scientists and Engineers, by Occupation and Highest Degree Received, United States, 1995.

1.9.   Cost per R&D Scientist or Engineer in R&D-Performing Companies, by Industry and Size of Company, United States, 1986–97.

Patents, High-Tech Trade, and High-Tech Company Formation

1.10.   Patents Granted by Date of Patent Grant (Granted January 1, 1963–December 31, 1998) (number of patents).

1.11.   Patents Granted by Date of Patent Grant (Granted January 1, 1963–December 31, 1998) (percent of patents).

1.12.   Ranked List of Some Organizations with 1,000 or More Patents, as Distributed by Year of Patent Grant, and as Distributed by Year of Patent Application (Granted January 1, 1969, through December 31, 1998).

1.13.   U.S. Trade in Advanced Technology Products, 1990–96.

1.14.   High-tech Companies Formed in the United States, 1980–94.

CHAPTER 2. AGRICULTURE

Agricultural Methods

2.1.   Commercial Fertilizer Use, United States, 1960–95.

2.2.   Changes in Irrigation System Acreage, United States, 1979–94.

2.3.   Irrigation Application Systems, by Type, United States, 1994.

2.4.   Use of Selected Biological and Cultural Pest Management Practices on Fruit, Vegetable, and Nut Crops, Major Producing States, United States, 1990s.

2.5.   Overall Pesticide Use on Selected U.S. Crops by Pesticide Type, 1964–95.

2.6.   National Use of Crop Residue Management Practices, United States, 1989–96.

Farm Equipment

2.7.   Farmers' Use of Communications Tools, United States, 1995–98.

2.8.   Domestic Farm Machinery Unit Sales, United States, 1986–96.

2.9.   Farm Machines, Selected Types: Units Shipped in the United States, 1987–96.

2.10.   Farm Machinery and Equipment: Number of Certain Kinds on Farms, and Tractor Horsepower, United States, Census Years, 1954–92.

Genetics and Farming

2.11.   Utility Patents Issued for Multicellular Organisms through 1994, United States.

2.12.   Field Test Permits Issued for Genetically Modified Plants through June 1993, United States.

CHAPTER 3. BUSINESS, MANUFACTURING, AND MATERIALS

Business Technology

3.1.   Manufacturers' Shipments of Business Appliances, United States, 1995–96.

3.2.   World Smart Card Usage by Global Region, 1995–2002.

3.3.   Number of Automated Teller Machines Deployed in the United States, 1990–97.

3.4.   Number of Monthly Transactions per Automated Teller Machine in the United States, 1988–97.

Manufacturing Technology

3.5.   U.S. Robot Population, 1982–97.

3.6.   Manufacturers' Use and Planned Use of Certain Advanced Technologies in the United States, 1988 and 1993.

3.7.   Soft Technology Use among Small and Large Plants, United States, 1997.

3.8.   Hard Technology Use among Small and Large Plants, United States, 1997.

Materials Technology

3.9.   Nonfuel Mineral Production in the United States, 1994–97.

3.10.   Refractories, Value of Shipments, United States, 1986–97.

3.11.   Changes in Use of Refractory Raw Materials, United States, 1970–94.

3.12.   U.S. Chemical Specialties Market, 1997, and Percent Change in 1998.

3.13.   Plastics: Production, Sales, and Captive Use, United States, 1996 and 1997.

3.14.   Total Resin Sales and Captive Use by Major Market, United States, 1992–97.

3.15.   Production of Thermosets and Thermoplastics, United States, 1987–97.

3.16.   World's Most Widely Used Commodity Plastics, 1998.

3.17.   Plastics Consumption by World Region, 1998.

3.18.   Composites Shipments Comparison, United States, 1987–97.

3.19.   Consumption of Oils and Fats for Industrial Use, United States, 1988–95.

3.20.   Industry Shipments and Number of Producers of Adhesives, United States, 1992.

3.21.   Adhesives Production by Major Class, United States, 1995.

CHAPTER 4. COMMUNICATIONS

Communication Cables

4.1.   Fiber Miles of Fiber Deployed by Local Operating Companies, United States, 1985–97.

4.2.   Competitive Access Fiber Systems, United States, 1990–97.

4.3.   Fiber Miles Installed—Interexchange Carriers, United States, 1985–97.

4.4.   Fiber Route Miles—Interexchange Carriers, United States, 1985–97.

4.5.   Transatlantic Cable Systems, 1956–98.

4.6.   Telephone Loops by State as of December 31, 1997.

4.7.   Statistics of Communications Common Carriers, Selected Operating Statistics, United States, 1984–97.

Communication Equipment

4.8.   Value of Shipments of Communication Equipment by Class of Product, United States, 1988–97.

4.9. Quantity and Value of Shipments of Communication Equipment: Modems, Telephone Sets, Other Data Communications Equipment, Video Teleconferencing Equipment, Voice/Call Message Processing Equipment, Facsimile Communication Equipment, United States, 1996–97.

4.10. Wireless Telephones—U.S. Sales to Dealers, 1993–96, and Estimated for 1984–92 and 1997–98.

4.11. Corded Telephones—U.S. Sales to Dealers, 1982–96, Estimated for 1997, and Projected for 1998.

4.12. Cordless Telephones—U.S. Sales to Dealers, 1980–96, Estimated for 1997, and Projected for 1998.

4.13. Fax Machines—U.S. Sales to Dealers, 1990–96, Estimated for 1997, and Projected for 1998.

Communication Services, Costs, and Usage

4.14. Average Monthly Residential Rates (in October of each year), United States, 1983–97.

4.15. Average Local Rates for Businesses with a Single Line in Urban Areas, United States (in October of each year), 1983–97.

4.16. Changes in the Price of Directly Dialed Five-Minute Long Distance Calls, United States, 1984 and 1998.

4.17. Telephone Service Expenditures, United States, 1980–97.

4.18. Area Code Assignments, United States, 1984–2000.

4.19. New Area Code Assignments by Year, 1984–98.

4.20. Telephone Numbers Assigned for 800 Service, United States, 1993–98.

4.21. Telephone Penetration in the United States, 1983–98.

4.22. Historical Telephone Penetration Estimates.

4.23. Dial Equipment Minutes, United States, 1980–97.

4.24. International Message Telephone Service for 1997.

4.25. International Service from United States to Foreign Points, 1980–97.

4.26. Cellular Telephone Subscribers, United States, 1984–98.

4.27. Cellular Telephone Service Average Monthly Bill, United States, 1987–98.

4.28. Paging Industry Numbers, United States, 1993–97.

4.29. Mobile Telephone Industry—U. S. National Penetration, 1993–97.

Broadcasting

4.30. Broadcast Stations Licensed in the United States as of May 31, 1998.

4.31. FCC Auctions Summary—Service Design, as of 1998.

4.32. FCC Auctions Summary—Auction Results, July 1994–February 1998.

4.33. Value of Shipments of Radio, Television, and Studio Equipment, United States, 1996–97.

CHAPTER 5. COMPUTERS

Computer Equipment

5.1. History of Microprocessors, 1971–98.

5.2. Computing Milestones, 1971 to 1995.

5.3. Intel Chip Sizes.

5.4. Computers, Value of Shipments, United States, 1989–97.

5.5. Computers, Quantity and Value of Shipments, United States, 1995–97.

5.6. Modems/Fax Modems—U.S. Sales to Dealers, 1993–96, Estimated for 1997, and Projected for 1998.

5.7. Personal Computers—U.S. Sales to Dealers, 1982-96, Estimated for 1997, and Projected for 1998.

5.8. Computer Printers—U.S. Sales to Dealers, 1993–96, Estimated for 1997, and Projected for 1998.

5.9. Value of Shipments of Semiconductors, Printed Circuit Boards, and Other Electronic Components by Class of Product, 1991–97.

5.10. Worldwide Market Size, Speech and Voice Recognition, 1993 and Forecast for 1998.

5.11. The U.S. Data Communications Market, 1996–97 and Projected for 1998.

Networks and the Internet

5.12.   Worldwide Intranet Market for 1996–2000.

5.13.   Worldwide Intranet Market by Category, 1996–2000.

5.14.   The Total Electronic Mail Market, United States, 1994–2000.

5.15.   Electronic Mail at Work (Usage), United States, 1994–2000.

5.16.   Electronic Mail at Home (Usage), United States, 1994–2000.

5.17.   Online Retail Revenues, United States and Canada, Projected 1997–2001.

Databases, Programming Languages, and Security

5.18.   Number of Database Records by Year, 1975–97.

5.19.   Growth in Number of Database Vendors, Producers, Database Entries, and Databases, 1975–97.

5.20.   Database Entries Classed by Form of Representation of Data, 1988–97.

5.21.   Media for Database Distribution/Access, 1989–97.

5.22.   Data Mining Market Forecast, 1996 and 2000.

5.23.   U.S. Java Software Market Size Projections, 1997–2000.

5.24.   1998 Security Revenue, United States, Projected.

CHAPTER 6. CONSTRUCTION AND INFRASTRUCTURE

Buildings and Components

6.1.   Building Technologies: Plans for Modernization and/or New Construction, United States, 1998.

6.2.   Building Types: Percent of Total Square Feet for Each, New Construction, United States, 1993–97.

6.3.   Buildings, Percent of Square Feet by Number of Stories, All Types, New Construction, United States, 1993–97.

6.4.   Value of Shipments of Alarm Systems and Fire Detection and Prevention Equipment, United States, 1996–97.

6.5.   Carbon Monoxide and Smoke Detectors, Unit Shipments, United States, 1996.

6.6.   U.S. Market Size for Air Conditioners, Heat Pumps, Dehumidifiers, and Gas Furnaces in 1996.

Infrastructure

6.7.   Highway Pavement Conditions, United States, 1994 and 1996.

6.8.   Funding for Highways, United States, 1980–97.

6.9.   Highway Mileage—Functional Systems and Urban/Rural, United States, 1995.

6.10.   Condition of U.S. Bridges, 1990–96.

6.11   Value of Some Public Utility and Infrastructure Construction, United States, 1990–95.

CHAPTER 7. CONSUMER PRODUCTS AND ENTERTAINMENT

Household Appliances and Other Consumer Products

7.1.   Quantity and Value of Shipments of Major Household Appliances, United States: Ranges, Ovens, Surface Cooking Units, Outdoor Cooking Equipment, 1991–97.

7.2.   Quantity and Value of Shipments of Major Household Appliances, United States: Refrigerators and Freezers, 1993–97.

7.3.   Quantity and Value of Shipments of Major Household Appliances, United States: Washers and Dryers, 1991–97.

7.4.   Quantity and Value of Shipments of Major Household Appliances, United States: Water Heaters and Dishwashers, 1991–97.

7.5.   U.S. Households Owning Particular Appliances or Consumer Electronics Products, 1994–96.

7.6.   Telephone Answering Devices—U.S. Sales to Dealers, 1982–96, Estimated for 1997, and Projected for 1998.

7.7.   Aftermarket Vehicle Security—Factory Sales, 1993–95, Estimated for 1990–92 and for 1996–92 and for 1996–97.

Entertainment

7.8.   Portable CD Equipment—U.S. Factory Sales, 1990–96, Estimated for 1997, and Projected for 1998.

7.9.   Compact Audio Systems—U.S. Factory Sales, 1980–96, Estimated for 1997, and Projected for 1998.

7.10.   Rack Audio Systems—U.S. Factory Sales, 1988–96, Estimated for 1997, and Projected for 1998.

7.11.   Separate Audio Component Systems—U.S. Factory Sales, 1979–96, Estimated for 1997, and Projected for 1998.

7.12.   Camcorders—U.S. Sales to Dealers, 1985–96, Estimated for 1997, and Projected for 1998.

7.13.   Total CD Players—U.S. Factory Sales, 1987–96, Estimated for 1983–86 and 1997, and Projected for 1998.

7.14.   Direct to Home (DTH) Satellite Systems, U.S. Sales to Dealers, 1986–89, Estimated for 1990–97, and Projected for 1998.

7.15.   Color TV Receivers—U.S. Sales to Dealers, 1954–96, Estimated for 1997, and Projected for 1998.

7.16.   Projection Television—U.S. Sales to Dealers, 1984–96, Estimated for 1997, and Projected for 1998.

7.17.   VCR Decks—U.S. Sales to Dealers, 1979–96, Estimated for 1997, and Projected for 1998.

7.18.   Laserdisc Players—U.S. Sales to Dealers, 1990–95, Estimated for 1985–89 and for 1996–97, and Projected for 1998.

7.19.   Home Radios—U.S. Factory Sales, 1950–96, Estimated for 1997, and Projected for 1998.

7.20.   The Computer Animation Industry, Market Size, Worldwide, 1991–96.

7.21.   The Computer Animation Industry, Total Use Analysis, Worldwide, 1996.

CHAPTER 8. EDUCATION AND LIBRARIES

Libraries and Technology

8.1.   Percentage of U.S. Public Libraries Offering Access to the Internet, 1996–97.

8.2.   Percentage of U.S. Public Libraries Connected to the Internet, 1996–97.

8.3.   Internet Access in U.S. Public Libraries Regionally, 1997.

8.4.   U.S. Public Libraries Hosting Their Own Web Sites, 1997.

8.5.   Percent of Schools with Library Media Centers and Percent of Pupils in Schools with Library Media Centers: Historical Summary, 1958–94.

8.6.   School Library Media Centers Offering Selected Services and Equipment, Public and Private, Elementary and Secondary, United States, 1993–94.

8.7.   School Library Media Center Expenditures for Selected Services and Equipment, Public and Private, Elementary and Secondary, United States, 1993–94.

8.8.   School Library Media Centers, United States: Number of Selected Items Held per 100 Students at End of 1992–93 School Year.

8.9.   College and University Library Hardware, Software, Audiovisual Materials, Microforms, and Machine-Readable Materials, 1991–92 and 1994–95.

Technology in Schools

8.10.   Distance Education in Higher Education Institutions, United States: Percent Distribution, by Current and Planned Use, 1995.

8.11.   Percent of U.S. Higher Education Institutions Using Various Technologies to Deliver Distance Education Courses, and Plans for Use by Level of Use, 1995.

8.12.   Percentage of Public Schools Having Access to the Internet in Fall 1994, 1995, 1996, 1997, and 1998, by School Characteristics, United States.

8.13.   Students Who Reported Using a Computer, by Location of Use and Grade, United States, 1984–96.

8.14.   Percentage of Students Who Used a Computer at School and/or Home, by Current Grade Level, Race-ethnicity, and Family Income, United States: 1984, 1989, 1993, and 1997.

8.15.   Percentage of Students Who Used a Computer at Home, by Purpose, Current Grade Level, Race-ethnicity, and Family Income, United States, 1997.

8.16.   Percent of Student Home Computer Users Using Specific Applications, by Selected Characteristics, United States, October 1997.

CHAPTER 9. ENERGY AND ENVIRONMENT

Energy

9.1.   U.S. Energy Consumption, by Type, 1949–97.

9.2.   U.S. Energy Consumption by Energy Source, 1992–97.

9.3.   U.S. and World Petroleum Consumption, 1949–96.

9.4.   Biomass Energy Consumption by Sector and Census Region, United States, 1992-97.

9.5.   Renewable Energy Consumption by Sector and Energy Source, United States, 1992–97.

9.6.   Housing Units Heated by Various Types of Energy, United States, 1950–95.

9.7.   U.S. Electric Generating Capacity from Renewable Sources, 1992–97.

9.8.   Electricity Generation from Renewable Energy by Energy Source, United States, 1992–97.

9.9.   Nuclear Power Plants, United States, Various Statistics, 1957-97.

9.10.   Motor Gasoline, All Types, Retail Price (Real), United States, 1978–97.

9.11.   Motor Vehicle Fuel Consumption Rate, United States, 1960–96.

9.12.   U.S. Energy Prices, by Type of Fuel, 1970–94.

9.13.   Shipments of Solar Collectors by Market Sector, End Use, and Type, United States, 1995–97.

9.14.   Shipments of Photovoltaic Modules and Cells by Market Sector, End Use, and Type, United States, 1995–97.

9.15.   Annual Photovoltaic and Solar Thermal Shipments, 1977–97.

9.16.   Number of Lights by Bulb Type by Room (Residential), United States, 1993.

Environment

9.17.   Number and Types of National Priorities List (NPL) Cleanup Sites in Construction Completion Stage as of September 30, 1996, United States.

9.18.   Cleanup Technologies Used at U.S. Superfund's 410 Construction Completion Sites as of September 30, 1996.

9.19.   Quantity of RCRA Hazardous Waste Generated, and Number of Hazardous Waste Generators, by State, 1993, 1995, and 1997.

9.20.   1997 CERCLA (Comprehensive Environmental Response, Compensation, and Liability Act) Priority List of Hazardous Substances.

9.21.   Shipments for Selected Industrial Air Pollution Control Equipment, 1992–97.

9.22.   Value of Shipments of Selected Air Pollution Control Equipment by End Use, United States, 1992–97.

9.23.   Approximate Quantities of Wood and Wood-Fiber Materials Recovered for Recycling in the United States, 1994.

9.24.   Estimated Emissions of Greenhouse Gases, 1985–96.

CHAPTER 10. THE MILITARY AND LAW ENFORCEMENT

Military

10.1.   U.S. Department of Defense FY 1999 Budget, Program Acquisition Costs with History for 1997 and 1998: Aircraft, Missiles, Vessels, Space Programs, and Other Programs.

10.2.   U.S. Defense Technology Area Plan, Resource Funding, 1997–2003.

10.3.   Nuclear Facts and Figures.

10.4.   U.S. Chemical Weapons Storage and Destruction.

10.5.   U.S. Defense Technology: Programmed Funding for Various Technologies, 1997–2003. Biomedical Defense Technology Objectives.

10.6.   Defense Technology: Programmed Funding for Various Technologies, 1997–2003. Ground and Sea Vehicles Technology Objectives.

10.7.   U.S. Defense Technology: Programmed Funding for Various Technologies, 1997–2003. Materials and Processes Objectives.

10.8.   U.S. Defense Technology: Programmed Funding for Various Technologies, 1997–2003. Air Platform Technology Objectives.

10.9.   U.S. Defense Technology: Programmed Funding for Various Technologies, 1997–2003. Chemical/Biological Defense and Nuclear Technology Objectives.

10.10. U.S. Defense Technology: Programmed Funding for Various Technologies, 1997–2003. Sensors, Electronics, and Battlespace Environment Objectives.

10.11. U.S. Defense Technology: Programmed Funding for Various Technologies, 1997–2003. Weapons Objectives.

10.12. U.S. Defense Technology: Programmed Funding for Various Technologies, 1997–2003. Information Superiority Objectives.

10.13. U.S. Joint Warfighting Capability Objectives Funding, Fiscal Year 1998.

10.14. U.S. Department of Defense Strategic Forces Highlights, 1992–2000.

Law Enforcement

10.15. Selected Summary Data for State and Local Law Enforcement Agencies with 100 or More Officers, by Type of Agency, United States, 1990 and 1993: Weapons and Body Armor.

10.16. Selected Summary Data for State and Local Law Enforcement Agencies with 100 or More Officers, by Type of Agency, United States, 1990 and 1993: 911 Systems, Computers and Information Systems, and Vehicles.

10.17. Crime Scene DNA Samples Received and Analyzed by U.S. Forensic Laboratories, 1995–98 and Projected for 1999.

10.18. U.S. Law Enforcement Agency Use of Mobile Data Terminals and Laptop Computers, 1997 and Estimated for 1999.

10.19. U.S. Law Enforcement Use of Wireless Voice and/or Data Security, 1997.

10.20. U.S. Law Enforcement Use of Computerized Crime Mapping, 1997 and 1998.

CHAPTER 11. MEDICINE AND BIOTECHNOLOGY

Medical Services and Procedures

11.1. Telemedicine Facilities by Clinical Service, United States, as of July 1998.

11.2. Top Ten Telemedicine Specialty Services, United States, as of July 1998.

11.3. Home Diagnostic Monitoring Revenues, 1994–97, and Projected for 1998–2000.

11.4. Medical Testing Chemicals, United States, 1994 and Forecast for 2000.

11.5. Number and Rate of Ambulatory and Inpatient Procedures by Selected Procedure Categories, United States, 1995.

11.6. Number and Rate of Ambulatory and Inpatient Surgical Procedures by Selected Procedure Categories, United States, 1995.

11.7. Number and Rate of Ambulatory and Inpatient Nonsurgical Procedures by Selected Procedure Categories, United States, 1995.

11.8. Annual Number of Organ and Tissue Transplants in the United States for All Ages and Regardless of Citizenship of Recipient, 1989–97, and Projected for 1998 and 1999.

11.9. Supply of Alternative Nonphysician Clinicians, United States, 1990 to 2015.

Drugs and Nutrition

11.10. Generic Substances Most Frequently Used at Ambulatory Care Visits, United States, 1995.

11.11. Drug Mentions by Therapeutic Classification, According to Ambulatory Care Setting, United States, 1995.

11.12. Drugs That Exceeded a Billion Dollars in Worldwide Sales in 1996.

11.13. The Top 100 Prescription Drugs by Worldwide Sales, 1995 and 1996.

11.14. Cancer and Cancer-related Treatments Worldwide, 1996.

11.15. Leading Therapeutic Classes—Dispensed Prescription Volume, United States, 1997.

11.16. Leading U.S. Rx Pharmaceutical Dollar Sales Volume Leaders, 1996–97.

11.17. U.S. Rx Pharmaceutical Prescription Leaders, 1996–97.

11.18. Value of Shipments of Pharmaceutical Preparations, Except Biologicals, United States, 1995–97.

11.19. Value of Product Class Shipments of Pharmaceutical Preparations, Except Biologicals, United States, 1986–97.

11.20. Nutraceuticals. Market Size, 1996.

11.21. Revenues in the U.S. Market for Pain Management, 1995 and 1996, and Forecast for 2000.

Medical Equipment

11.22. Number of Persons Using Assistive Technology Devices, by Age of Person and Type of Device, United States, 1994.

11.23. Electromedical and Irradiation Equipment, Quantity and Value of Shipments, United States, 1996–97.

Biotechnology

11.24. Biotechnology Market Areas: Participation of Biotechnology Companies by Primary Focus, United States, 1996.

11.25. U.S. Biotechnology Firms, 1997.

11.26. 1997 U.S. Biopharmaceutical Facts and Figures.

CHAPTER 12. SPACE

Global Positioning Systems and Space Commercialization

12.1. U.S. Global Positioning Satellite Market, 1992 and 1996, and Forecast for 2001.

12.2. Global Positioning System Users, United States, 1995.

12.3. Global Positioning Satellite Market Projections Worldwide, 1993–2000.

12.4. U.S. Commercial Space Revenues, 1990–99.

12.5. U.S. and World Satellite Revenues, 1997.

Space Debris

12.6. Number of Cataloged Space Objects in Orbit.

12.7. Space Debris Composition

Spacecraft and Space Programs

12.8. U.S. Department of Defense Space Platforms Goals, Payoffs, and Time Frames: Space Vehicles, 2000 and 2005.

12.9. U.S. Spacecraft Parameters, 1990–99.

12.10. Space Shuttle Flights, 1981–98.

12.11. Hubble Space Telescope Statistics as of April 1999.

12.12. Worldwide Successful Space Launches, 1957–97.

CHAPTER 13. TRANSPORTATION

Transportation System

13.1. Major Elements of the U.S. Transportation System: 1996.

Air Transportation

13.2. Annual Shipments of New U.S. Manufactured General Aviation Aircraft by Units Shipped, Number of Companies Reporting, and Factory Net Billings, 1946–98.

13.3. General Aviation Aircraft Shipments by Type of Aircraft, U.S.-Manufactured, 1962–98.

13.4. Average Age of U.S. General Aviation Fleet, 1996 and 1997.

13.5. Aircraft Reported in Operation by Air Carriers, Sorted by Manufacturer, and Model, 1984–96.

13.6. U.S. FAA Air Route Facilities and Services, 1972–97.

13.7. Airports by Geographic Area: U.S. Civil and Joint Use Airports, Heliports, STOLports, and Seaplane Bases on Record by Type of Ownership, 1997.

13.8. U.S. Aerospace Industry Sales by Product Group in Current Dollars, 1981–98.

13.9. U.S. Aerospace Industry Sales by Product Group in Constant Dollars, 1981–98

Rail and Urban Transport

13.10. Physical Condition of U.S. Transit Rail Systems, 1984 and 1992.

13.11. Average Age of Urban Transit Vehicles, United States, 1985 and 1995.

13.12. Growth in U.S. Urban Transit Service, 1985 to 1995.

Cars and Trucks

13.13. A Typical U.S. Family Vehicle, Material Content and Total Weight, 1976, 1986, and 1996.

13.14. Material Usage per Passenger Car Built in the United States, 1980 and 1993.

13.15. Sales, Market Shares, and Fuel Economies of New Domestic (U.S.) and Import Light Trucks, 1980–96.

13.16. Average Annual Growth Rate of Passenger Cars and Their Use, In Relation to Population, United States, Canada, European Union, and Japan, 1970–90.

13.17. U.S. Cars and Light Trucks with Airbags, 1990–95.

13.18. Number of Vehicles with Airbags as of September 1, 1998.

Alternative Fuels, Fuel Efficiency, and Vehicle Emissions

13.19. Alternative Fuel Vehicles by Fuel Type, United States, 1992–99.

13.20. Energy Intensiveness of Passenger Modes, United States, 1960–96.

13.21. U.S. Energy Consumption by the Transportation Sector, 1955–96.

13.22. Estimated Consumption of Vehicle Fuels in the United States: 1992–97.

13.23. Sources of Nonpetroleum Transportation Energy, United States, 1981–95.

13.24. U.S. Transportation-related Air Emissions, 1970–95.

Preface

This volume tells stories about the tools we humans use, but unlike myths, ballads, and sagas, these narratives are told in numbers rather than words. Although myriad sources for technology statistics exist, no work gathers a wide and representative range of them in one place. This book fills that void, offering statistical overviews of technologies related to

- Agriculture
- Biotechnology
- Business
- Communications
- Computers
- Construction and infrastructure
- Consumer products and entertainment
- Education and libraries
- Energy
- Environment
- Law enforcement
- Manufacturing
- Materials
- Medicine
- Military
- Space
- Transportation

You will also find general indicators that help assess how technologically advanced and competitive we are as a society.

To gain a well-rounded picture of the current and past state of the art, one must examine a variety of measures. Thus, you will find about 260 tables and figures that examine

- Access to technologies
- Budgets
- Capacities
- Consumption
- Costs
- Employment
- Expenditures
- Funding sources

- Incidence of procedures
- Number of companies
- Number of products on the market
- Number of units existing
- Patents
- Penetration
- Physical sizes and conditions
- Popularity
- Prices
- Production
- Sales
- Shipments
- Trade
- Usage
- Values

Many items comprise time series so that you can track performance. Some include forecasts as well.

The tables and figures will tell you what technologies exist, which are important, how they contribute to the economy, and how they stack up against others. You will be able to track how the technologies have changed—how usage has varied, and how current prices compare with previous ones.

You will also gain insight into how we adopt and discard our technologies. The tables tell of meteoric rises (electronic mail use), sometimes preceding steep falls (sales of Zantac and amoxicillin); slow growth (camcorders); and good old steady persistence (refrigerators). The reasons behind the numbers fall outside the scope of this book; I hope you will be intrigued enough to dig further into the wherefores and whys.

Perhaps most exciting of all, you can watch new technologies emerge—some slowly (alternative-fueled vehicles), others bursting on the scene (electronic commerce). The numbers you will find here will serve as baselines as you watch these technologies flourish or flame out.

WHO SHOULD USE THIS BOOK

This book is addressed to anyone who desires to assess the current state of a technology, estimate where it is going, or track its development over time. Market research professionals, strategic planners, policy makers and their advisors, venture capitalists, bankers, manufacturers, engineers, students, and consumers will all find valuable information here. The time series showing how technology has developed will particularly appeal to sociologists, historians, writers, librarians, teachers, and movie makers.

SELECTION CRITERIA AND EDITING

In selecting statistics for this book, I have embraced a broad two-pronged definition of technology, including not just devices, but processes, methods, and principles for achieving practical ends. Thus, it's possible to speak of "DNA technology" and "pest scouting" as technologies without reference to any particular tools or implements. At the same time, I have included many technologies you *can* see and touch.

Even though high-tech fascinates us, many low-tech devices and methods still pervade our lives, so I have included a range of high-, medium-, and low-tech subjects. I have also attempted to cover many of the hot topics fulminating within each technology group, such as:

- Farmers' use of "green" technologies and high-tech communications tools
- Smart cards, which allow us to carry large volumes of personal information with us
- High technology in manufacturing, such as robots and computer-controlled machines
- Wireless communications
- The explosion in area codes
- Sales of laptop computers
- The intranet market
- Electronic commerce
- Smart buildings
- Vehicle security systems
- The computer animation industry
- Access to the Internet in schools and libraries
- Students' access to computers and other technology
- Renewable energy
- Environmental cleanup technologies
- Military use of high-tech equipment
- DNA evidence in the courts
- Telemedicine

- Alternative medicine
- Antidepressant drug sales
- Activities in the biotechnology industry
- Global positioning satellites
- Orbital debris
- Alternative-fueled vehicles
- The number of cars on the road
- Transportation-related air emissions

Wherever possible, I have chosen statistics that illuminate the state of the art or that track trends. At writing time, figures for some subjects, particularly emerging technologies, were available primarily or solely as projections. I have included these numbers because of the importance of the technologies and avoided projections whenever actual measurements were available.

Where sources or methodologies were obviously suspect, I avoided using them. However, I have not been able to check the methodology of every survey cited in the book meticulously, so errors may exist.

I have attempted to be as complete as possible when citing sources, so you will find many references to specific page or table numbers. However, where data have been merged from several years of a publication or when information has been retrieved from a Web site, such specificity has not always been posible. In those cases, you will sometimes see "various years" or the date I retrieved the Web-based data.

I have also attempted to provide data current as of the time of writing. You will note, however, that some of the statistics in this book cover dates no later than the early or mid-nineties. The age of those numbers stems from two realities:

- Taking and collating surveys require time and depend both on the nature of the surveys and the staffing levels of the organizations taking them. Results can be published long after the questions were asked.
- Some surveys are taken only once, or infrequently.

However, to help alleviate this problem, I have cited the source so that you can seek the absolute latest statistics available. Also, in many cases, the numbers change so slowly (miles of highway, for instance) that older and newer numbers differ little.

Most data relate to the United States. However, a few tables treat other countries, regions, or the world as a whole.

I have consulted both public and private sources, so you will find a well-rounded mix of associations, market research firms, academics, trade publications, and government agencies counting different things in different ways.

Many tables have been lightly edited for clarity. However, most footnotes have been preserved. In some cases, I have merged data from multiple sources, usually so that a time series could be lengthened; in others, the producer has premerged data from various sources.

ORGANIZATION

The book is divided into broad subject areas like energy, agriculture, and so on; the broad subject areas are then broken down into subcategories. In some cases, I have combined two or more topics with similar characteristics to improve the flow of the book. While the basic categories grouping the tables are useful, the interdisciplinary nature of some data makes for unavoidable overlap. For example, use of energy for transportation logically belongs in both the energy and the transportation chapters; personal computers can apply to business, computers, and consumer products. Rather than repeat these tables, I have placed them in the chapters where they seem to make the most sense. If you can't find what you need, use the index to help you zero in on specific topics.

Each chapter is preceded by a brief introduction that explains the importance of the technology and provides historical and current context of its use. Most introductions also list major sub-fields or categories of the technology, important issues to consider, and hot topics. This discussion is followed by a short description of the tables in that chapter and what data you can expect to find and anything unusual or noteworthy about that data.

A short glossary defines some of the uncommon terms used in multiple tables or in introductory material. When terms are used only once, you will find the definition within the table itself.

READING THE DATA

Ralph Waldo Emerson probably didn't have the U.S. Census Bureau in mind when he said, "A foolish consistency is the hobgoblin of little minds," but the agency has taken the great man's words to heart. Consistency is not the strong suit of either the Census Bureau or any other federal agency that publishes statistics. Here are some issues to keep in mind as you examine the data:

- The Census Bureau constantly changes definitions. One year a certain product class includes certain types of equipment, and another year it doesn't. These changes account for some instances of "NA"—not available—but they also affect counts.
- The bureau and other agencies constantly change what and how they count. For a while, manufacturing technologies were being tracked; now they're not. The Bureau of Justice Statistics used to ask state and local law enforcement agencies about the technologies they use; now they don't.
- Many technologies of interest to large numbers of people aren't tracked by the government at all. Try to find something about alternative medicine or nanotechnology in the spaghetti of government statistics. By the time you read these words, those subjects may have inspired new surveys or special studies, but right now, the government seems to be silent on their behalf.
- The government constantly revises older figures, sometimes without indicating that such figures have changed. If you look at two versions of the same table, one for 1995–96 and one for 1996–97, you may notice that 1994's numbers have changed from the first table to the second. That's not so bad when an "r" appears next to the number to show that it's been revised, but that magic little "r" doesn't always appear. Usually the "r" is used when figures have been revised more than a certain percentage, but I have come across cases that deviate from that standard.
- The government mixes figures gained through different methodologies in the same table, making it appear that the numbers for various years are comparable. In fact, numbers for various years are seldom comparable.
- The government loves to round figures. Usually it tells you that it's rounding, but not always. One problem stemming from this rounding is that one table may take a figure out to four decimal places, and another table for succeeding years may round to two places. Time series that merge both tables may appear inconsistent.
- Agencies are constantly adding and removing documents and data from their Web sites. What was there a month ago may be gone now, and even if something similar appears in its place, it may offer the data sliced a different way. Hunting statistics is like pursuing a changeling, so expect the unexpected when you read the data and attempt to consult the source.

However, the Census Bureau *does* use the same *type* of product class codes in all the Current Industrial Reports cited in this book. These product classes are based on the 60-year-old SIC (Standard Industrial Classification) codes. The new system, NAICS, or North American Industry Classification System, was approved in 1997, published in 1998, and will be phased in by various government agencies between 1999 and 2004. NAICS, developed and agreed upon by the United States, Canada, and Mexico, completely revises SIC, adding new sectors, subsectors, and product codes that more accurately reflect today's economies. While some NAICS codes will map directly to SIC codes, most will not. Three hundred and fifty-eight industries are completely new, 390 are revised, and 422 remain more or less the same. However, the numbering system is entirely different, even for those businesses that remain comparable under both schemes.

Despite these caveats, you will find all sorts of interesting things to do with the data besides look up that one number you need. You can contrast market size for a consumer product with the money spent on a weapon or sales of life-saving drugs. If you want to know what's on the cutting edge, check out military technologies, which often seep into commercial life down the line. Watch the biotechnology industry change everything about our world from medical treatments to environmental cleanup. Look for patterns and anomalies.

Acknowledgments

Every book owes its birth and quality to many people whose names don't end up on the cover but whose work and encouragement make it special. *The Statistical Handbook of Technology* wouldn't exist without the terrific people at Oryx Press, the generous information providers who contributed to its pages, and my much-loved and appreciated personal cheering section.

At Oryx Press, a thousand thank yous to all, including

Jake Goldberg, who believed in and fought for me and made my job a whole lot easier.

Sean Tape, eagle-eyed editor extraordinaire and class act who can get you to work around the clock and feel grateful for the privilege.

Linda Vespa, Sandi Bourelle, and Chris Crites, who endured more author misformatting than any production department should have to (and with smiles, yet)

Anne Thompson and John Wagner, helpful execs.

Mary Swistara (and her little postcards too).

Linda Gorman, Angelea duMont, and Brian Johnson, publicity wizzes.

To many generous contributors, thank you for rounding out the book with dynamite information:

Jonathan Avins at Strategic Analysis, Inc.

Jeff Burnstein at the Robotics Industry Association.

Rick Bush at the American Farm Bureau.

Scott Dempster at Find/SVP.

Mark Dibner at the Institute for Biotechnology Information.

Jennifer Dills and Melva Garrison at the Society of the Plastics Industry, Inc.

Nancy Duckwitz and Donna Wauhop at IMS Health.

Rich Dzierwa and Randy Skazny at *Appliance* magazine.

Scott Edmunds at *Buildings* magazine.

Ori Epstein and Stephanie Bosson at Forrester Research.

Kris Estes at the General Aviation Manufacturers Association.

Lisa Portner Fasold at the Consumer Electronics Manufacturers Association (CEMA).

Aaron Fisher at *Data Communications* magazine. All data from *Data Communications* magazine, December 1997. Copyright 1998. The McGraw-Hill Companies, Inc. All rights reserved.

Liz Foster at Faulkner & Gray.

Melanie Glover at Crain Communications, Inc.

Kristi Grahl at *Ceramic Industry* magazine.

Suzanne Grappi at *PC Magazine*.

Bob Harter and Amy Goulla at ITT Systems and Sciences.

Richard H. Hauboldt and Mary B. Wendt at Milliman & Robertson, Inc.

Kathy Huston at Engel Publishing Partners.

Krysten Jenci at the International Trade Administration.

Andy Johnson at AISC Marketing, Inc.

Dave Johnson at the Center for Defense Information.

Lee Kurnoff at Zona Research.

Eric Linak at SRI Consulting.

J. Andrew Magpantay and Sally Benson at the American Library Association.

John Oberteuffer at Voice Information Associates, Inc.

Ross Pecoraro, Keith Hammond, and Nick Mariottini at Frost & Sullivan.

David Peyton at the National Association of Manufacturers.

Sarah Reardon at the Electronic Messaging Association.

Robi Roncarelli at PIXEL—the computer animation news people.

Tara Rummell at Datamonitor.

Clare M. Searby at the FBI.

Jeff Shear and John Kershman at Shear/Kershman Laboratories, Inc.

Steve Short at Intel Corp.

Tom D. Snyder at the National Center for Education Statistics.

Dr. Paul M. Swamidass at the Thomas Walter Center for Technology Management, Auburn University.

Mike Swiek at the U.S. GPS Industry Council.

David Vadas and David Napier at the Aerospace Industries Association of America.

Ada Walker at the American Medical Association.

Peter Ward at the Meta Group.

Kim Wells at the International Trade Administration.

Dr. Martha E. Williams at the University of Illinois.

Eileen Zix at the Freedonia Group.

To my personal people (and cat), thank you for being there, and especially for making me laugh:

Kathy and Scott Watters, Jim Cornelius, Barbara Javor, Howard Resnick, Mark Marcus, Bob Horacek, Patricia Barnes-Svarney, John Calderon, Gary and Barb Cooper, Ken Wong, Trung Huynh, Julie Gesin, John Bryans, Rick Benzel, Barbara Quint, Seth Potter, Jack Flannery, my parents, my sister, and Despina.

Thank you also to Ruth Balkin, Bernie Born, Linda Cooper, Susan Detwiler, Karen Ann Drebes, Josh Duberman, Martin Goffman, Susan Golden, Amelia Kassel, Alex Kramer, Sheri Lanza, John Levis, Fred Loewen, Ann Potter, Allan Rypka, Stuart Sandow, Cindy Shamel, and Linda Stevenson for help brainstorming about famous lawmen and offices of the 1950s.

Grazie mille to my muses: J. Michael Straczynski, Jerry Seinfeld, and the memory of Laura Nyro.

Introduction

Human beings employ technology to amplify their natural abilities. Transportation technologies expand the capabilities of our feet; communications devices project our voices; military technologies intensify what our fists can do. We are not the only animals to employ technology—beavers construct dams, ants excavate tunnels, birds make nests—but we are the only ones who build on our knowledge to devise increasingly complex ways of doing things.

Of course, we experience fits and starts in our quest for progress. As all inventors know, not every technology sells, even if it works brilliantly, and even if it's technically superior to competing technologies. (Look at Beta vs. VHS videotape formats, for example.) Many factors influence whether and how widely technologies are adopted, such as

- *Cost and cost-effectiveness.* Can I afford it, and is the cost worth it? Can I afford not to adopt this technology?
- *Ease of use.* Will this technology cause such a learning curve or take so much of my time to administer that it's not worth the trouble?

- *Ease and cost of manufacture.* Is the technology profitable to make, and can I find enough trained people to staff my manufacturing operation?
- *Government regulations.* Does the government say I have to adopt this technology? Does it provide incentives that make me want to adopt it even if I don't have to?
- *Political pressures.* Are powerful lobbies pushing for the use or banning of this technology?
- *Effectiveness.* Does it work?
- *Priorities.* Even if it works, do I have other things I need to do that are more important?
- *Usefulness.* Does it make a difference?

Once adopted, even successful technologies may be overshadowed or killed by new kids in town packing shinier, faster, cheaper, more capacious, more useful, or just plain more practical methods and devices. But that's progress.

Chapter 1. Basic Technology Indicators

People always seek to compare one society with another, sometimes out of curiosity, often with competitive fervor, but usually to assess where they stand and where to put resources. We can gauge the technological status of a society by looking at certain facts about it. But to do so, we need to define what "technologically advanced" means. It is not an easy concept to pin down. If Merlin could magically manipulate objects and events, was Camelot technologically advanced? Was ancient China, with fireworks and acupuncture, or ancient Rome, with its aqueducts, or ancient Egypt, with its pyramids, technologically advanced? More recently, was Scotland more advanced than the rest of the world because its scientists produced the first cloned animal (Dolly, the sheep), and do we consider it to have fallen behind just because other countries announced new cloning successes?

One way of defining technological advancement is to designate certain technologies as critical to national goals, then measure progress in those areas. Another is to emphasize those technologies that raise standards of living by making people more comfortable and healthier. Or, you could count inventions, or technical papers published in journals and proceedings, or scientists and engineers. Perhaps you see technological advancement as the ability to fulfill human needs without damaging the planet and its other inhabitants. Maybe technological advancement combines all these measurements.

Table 1.1 lists technologies considered critical to U.S. national security and economic prosperity. Congress has mandated that the list be re-examined and reported on at regular intervals so that research and development efforts can be prioritized and progress can be measured. Even though Table 1.1 dates from 1995, the list has changed little.

Highlights of the list include

- Energy efficiency for buildings and transportation systems, including developing noninternal combustion engines, and renewable energy development

- Interoperability of computer systems, a lack of which contributes to the daily frustration of millions
- Better biotechnology processes and products like gene therapy and diagnostics
- Improved pathogen detection and reduction in foods
- Cost-effective and widespread nanofabrication
- Stronger and more versatile materials, like composites for aircraft and automobile engines, and high-durability materials for highways and infrastructure
- Spacecraft power systems that result in lower overall weight and deeper space missions
- Improved aircraft engines that are safer, more powerful, and less polluting
- Better pollution control systems and better environmental remediation techniques

This chapter contains some U.S. government attempts to assess how scientifically and technologically advanced we are in general. The tables in the rest of the book can be used to round out the picture in specific areas.

Other information in this chapter includes

- A comparison of industrial research and development funding by the federal government and private companies from 1953 through 1997 (Table 1.2). Compared to the government, private industry has increased its contribution more dramatically over that time, by a factor of 11 as compared with 3 for the government (constant dollars).
- A comparison of industrial research-and-development funding, by industry and company size, from 1986 to 1997 (Table 1.3). Computers, aircraft and missiles, drugs and medicines, and other transportation received the most. Table 1.4 shows federally funded industrial R & D for the same period.

- The number of scientists and engineers in research-and-development-performing companies from 1986 through 1998, by industry and company size (Table 1.6). Table 1.7 shows the employment status of scientists and engineers by occupation and highest degree received. The figures are encouraging except for physical scientists who received degrees other than bachelor's, master's, or Ph.D. Other than that, employment rates vary from 97% to 100%. Table 1.8 shows the numbers of employed scientists and engineers by occupation and degree. Computer scientists are the most numerous, with electrical engineers second.

- The numbers and percent of patents granted in various countries and by company (Tables 1.10 through 1.12). Table 1.12 shows that General Electric Co. has been granted more patents than any other company (over 25,000).
- U.S. trade in advanced technology products, 1990 through 1996 (Table 1.13).
- The huge jump in nuclear technology imports in 1996 is correct and due to restructuring of the definition of "nuclear technology."
- Table 1.14 takes a look at high-tech company formation in the United States between 1980 and 1994, though the first row indicates total formations since 1960.

1.1. United States National Critical Technologies.

Technology Area	Technology Sub-Area	Specific Technologies	Sample Applications to Which Specific Technologies Contribute
Energy			
Energy efficiency	Building technologies	Superwindows Modular utility components Energy-efficient lighting/appliances Advanced building management	Reduced demand growth in energy sector by increasing energy efficiency of buildings, industry
	Noninternal combustion propulsion systems		Improved combustion systems for automobiles
Storage, conditioning, distribution, and transmission	Advanced batteries	Lead-acid, lithium, aluminum iron, sodium metal chloride, sodium sulfur, zinc-bromine, iron-air, zinc chloride, iron-chromium, zinc-ferrocyanide Li-Fe-S	Power sources for electric vehicles with consumer appeal; even out intermittent renewables generation for greater feasibility, appeal; area and frequency regulation, spinning reserve, peak shaving, and power quality
	Power electronics	High-power solid state switches Utility electronics	Greater stability, lower losses, faster switching of electric grid, conditioning of current from nonphase-locked intermittent sources (renewables), high voltage direct current converter stations, and real time systems control
	Capacitors		Electric vehicle energy storage, high pulse repetition frequency systems (e.g., elec. guns)
Improved generation	Gas turbines	Combuster design	Highly efficient energy generation with minimum pollution
		High-temperature materials	
	Fuel cells	Proton exchange membrane Phosphoric acid fuel cells Molten carbonate fuel cells Solid oxide fuel cells	Commercial development of fuel cell technology for distributed generation, transport Using hydrogen and hydrocarbon fuels
	Next generation nuclear reactors	Light-water Inherently stable reactor technology	Increased safety margins, reduced relative costs through passive response to non-normal conditions, simplified systems, inherently improved thermal management, tolerance
	Power supplies	High energy-density, low mass supplies	Improved weapons systems/reconnaissance vehicles, electronic warfare systems
		Pulse supplies	Simulating nuclear weapons effects

1.1. United States National Critical Technologies *(continued).*

Technology Area	Technology Sub-Area	Specific Technologies	Sample Applications to Which Specific Technologies Contribute
	Renewable energy	Solar thermal Photovoltaic Wind turbines	Mass production of devices for direct sunlight conversion and competitive prices Improved performance in a wider range of wind resource sites
		Biomass fuels	Conversion to synthesis gas or oil for production of electric power, liquid/gaseous fuels
Environmental quality			
Monitoring and assessment	Integrated environment monitoring	Sensors Software	Global climate/ocean observing systems; battlefield identification of dangerous chemical and biological agents; proliferation monitoring
		Networking Simulation	
	Remote assessment of biosystems	Ground truth biomarkers and pattern recognition	Forestry and fisheries management, crop yield prediction
Pollution control	Pollution control	Physical separation Component separation Chemical transformation Biological agent separation Waste elimination	Increase fraction of waste stream recovered for recycling and reuse by increasing efficiency and lowering cost with which components of stream can be separated
Remediation and restoration	Remediation and restoration	Soil washing Thermal desorption Composting Electrochemical separation Super-critical water oxidation Recovery of spilled oil and other hazardous substances	Increase efficiency of, reduce cost and cycles required to remediate sites, especially those with multiple contaminants, habit restoration
	Bioremediation	Microbial metabolism of organic pollutants Sequestration of heavy materials	
	Nuclear waste storage/ disposal	Storage, treatment, separation	Increased safety/decreased uncertainty for storage and containment of nuclear waste
Information and communication			
Components	High-density data storage	Thin-film recording head transducers Recording media High-density RAM Format compatibility/standards for optical storage	One-gigabyte micro-floppies Storage requirements for high-definition displays, high-performance computing Long-term commercial viability of media, storage requirements for high-definition displays
		Magneto-optical storage	Eraseable data storage with conventional optical disk data density
		Holographic optical elements Parallel data storage controllers	Improved access time, reduced weight, cost Meet storage requirements of massively parallel computers
	High-definition displays	Lithography Circuitry patterning Glass sheet production Thin-film techniques Holographic displays	Competitive high-definition display manufacturing capability supporting (e.g.) flat panel, HDTV, military systems Displays for military and commercial systems
Communications	High-resolution scanning Data compression	High-yield detector arrays Chaotic graphics compression	Search engines on optical storage High data-transfer requirement applications, e.g., HDTV
	Signal conditioning and validation		Support high-performance, arrayed/ multispectral sensor applications: target recognition, munitions guidance, computer-integrated manufacturing, environmental monitoring
	Telecom/data routing	Broadband switching	Full utilization of fiber-optic capacity, especially in support of national information infrastructure

1.1. United States National Critical Technologies *(continued).*

Technology Area	Technology Sub-Area	Specific Technologies	Sample Applications to Which Specific Technologies Contribute
		Programmable radios	All-spectrum/format radios for military interoperability, reduced costs
		Wireless communications Cable Fiber	Improved economics, technical characteristics to support maximum access to national information infrastructure and robust military communications system
		Satellite-ground communication protocols	Improved space-ground interoperability
Computing systems	Interoperability	Data interchange standards	Full standardized multimedia data exchange on national information infrastructure or other systems
		Product data exchange	Full exchange of product characteristic data between systems
	Parallel processing	Multiple Instruct, Multiple Data Stream (MIMD) Single Instruction, Multiple Data Stream (SIMD) Very Large Instruction Word (VLIW) Systolic arrays Specialized parallel coprocessors Hypercubes Parallel data storage architecture Mobile computing systems	Parallel-processing hardware and software to support high-performance computing activities in prediction and analysis of complex systems, management, simulation and design of complex systems/processes
Information management	Data fusion	Multi-spectral data processing Distributed array processing	Advanced targeting and guidance, environmental monitoring, machine vision, intelligence capabilities, anti-stealth techniques
	Large-scale information systems	Very large database management tools Real-time largescale information retrieval tools	Financial systems, electrical power grids, battlefield surveillance systems, fusion of global ocean- and climate-observing data
	Health systems and services	Integrated systems of electronic patient records and decision support systems User interfaces	Patient management and outcome studies, telemedicine, virtual medical groups, evaluation of preventive interventions, health resource allocation
	Integrated navigation systems	Electronic charts data information systems On-the-fly DGPS positioning Real-time environmental information systems	Improved ship navigation capabilities, navigation systems for air traffic control, intelligent transportation systems
Intelligent complex adaptive systems	Autonomous robotic devices	Sensors, signal processing, software, robotics	Automatic sentries, hazardous materials handling, mine detection and removal, counterterrorist ops, law enforcement, environmental remediation, public disaster responses
	Artificial intelligence	Knowledge representation Computer-based reasoning methods Machine learning methods	Artificial intelligence systems for military and commercial/manufacturing use
Sensors	Physical devices	Microsensors	Integrated sensing/signal processing, expanded in-situ monitoring, integrated systems
		Biosensors	Diagnostics, environ/exposure monitoring, nano-controls, biological weapons use detection
		Chemical sensors	Process control, environmental monitoring, chemical weapons use detection

1.1. United States National Critical Technologies *(continued).*

Technology Area	Technology Sub-Area	Specific Technologies	Sample Applications to Which Specific Technologies Contribute
		Passive thermal imagers	Battlefield night vision, medical diagnostics
		IRST systems (point source passive thermal imaging)	Passive targeting systems, identification, kill assessment
		High-yield, high-density photo/ infrared detectors	HDTV, scanning, medical imaging, environmental monitoring, night vision systems, machine vision, advanced targeting and precision-guided munitions
		Multi-spectral integrated sensors	
		Infrared/radar sensors	
	Integrated signal processing		Improved miniaturized sensors with sensor and processing circuitry on same chip
Software and toolkits	Education/training software	Military training	Battle simulation and training
		Multimedia authoring tools	More effective education and lifelong training
	Network and system software	Navigation and resource discovery tools	High-utility, easy-access national information infrastructure via tools for organizing and finding information
		Directories	
		Registries	
		Transparent embedding software	Integrated heterogeneous concurrently operating systems in distributed environment
		Operating systems	
		Run-time execution systems	
		Programming languages	
		Interpreters	
		Compilers	
	Modeling and simulation software	Virtual battlefields & weapons systems	Improved effectiveness, economics of training, "virtual prototyping"
		Atmospheric and global systems	Improved weather prediction, knowledge of global change phenomena
		Computational fluid dynamics	Better aero/hydro-dynamic design, combustion and industrial process design
		Agriculture	Sustainable agricultural practices, optimized yields
		Cellular/biomolecular function	
		Computational physics	
		Computational chemistry	Rational design of new chemicals and pharmaceuticals; countermeasures against biological warfare
		Economic systems	
		Numerical control simulation	Visualization of machining process: design for minimal waste, computer-integrated manufacturing
		Manufacturing	"Virtual computer-integrated manufacturing" of products: allow rapid prototyping, reduced startup costs
	Software engineering tools	CASE (Computer-aided software engineering) tools	Increased efficiency of software authoring, better software maintainability
		User interface design tools	Contain software costs
		Software testing tools	Increased reliability, reduced cost of software, especially in complex systems
		Interface card design tools	Improved design capabilities for complex interface cards
	Pattern recognition	Virtual reality software	
		Natural language recognition (speech, handwriting)	
		Multimedia operating systems	
	Software production	Rapid prototyping	Reduced development costs, fewer development cycle iterations

1.1. United States National Critical Technologies *(continued).*

Technology Area	Technology Sub-Area	Specific Technologies	Sample Applications to Which Specific Technologies Contribute
		Modular/object-oriented programming	
Living systems			
Biotechnology	Bioprocessing	Drug, chemical, enzyme production Mineral extraction	Full commercialization of biotech by increased efficiency, reduced cost of cell bioreactors/bioprocessing; aquaculture
	Monoclonal antibody production	Diagnostics, therapeutics, separation processes, enzyme-like catalysts	Improved diagnostic capability for diseases and disorders; improved treatment of variety of diseases
	Protein engineering	Protein sequencing Protein folding	Site-specific binding/blocking Achieving the right shape, not just the right sequence
		Rationally designed drugs Structural biology Tailored protein catalysts	Improved potency and specificity Shaping molecules High selectivity; specificity catalysts: more efficient synthesis flow cost, water minimization
		Molecular electronics Biomolecular materials Molecular motors	Ultra-high density/low-power memories
	Recombinant DNA technologies	Gene mapping/sequencing Gene therapy/replacement	Diagnostics, prevention and therapy for diseases in humans, identification/ manipulation of specific genes in plants and animals
		Antisense Agricultural species modification	Turning off specific protein production Plant and livestock species with designed traits, plant pest/pathogen resistance
		Transgenic cell cultures Therapeutic agents	Modified cells to act as bioreactors
	Vaccines	Infectious disease AIDS vaccine Cancer	Enhanced immunogenicity Contain spread of AIDS virus
Medical technologies	Health information systems and services		
	Biocompatible materials	Microcapsules Replacement materials	Slow, targeted release of medication More durable, more effective materials for replacement of natural tissue or bone
		Material surface characteristics	Implantable devices, sensors, and electrodes
		Tailoring immune system response Tissue engineering	Artificial blood, skin, tendons, and ligaments
	Functional diagnostic imaging	High-field spectroscopic MRI Optical coherence tomography Artificial intelligence diagnostics support systems	Electronic biopsy Biochemical functional assessment
	Bacterial/viral detection and screening	Media, environments	Food and water quality assurance, antibiotic selection, infection control
		Epidemiology statistical techniques	Surveillance of new disease and spread of disease
	Medical devices & equipment	Patient monitoring equipment Biosensors	Pressure, electrophysiology, blood chemistry
		Functional electrical stimulation devices	Functional restoration, including hearing, breathing, continence, and mobility
		Critical care instrumentation	Cardiopulmonary bypass pumps, respirators
		Blood chemistry auto-analyzers Decontamination and sterilization	

1.1. United States National Critical Technologies *(continued).*

Technology Area	Technology Sub-Area	Specific Technologies	Sample Applications to Which Specific Technologies Contribute
		Endoscopic/laparoscopic surgical equipment	
Agriculture and food technologies	Sustainable agriculture production	Genetic resource identification, preservation, and utilization	Improved biological efficiency of agricultural organisms, biological pest control, sustainable agricultural systems, soil and water conservation
		Biological and engineering criteria	
		Ecosystem management	
	Food safety assurance	Pathogen detection, isolation, reduction/elimination	Food safety
		Drug and residue detection	
		Waste management	Byproduct recover and value-added product development
	Aquaculture and fisheries		Sustained population management and production, food quality assurance
Human systems	Advanced human-machine interfaces	Psychophysiology of learning	Education, training, performance enhancement
		Mental workload assessment	
		Neurotechnologies	
		Software for advanced human-computer interfaces	Human-machine interfaces more accurately simulating natural experience for increased ease of use and utility
		Sensors	
		Signal processing	
Manufacturing			
Discrete product manufacturing	Computer-integrated manufacturing support software	Computer-aided design	Improved solid object representation design
		Computer-aided engineering	Improved interactive design/ performance analysis of product
		Process, machine performance databases	Ensure consistency with high skills production and work systems
	Equipment interoperability	Group technology	Improved factory, production efficiency by identifying job/item similarities
		Computer-aided process planning	Optimized factory routing, production lead times with variant, generative computer-aided process planning
		Data-driven management information systems	Capabilities to effectively design, manage computer-integrated manufacturing
		Factory scheduling tools	Enhanced forecasting and scheduling capabilities for improved efficiency
	Intelligent processing equipment	Sensors	Process monitoring and control; automated design of large-scale interface cards
		Next-generation controller	
	Robotics		Automated construction; demining
	Automation systems for facilities operations	Computer-aided production cycle management	
		Building automation systems	
	Net shape processing	Hot isostatic pressing	More economic/technically feasible finishing of nontraditional materials, e.g., ceramics, superalloys
		Metal injection molding	
		Superplastic forming	
		Liquid transfer molding of polymer matrix composites	Low-cost manufacturing of large parts
	Rapid solidification processing	Spray forming	Uniform microstructure alloys, especially for hard-to-shape alloys
		Gas atomization	
Continuous materials	Catalysts	Tailored protein catalysts	
		Shape selective catalysts	Fuels and chemical production
		Catalysts by design	Increased-efficiency thermal processing, pollution control catalysts
		Biomimetic catalysts	Carbon dioxide utilization

1.1. United States National Critical Technologies *(continued)*.

Technology Area	Technology Sub-Area	Specific Technologies	Sample Applications to Which Specific Technologies Contribute
	Surface treatments	Laser hardening	Selective surface hardening in medical implants, high-perf bearings, extrusion dies
		Thin films	
		Grinding and machining of ceramics	Reduced cost and increased reliability of parts
		Ceramic coatings	
	Ultrapure refining methods	Various refining methods	Stronger polymer matrix composites and ceramics
		Electron beam processing	High-purity alloys for jet turbines
	Pollution avoidance	Process design strategies	Achieve lower levels of environmental loading
		Improved processes	Increase resource efficiency of process industries, e.g., petroleum refining, metals, etc.
		Design for the environment	Lifecycle approach to design of products and processes
		Industrial ecology	Multidimensional production and recycling reuse systems
	Predictive process control	Sensors, data processing	
Micro/nanofabrication and machining	Microdevice manufacturing technologies	Silicon machining	Better quality, expanded capability, lower-cost micromachines for sensors, controls
	Semiconductor manufacturing	X-ray lithography	Miniaturization and cost containment for chip manufacture
		Microwave plasma processing	
		Electron/ion micro-beams	
		Artificially structured materials	
		Laser-assisted processing	
		Metrology	
		Design testing	
	Semiconductor integration technologies	Integrated packaging	Easier/lower-cost integration/ interconnection of multiple chips
		Multichip modules	
	Artificial structuring methods	Chemical vapor & sputter deposition	Thin films, high-speed microelectronics
		Molecular and chemical beam expitaxy	High-performance alloys, ceramics, coatings (e.g., diamond) for improved wear, thermal, and radiation-resistance characteristics
		Spin-on deposition	
		Vacuum evaporation	
Materials			
Materials	Alloys	Lightweight structural alloys	Lighter, stiffer airframes, automobile frames
		Intermetallic alloys	High-temp structural applications in aerospace
	Ceramic materials	Ceramic interface card packaging	Improved interconnection/data rates for multi-interface card assemblies
		Ceramic matrix composites	
		Ceramic coatings	Improved wear characteristics in high-speed moving parts: turbine engines, cutting tools
	Composites	Polymer matrix composites	Economical, bulk polymer matrix composites for jet engines, autos
		Ceramic matrix composites	High-performance aircraft engines, auto engines, advanced armor
		Metal-matrix composites	Space applications for null-outgassing, radiation-resistant important, hypersonic aircraft
		Carbon-carbon composites	Ultra-high temperature aerospace applications (rocket nozzles, thermal surfaces)
		Molecular-scale composites	
	Electronic materials	Gallium arsenide	Superior, economic gallium arsenide–based interface cards
		Thin-film dielectric materials	Miniaturization of microwave communication devices
	Photonic materials	Semiconductor lasers/laser arrays	Laser/detector technologies to support communications and optical data processing
		Advanced detectors	

1.1. United States National Critical Technologies *(continued)*.

Technology Area	Technology Sub-Area	Specific Technologies	Sample Applications to Which Specific Technologies Contribute
	High energy-density materials	Advanced solid and liquid propellants	Space launch vehicles, longer-range artillery
		Explosives	Improved conventional weapons
	Highway/infrastructure	Paving	Higher-durability, more economical materials to maintain surface transport infrastructure
		Repair materials	
		Polymer matrix composites	Retrofit reinforcement for earthquake damage prevention
	Biocompatible materials		
	Stealth materials	Radar-absorbing materials/coatings	Increased abilities for military aircraft to operate unobserved by radar
	Superconductors	High-temperature superconductors	Sensors, low-power electronics, power transmission, energy storage, powerful magnets for research, medical diagnostics, maglev technology
		Advanced low-temperature superconductors	
Structures	Aircraft structures	Computational structural mechanics	Enhance productivity of prototype development
		Hot structures	Hypersonic cruise/reentry functions
		Actively cooled structures	
		Materials life behavior prediction	Improved safety, design for lifecycle
		Life extension technologies	
		Stealth structures	Enhanced survivability, effectiveness of military aircraft through reduced observability
		Infrastructure materials	
Transportation			
Aerodynamics	Aircraft aerodynamics	Turbulence prediction and control	Engine efficiency, aircraft stability improvements
		Laminar flow control	
		Noise control	Reduced environmental impact aircraft, low-observable aircraft
		Hypersonic/aero-assist/wave-rider designs	Long-range, high-speed aircraft
	Surface vehicle aerodynamics	Low-drag designs	More efficient, lower-emissions vehicles
Avionics and controls	Aircraft/spacecraft avionics	Glass cockpit technologies	Reduce pilot workload
		Foul weather flying	
		Fly-by-light technologies	Reduced weight, increased reliability and stability control
		Power-by-wire technologies	
		Highly integrated engine/airframe controls	High-speed (greater than Mach 3) applications
		Space-qualified computers	Better performance, autonomous operations, improved reliability
	Surface transport controls	Advanced engine controls	More efficient, lower-emissions engines
Propulsion and power	Aircraft turbines	High-thrust turbines	Sustained supersonic cruise without afterburners
		High-efficiency turbines	Better specific fuel consumption
		Multi-fuel/hybrid engines	More powerful engines with better fuel economy and lower nitrogen oxide emissions
		Advanced engine components	Increased safety, quality assurance, lower manufacturing cost
	Spacecraft power systems	High-efficiency solar cells	Lower spacecraft weight
		High-power compact sources	Deep space mission requirements
		Remote power transmission	In space/space-to-ground power transmission
		Batteries	Lower spacecraft weight
	Electrically powered vehicles	Energy storage technologies	Reduce emissions/secondary pollutant concentrations in urban areas
Systems integration	Intelligent transportation systems	Sensors, networks, software, satellite navigation	Add information and control infrastructure to existing physical transport infrastructure to increase safety, capacity, driver convenience, and reduce emissions, fuel consumption, congestion

1.1. United States National Critical Technologies *(continued).*

Technology Area	Technology Sub-Area	Specific Technologies	Sample Applications to Which Specific Technologies Contribute
	Aircraft/spacecraft integration	Satellite assembly Advanced fabrication, lay-up, and joining technologies Multidisciplinary aircraft design techniques	Industrywide, common interface standards Successful adaptation to new materials, reduced costs of design, prototyping
Human interface	Human factors engineering	Ergonomics Long-duration flight countermeasures High-gravity countermeasures	Improved automobile control Improve safety and situational awareness/ counter aircrew fatigue Improve pilot function under high-gravity maneuvers
	Spacecraft life support	Closed/mostly closed extended life support systems	Reduced space logistics costs

Source: National Critical Technologies Review Group. *National Critical Technologies Report.* Washington, DC: White House Office of Science and Technology, March 1995. Appendix A. http://www.whitehouse.gov/WH/EOP/OSTP/CTIformatted/ Also at http://www.nsf.gov/sbe/srs/seind96/chap_6.pdf. For private sector views on which technologies are critical to the United States, see Steven W. Popper et al. *New Forces at Work: Industry Views Critical Technologies.* Santa Monica, CA: RAND, 1998. Appendix B. This study builds on the 1995 report cited above.

RESEARCH AND DEVELOPMENT

1.2. Trends in Industrial R&D Performance, by Source of Funds, in Current and Constant Dollars, United States, 1953–97 (millions of dollars).

Year	Total R & D Current dollars	Total R & D Constant 1992 dollars	Federal Current dollars	Federal Constant 1992 dollars	Company [1] Current dollars	Company [1] Constant 1992 dollars
1953	$3,630	$17,988	$1,430	$7,086	$2,200	$10,902
1954	$4,070	$19,941	$1,750	$8,574	$2,320	$11,367
1955	$4,640	$22,372	$2,180	$10,511	$2,460	$11,861
1956	$6,605	$30,764	$3,328	$15,501	$3,277	$15,263
1957	$7,731	$34,856	$4,335	$19,545	$3,396	$15,311
1958	$8,389	$36,940	$4,759	$20,956	$3,630	$15,984
1959	$9,618	$41,908	$5,635	$24,553	$3,983	$17,355
1960	$10,509	$45,161	$6,081	$26,132	$4,428	$19,029
1961	$10,908	$46,338	$6,240	$26,508	$4,668	$19,830
1962	$11,464	$48,087	$6,434	$26,988	$5,029	$21,095
1963	$12,630	$52,363	$7,270	$30,141	$5,360	$22,222
1964	$13,512	$55,196	$7,720	$31,536	$5,792	$23,660
1965	$14,185	$56,831	$7,740	$31,010	$6,445	$25,821
1966	$15,548	$60,569	$8,332	$32,458	$7,216	$28,111
1967	$16,385	$61,854	$8,365	$31,578	$8,020	$30,276
1968	$17,429	$63,057	$8,560	$30,970	$8,869	$32,088
1969	$18,308	$63,262	$8,451	$29,202	$9,857	$34,060
1970	$18,067	$59,275	$7,779	$25,522	$10,288	$33,753
1971	$18,320	$57,143	$7,666	$23,911	$10,654	$33,231
1972	$19,552	$58,504	$8,017	$23,989	$11,535	$34,515
1973	$21,249	$60,195	$8,145	$23,074	$13,104	$37,122
1974	$22,887	$59,493	$8,220	$21,367	$14,667	$38,126
1975	$24,187	$57,465	$8,605	$20,444	$15,582	$37,021
1976	$26,997	$60,599	$9,561	$21,461	$17,436	$39,138
1977	$29,825	$62,882	$10,485	$22,106	$19,340	$40,776
1978	$33,304	$65,443	$11,189	$21,987	$22,115	$43,456
1979	$38,226	$69,212	$12,518	$22,665	$25,708	$46,547
1980	$44,505	$73,769	$14,029	$23,254	$30,476	$50,515
1981	$51,810	$78,488	$16,382	$24,817	$35,428	$53,671
1982	$58,650	$83,583	$18,545	$26,429	$40,105	$57,154
1983	$65,268	$89,213	$20,680	$28,267	$44,588	$60,946
1984	$74,800	$98,525	$23,396	$30,817	$51,404	$67,708
1985	$84,239	$107,270	$27,196	$34,631	$57,043	$72,638
1986	$87,823	$108,989	$27,891	$34,613	$59,932	$74,376
1987	$92,155	$110,950	$30,752	$37,024	$61,403	$73,926
1988	$97,015	$112,690	$30,343	$35,246	$66,672	$77,445
1989	$102,055	$113,748	$28,554	$31,826	$73,501	$81,923
1990	$109,727	$117,230	$28,125	$30,048	$81,602	$87,182
1991	$116,952	$120,173	$26,372	$27,098	$90,580	$93,074
1992	$119,110	$119,110	$24,722	$24,722	$94,388	$94,388
1993	$117,400	$114,380	$22,809	$22,222	$94,591	$92,158
1994	$119,595	$113,802	$22,463	$214,375	$97,131	$92,426
1995	$132,103	$122,590	$23,451	$21,762	$108,652	$100,828
1996	$144,667	$131,265	$23,653	$21,462	$121,015	$109,804
1997	$157,539	$140,159	$23,928	$21,288	$133,611	$118,871

Notes:

[1]Company funds include funds for industrial R&D work performed within company facilities from all sources except the federal government. The funds predominately are the companies' own, but also include funds from outside organizations such as other companies, research institutions, universities and colleges, nonprofit organizations, and state governments. Company-financed R&D not performed within the company is excluded. Gross domestic product implicit price deflators were used to convert current dollars to constant (1992) dollars. As a result of a new sample design, statistics for 1988–91 have been revised since originally published. These statistics better reflect R&D performance among firms in the nonmanufacturing industries and small firms in all industries. As a result of the new sample design, statistics for 1991 and later years are not directly comparable with statistics for 1990 and earlier years.

Source: National Science Foundation, Division of Science Resources Studies. *Research and Development in Industry: 1997 (Early Release Tables).* Arlington, VA: National Science Foundation, 1999. Table A-1. http://www.nsf.gov/sbe/srs/indus/start.htm

1.3. Total (Company, Federal, and Other) Funds for Industrial R&D Performance, by Industry and Size of Company, United States, 1986–97 (millions of dollars).

Industry and size of company

Distribution by industry:

	1986	1987	1988	1989	1990	1991	1992	1993	1994	1995	1996	1997
All Industries	$87,823	$92,155	$97,015	$102,055	$109,727	$116,952	$119,110	$117,400	$119,595	$132,103	$144,667	$157,539
Manufacturing	$80,377	$84,311	$86,503	$88,024	$88,934	$88,506	$90,177	$86,569	$90,749	$100,067	$111,864	$121,025
Food, kindred, and tobacco products	D	$1,206	D	D	D	$1,277	$1,386	$1,345	$1,476	$1,566	$1,564	$1,787
Textiles and apparel	D	$137	D	D	D	D	D	D	D	$1,566	D	D
Lumber, wood products, and furniture	$144	D	D	$192	$216	D	D	D	D	D	D	$348
Paper and allied products	D	D	D	$879	$1,059	D	D	D	D	D	D	D
Chemicals and allied products	$8,843	$9,635	$11,067	$12,069	$13,291	$14,648	$15,381	D	D	$17,547	D	D
Industrial chemicals	$3,552	$3,716	$4,172	$4,451	$5,010	$5,390	$5,165	D	D	D	D	D
Drugs and medicines	$3,658	D	$4,906	D	D	D	$7,944	$9,146	$9,633	$10,215	$9,773	$11,589
Other chemicals	$1,633	D	$1,989	D	D	D	$2,272	D	D	D	$2,505	D
Petroleum refining and extraction	D	$1,897	$1,997	$2,180	$2,306	$2,498	$2,277	$2,152	$1,950	$1,760	$1,654	D
Rubber products	D	D	D	D	D	D	D	D	D	D	D	D
Stone, clay, and glass products	$950	$995	D	D	D	D	$522	$538	$591	$448	$468	$608
Primary metals	D	$730	$637	$686	$739	$714	D	$669	$690	$593	D	$988
Ferrous metals and products	D	D	$253	D	D	D	D	$289	D	D	D	D
Nonferrous metals and products	$458	D	$384	D	D	D	D	$380	D	D	D	D
Fabricated metal products	$895	$783	$881	$904	$939	$974	$1,017	$1,158	$1,111	$1,023	D	$1,798
Machinery	D	D	D	D	$14,446	$14,775	$14,938	$8,381	$8,110	D	$13,455	$18,499
Office, computing, and accounting machines	D	D	D	D	D	D	D	$4,950	$4,106	D	D	$12,840
Other machinery, except electrical	$2,396	$2,428	$2,682	$2,729	D	D	D	$3,431	$4,004	$5,041	D	$5,659
Electrical equipment	$14,980	$15,848	$14,128	$13,318	$13,400	$13,415	13,360	$13,349	$15,338	$18,751	$22,498	$24,585
Radio and TV receiving equipment	$133	$139	$149	$96	$114	D	D	D	D	D	D	D
Communication equipment	$9,669	$10,184	$8,427	$7,071	$5,928	$4,787	$3,567	$5,311	$6,032	D	D	D
Electronic components	D	$4,286	$4,133	$4,025	$3,914	D	D	D	D	D	D	D
Other electrical equipment	D	$1,239	$1,419	$2,126	$3,444	D	D	D	D	D	D	$4,909
Transportation equipment	$31,275	$34,246	$34,775	$33,859	$31,361	$27,428	$27,494	$27,258	$28,087	$32,441	$32,737	$31,993
Motor vehicles and motor vehicles equipment	D	D	D	D	D	D	D	$11,718	D	D	D	D
Other transportation equipment	D	D	D	D	D	D	D	$483	D	D	D	D
Aircraft and missiles	$21,050	$24,458	$24,168	$22,331	$20,635	$16,629	$17,158	$15,056	$14,260	$16,951	$16,224	$16,296
Professional and scientific instruments	$5,103	$5,222	$5,530	$5,992	$7,055	$8,705	$9,542	$10,119	$11,441	$11,976	$12,149	$13,458
Scientific and mechanical measuring instruments	D	D	$1,959	$2,366	$3,346	D	$5,156	$5,681	$6,952	$7,146	D	$8,135
Optical, surgical, photographic, and other instruments	D	D	$3,571	$3,626	$3,709	D	$4,386	$4,438	$4,489	$4,831	D	$5,323
Other manufacturing industries[1]	$382	D	D	D	D	D	D	D	D	D	D	$2,798
Distribution by industry:												
Nonmanufacturing[2]	$7,446	$7,844	$10,513	$14,031	$20,793	$28,446	$28,933	$30,831	$28,846	$32,036	$32,803	$36,514
Transportation and utilities	NA	NA	NA	NA	NA	NA	NA	NA	NA	$5,435	$4,678	$3,013
Communications	NA	NA	NA	NA	NA	NA	NA	NA	NA	D	D	D
Telephone communications	NA	NA	NA	NA	NA	NA	NA	NA	NA	D	D	D
Other communications	NA	NA	NA	NA	NA	NA	NA	NA	NA	$59	$73	$58
Electric, gas, and sanitary services	NA	NA	NA	NA	NA	NA	NA	NA	NA	$440	$352	D

1.3. Total (Company, Federal, and Other) Funds for Industrial R&D Performance, by Industry and Size of Company, United States, 1986–97 (millions of dollars) (continued).

	1986	1987	1988	1989	1990	1991	1992	1993	1994	1995	1996	1997
Other transportation and utilities	NA	NA	NA	NA	NA	NA	NA	NA	NA	D	D	D
Trade	NA	NA	NA	NA	NA	NA	NA	NA	NA	D	$6,389	D
Finance, insurance, and real estate	NA	NA	NA	NA	NA	NA	NA	NA	NA	D	D	D
Services	NA	NA	NA	NA	NA	NA	NA	NA	NA	$17,876	$19,022	$22,400
Business services	NA	NA	NA	NA	NA	NA	NA	NA	NA	$9,293	$10,641	$11,960
Computer and data processing services	NA	NA	NA	NA	NA	NA	NA	NA	NA	$9,059	D	$11,706
Other business services	NA	NA	NA	NA	NA	NA	NA	NA	NA	$234	D	$254
Health services	NA	NA	NA	NA	NA	NA	NA	NA	NA	$756	D	D
Offices and clinics of medical doctors, hospitals, medical and dental labs	NA	NA	NA	NA	NA	NA	NA	NA	NA	D	$715 S	D
Other health services	NA	NA	NA	NA	NA	NA	NA	NA	NA	D	D	$49
Engineering and management services	NA	NA	NA	NA	NA	NA	NA	NA	NA	$7,662	$7,318	$9,290
Engineering, architectural, and surveying	NA	NA	NA	NA	NA	NA	NA	NA	NA	$2,933	$1,660	$2,030
Research, development, and testing, hospitals, medical, and dental labs	NA	NA	NA	NA	NA	NA	NA	NA	NA	D	$5,484	D
Other engineering and management services	NA	NA	NA	NA	NA	NA	NA	NA	NA	D	$173	D
Other services	NA	NA	NA	NA	NA	NA	NA	NA	NA	$165	D	D
Other nonmanufacturing industries[1]	NA	NA	NA	NA	NA	NA	NA	NA	NA	$413	D	$1,618
Distribution by size of company: [Number of employees]												
Total	$87,823	$92,155	$97,015	$102,055	$109,727	$116,952	$119,110	$117,400	$119,595	$132,103	$144,667	$157,539
Fewer than 500	$7,071	$7,163	S	$7,809	S	$13,172	$13,557	$14,620	$13,966	$16,662	$20,249	$24,063
500 to 999	$1,902	$1,725	$1,669	$1,825	$2,154	$8,000	$7,958	$3,230	$3,608	$4,693	$4,637	$4,966
1,000 to 4,999	$7,472	$7,262	$7,622	$7,881	$8,411	$10,453	$11,886	$13,334	$14,617	$16,960	$18,273	$19,590
5,000 to 9,999	$4,251	$4,501	$5,245	$5,756	$6,746	$8,049	$8,258	$9,135	$8,912	$9,532	$11,537	$14,266
10,000 to 24,999	$10,493	$12,043	$11,506	$10,450	$12,486	$15,770	$15,744	$15,421	$15,972	$17,071	$20,164	$21,510
25,000 or more	$56,991	$59,461	$63,694	$68,335	$71,030	$61,508	$61,707	$61,659	$62,519	$67,185	$69,808	$73,144

Notes:

[1]Beginning in 1996 manufacturing companies with 50 or fewer employees and nonmanufacturing companies with 15 or fewer employees were sampled separately without regard to industry classification to minimize year-to-year variation in survey estimates. Estimates for manufacturing companies in this group are combined with those for companies in "Other manufacturing industries." Estimates for nonmanufacturing companies in this group are combined with those for companies in "Other nonmanufacturing industries." As a result, statistics for "Other manufacturing industries" and for "Other nonmanufacturing industries" for 1996 are not comparable with statistics for prior years. [2]Until 1995, data were not broken down below this level of detail for most nonmanufacturing industries.

D = Data have been withheld to avoid disclosing operations of individual companies.

S = Indicates imputation of more than 50 percent. For years prior to 1993, data have been withheld. When a company returns an incomplete form, missing information is imputed, i.e., ascribed, based on the information that is provided, prior-year submissions, and information provided by other companies within the industry. When a company fails to return a form, information is imputed from prior years, adjusted to reflect changes noted among responding companies. If no previous year information is available for the company, data are imputed from other companies in the same industry.

NA = Not available.

As a result of a new sample design, statistics for 1988–91 have been revised since originally published. These statistics now better reflect R&D performance among firms in the nonmanufacturing industries and small firms in all industries. As a result of the new sample design, statistics for 1991 and later years are not directly comparable with statistics for 1990 and earlier years.

Source: National Science Foundation, Division of Science Resources Studies. *Research and Development in Industry: 1997 (Early Release Tables).* Arlington, VA: National Science Foundation, 1999. Table A-3. http://www.nsf.gov/sbe/srs/indus/start.htm

1.4. Federal Funds for Industrial R&D Performance, by Industry and Size of Company, United States, 1986–97 (millions of dollars).

Industry and size of company	1986	1987	1988	1989	1990	1991	1992	1993	1994	1995	1996	1997
Distribution by industry:												
All Industries	$27,891	$30,752	$30,343	$28,554	$28,125	$26,372	$24,722	$22,809	$22,463	$23,451	$23,653	$23,928
Manufacturing	$25,185	$28,052	$27,088	$24,826	$23,683	$20,867	$19,152	$16,669	$17,373	$18,831	$20,020	$19,823
Food, kindred, and tobacco products	D	2	D	D	D	0	0	0	0	0	0	0
Textiles and apparel	D	D	D	D	D	S	D	D	D	D	D	D
Lumber, wood products, and furniture	0	0	D	0	0	D	D	D	D	D	D	0
Paper and allied products	D	D	D	0	0	D	S	D	D	D	D	D
Chemicals and allied products	$179	$190	$238	$126	$123	$209	S	D	D	$210	D	D
Industrial chemicals	$178	$185	$232	$111	$109	$165	S	D	D	D	D	D
Drugs and medicines	$1	D	6	D	D	D	S	$15	$8	$14	$3	$3
Other chemicals	0	D	0	D	D	D	S	D	D	D	0	D
Petroleum refining and extraction	D	$14	$22	S	S	$11	$9	$14	$10	$6	$24	D
Rubber products	D	D	D	D	D	D	D	D	D	D	D	D
Stone, clay, and glass products	$9	$10	$17	$22	D	D	D	$9	$38	$6	$5	$2
Primary metals	D	$19	$17	D	D	$8	S	$23	$17	$13	D	$221
Ferrous metals and products	D	D	$1	D	D	$1	D	$17	D	D	D	D
Nonferrous metals and products	$8	D	$16	D	D	$7	D	$6	D	D	D	D
Fabricated metal products	$95	$150	$163	$178	$203	$226	$294	$222	$243	$86	D	$129
Machinery	D	D	D	D	$871	$1,055	$1,035	$86	$99	D	$117	$106
Office, computing, and accounting machines	D	D	D	D	D	D	D	$33	28	D	D	$53
Other machinery, except electrical	$75	$44	$101	$112	D	D	D	$53	$71	$64	D	$53
Electrical equipment	$5,213	$5,399	$4,153	$3,743	$4,133	$4,550	$3,844	$1,667	$1,801	$1,690	$2,143	$1,839
Radio and TV receiving equipment	0	0	0	0	0	0	D	D	D	D	D	D
Communication equipment	$4,552	$4,729	$3,630	$2,911	$2,344	D	D	D	D	D	D	D
Electronic components	D	$656	$449	$369	$418	D	$247	$206	$162	D	D	D
Other electrical equipment	D	$14	$74	$463	$1,371	D	D	D	D	D	D	$477
Transportation equipment	$17,708	$20,784	$20,865	$19,262	$17,097	$12,570	$11,202	$10,617	$10,392	$13,130	$12,202	$12,251
Motor vehicles and motor vehicles equipment	D	D	D	D	D	D	D	D	D	D	D	D
Other transportation equipment	D	D	D	D	D	D	S	D	D	D	D	D
Aircraft and missiles	$14,984	$18,519	$18,402	$16,828	$15,248	$11,096	S	$9,372	$8,794	$11,462	$10,515	$10,619
Professional and scientific instruments	$351	$272	$191	$263	$737	$1,865	$2,221	$2,577	$3,384	$3,460 S	$3,942	$4,499
Scientific and mechanical measuring instruments	D	D	S	S	S	D	$2,143	$2,484	$3,266	$3,358 S	D	$4,416
Optical, surgical, photographic, and other instruments	D	D	$95	$101	$87	D	$78	$92	$118	$102	D	$84
Other manufacturing industries[1]	$2	D	D	D	D	D	$61	D	D	D	D	$156
Distribution by industry:												
Nonmanufacturing[2]	$2,706	$2,700	$3,256	$3,729	$4,442	$5,505	$5,570	$6,140	$5,090	$4,620	$3,633	$4,105
Transportation and utilities	NA	NA	NA	NA	NA	NA	NA	NA	NA	$252	$186	$200
Communications	NA	NA	NA	NA	NA	NA	NA	NA	NA	D	D	D
Telephone communications	NA	NA	NA	NA	NA	NA	NA	NA	NA	D	D	D
Other communications	NA	NA	NA	NA	NA	NA	NA	NA	NA	0	0	0

1.4. Federal Funds for Industrial R&D Performance, by Industry and Size of Company, United States, 1986–97 (millions of dollars) *(continued)*.

Industry and size of company	1986	1987	1988	1989	1990	1991	1992	1993	1994	1995	1996	1997
Electric, gas, and sanitary services	NA	NA	NA	NA	NA	NA	NA	NA	NA	$93	$42	D
Other transportation and utilities	NA	NA	NA	NA	NA	NA	NA	NA	NA	D	D	D
Trade	NA	NA	NA	NA	NA	NA	NA	NA	NA	D	D	D
Finance, insurance, and real estate	NA	NA	NA	NA	NA	NA	NA	NA	NA	D	D	D
Services	NA	NA	NA	NA	NA	NA	NA	NA	NA	$4,270	$3,118	$3,805
Business services	NA	NA	NA	NA	NA	NA	NA	NA	NA	$612	$361	$400
Computer and data processing services	NA	NA	NA	NA	NA	NA	NA	NA	NA	$514	D	$389
Other business services	NA	NA	NA	NA	NA	NA	NA	NA	NA	$98	D	$11
Health services	NA	NA	NA	NA	NA	NA	NA	NA	NA	$4	D	D
Offices and clinics of medical doctors, hospitals, medical and dental labs	NA	NA	NA	NA	NA	NA	NA	NA	NA	D	$3	$4
Other health services	NA	NA	NA	NA	NA	NA	NA	NA	NA	D	D	D
Engineering and management services	NA	NA	NA	NA	NA	NA	NA	NA	NA	$3,650	$2,746	$3,381 S
Engineering, architectural, and surveying	NA	NA	NA	NA	NA	NA	NA	NA	NA	$1,883	$994	$1,258
Research, development, and testing	NA	NA	NA	NA	NA	NA	NA	NA	NA	D	$1,708	D
Other engineering and management services	NA	NA	NA	NA	NA	NA	NA	NA	NA	D	$45	D
Other services	NA	NA	NA	NA	NA	NA	NA	NA	NA	$4	D	D
Other nonmanufacturing industries[1]	NA	NA	NA	NA	NA	NA	NA	NA	NA	$11	D	$77
Distribution by size of company:												
[Number of employees]												
Total	$27,891	$30,752	$30,343	$28,554	$28,125	$26,372	$24,722	$22,809	$22,463	$23,451	$23,653	$23,928
Fewer than 500	$868	$963	$816	$901	$895	$1,887	$2,025	$1,614	$1,164	$1,978	$2,301	$2,209
500 to 999	$137	$115	$131	$117	S	$181	$151	$182	$182	$225	$219	$376
1,000 to 4,999	$1,229	$981	$1,093	$958	$881	$1,050	S	$1,115	$1,083	$798	$512	$540
5,000 to 9,999	$796	$748	$864	$740	$257	$816	$763	$764	$825	$243	$468	$612
10,000 to 24,999	$2,004	$2,362	$1,705	$1,129	$1,526	$3,373	$3,416	$2,816	$2,348	$1,946	$1,031	$913
25,000 or more	$23,213	$25,583	$25,734	$24,709	$24,436	$19,065	$17,346	$16,319	$16,862	$18,261	$19,122	$19,277

Notes:

[1] Beginning in 1996 manufacturing companies with 50 or fewer employees and nonmanufacturing companies with 15 or fewer employees were sampled separately without regard to industry classification to minimize year-to-year variation in survey estimates. Estimates for manufacturing companies in this group are combined with those for companies in "Other manufacturing industries." Estimates for nonmanufacturing companies in this group are combined with those for companies in "Other nonmanufacturing industries." As a result, statistics for "Other manufacturing industries" and for "Other nonmanufacturing industries" for 1996 are not comparable with statistics for prior years.

[2] Until 1995, data were not broken down below this level of detail for most nonmanufacturing industries.

D = Data have been withheld to avoid disclosing operations of individual companies.

S = Indicates imputation of more than 50 percent. For years prior to 1993, data have been withheld. When a company returns an incomplete form, missing information is imputed, i.e., ascribed, based on the information that is provided, prior-year submissions, and information provided by other companies within the industry. When a company fails to return a form, information is imputed from prior years, adjusted to reflect changes noted among responding companies. If no previous year information is available for the company, data are imputed from other companies in the same industry.

NA = Not available.

As a result of a new sample design, statistics for 1988–91 have been revised since originally published. These statistics now better reflect R&D performance among firms in the nonmanufacturing industries and small firms in all industries. As a result of the new sample design, statistics for 1991 and later years are not directly comparable with statistics for 1990 and earlier years.

Source: National Science Foundation, Division of Science Resources Studies. *Research and Development in Industry: 1997 (Early Release Tables).* Arlington, VA: National Science Foundation, 1999. Table A-13. http://www.nsf.gov/sbe/srs/indus/start.htm

1.5. Total (Company, Federal, and Other) Funds for Performance of Basic Research, Applied Research, and Development, in Current and in Constant Dollars, United States, 1953–97 (millions of dollars).

Year	Total		Basic research		Applied research		Development	
	Current dollars	Constant 1992 dollars	Current dollars	Constant 1992 dollars	Current dollars	Constant 1992 dollars	Current dollars	Constant 1992 dollars
1953	$3,630	$17,988	$151	$748	$726	$3,598	$2,753	$13,642
1954	$4,070	$19,941	$166	$813	$814	$3,988	$3,090	$15,140
1955	$4,640	$22,372	$189	$911	$928	$4,474	$3,523	$16,986
1956	$6,605	$30,764	$253	$1,178	$1,268	$5,906	$5,084	$23,680
1957	$7,731	$34,856	$271	$1,222	$1,670	$7,529	$5,790	$26,105
1958	$8,389	$36,940	$295	$1,299	$1,911	$8,415	$6,183	$27,226
1959	$9,618	$41,908	$320	$1,394	$1,991	$8,675	$7,307	$31,839
1960	$10,509	$45,161	$376	$1,616	$2,029	$8,719	$8,104	$34,826
1961	$10,908	$46,338	$395	$1,678	$1,977	$8,398	$8,536	$36,262
1962	$11,464	$48,087	$488	$2,047	$2,449	$10,273	$8,527	$35,768
1963	$12,630	$52,363	$522	$2,164	$2,457	$10,187	$9,651	$40,012
1964	$13,512	$55,196	$549	$2,243	$2,600	$10,621	$10,363	$42,333
1965	$14,185	$56,831	$592	$2,372	$2,658	$10,649	$10,935	$43,810
1966	$15,548	$60,569	$624	$2,431	$2,843	$11,075	$12,081	$47,063
1967	$16,385	$61,854	$629	$2,374	$2,915	$11,004	$12,841	$48,475
1968	$17,429	$63,057	$642	$2,323	$3,124	$11,302	$13,663	$49,432
1969	$18,308	$63,262	$618	$2,135	$3,287	$11,358	$14,403	$49,768
1970	$18,067	$59,275	$602	$1,975	$3,427	$11,243	$14,038	$46,056
1971	$18,320	$57,143	$590	$1,840	$3,415	$10,652	$14,315	$44,651
1972	$19,552	$58,504	$593	$1,774	$3,514	$10,515	$15,445	$46,215
1973	$21,249	$60,195	$631	$1,788	$3,825	$10,836	$16,793	$47,572
1974	$22,887	$59,493	$699	$1,817	$4,288	$11,146	$17,900	$46,530
1975	$24,187	$57,465	$730	$1,734	$4,570	$10,858	$18,887	$44,873
1976	$26,997	$60,599	$819	$1,838	$5,112	$11,475	$21,066	$47,286
1977	$29,825	$62,882	$911	$1,921	$5,636	$11,883	$23,278	$49,079
1978	$33,304	$65,443	$1,035	$2,034	$6,300	$12,380	$25,969	$51,030
1979	$38,226	$69,212	$1,158	$2,097	$7,225	$13,082	$29,843	$54,034
1980	$44,505	$73,769	$1,325	$2,196	$8,450	$14,006	$34,730	$57,567
1981	$51,810	$78,488	$1,614	$2,445	$10,699	$16,208	$39,497	$59,835
1982	$58,650	$83,583	$1,904	$2,713	$12,323	$17,562	$44,423	$63,308
1983	$65,268	$89,213	$2,223	$3,039	$13,927	$19,036	$49,118	$67,138
1984	$74,800	$98,525	$2,608	$3,435	$15,765	$20,765	$56,427	$74,324
1985	$84,239	$107,270	$2,862	$3,644	$18,255	$23,246	$63,122	$80,379
1986	$87,823	$108,989	$4,047	$5,022	$19,759	$24,521	$64,017	$79,445
1987	$92,155	$110,950	$4,324	$5,206	$19,813	$23,854	$68,018	$81,890
1988	$97,015	$112,690	$4,500	$5,227	$20,748	$24,100	$71,767	$83,363
1989	$102,055	$113,748	$5,216	$5,814	$22,691	$25,291	$74,148	$82,644
1990	$109,727	$117,230	$5,128	$5,479	$24,785	$26,480	$79,814	$85,271
1991	$116,952	$120,173	$7,837	$8,053	$27,446	$28,202	$81,669	$83,918
1992	$119,110	$119,110	$7,002	$7,002	$26,168	$26,168	$85,940	$85,940
1993	$117,400	$114,380	$6,919	$6,741	$24,686	$24,051	$85,796	$83,589
1994	$119,595	$113,802	$7,017	$6,677	$23,490	$22,352	$89,088	$84,773
1995	$132,103	$122,590	$6,099	$5,660	$27,454	$25,477	$98,552	$91,455
1996	$144,667	$131,265	$8,207	$7,447	$29,241	$26,532	$107,218	$97,285
1997	$157,539	$140,159	$10,419	$9,270	$32,642	$29,041	$114,478	$101,849

Notes:

The character-of-work estimation procedure, that is, how "basic research," "applied research," and "development" are defined, was revised for 1986 and later years; hence, these data are not directly comparable with data for 1985 and earlier years. The 1992 gross domestic product (GDP) implicit price deflators were used to convert current dollars to constant dollars. As a result of a new sample design, statistics for 1988-1991 have been revised since originally published. The statistics now better reflect R&D performance among firms in the nonmanufacturing industries and small firms in all industries. As a result of the new sample design, statistics for 1991 and later years are not directly comparable to statistics for 1990 and earlier years.

Source: National Science Foundation, Division of Science Resources Studies. *Research and Development in Industry: 1997 (Early Release Tables).* Arlington, VA: National Science Foundation, 1999. Table A-26. http://www.nsf.gov/sbe/srs/indus/start.htm

1.6. Number of Full-Time-Equivalent (FTE) R&D Scientists and Engineers in R&D-Performing Companies, by Industry and Size of Company, United States, 1986–98 (in thousands).

Industry and size of company	1986	1987	1988	1989	1990	1991	1992	1993	1994	1995	1996	1997	Projected 1998
Total (January)	671.0	695.8	708.6	722.5	743.6	773.4	779.3	764.7	768.5	746.1	832.8	885.7	951.5
Total (Annual average)	683.4	702.2	715.6	733.0	758.5	776.4	772.0	766.6	758.8	789.4	859.3	918.6	NA
Distribution by industry:													
All Industries	671.0	695.8	708.6	722.5	743.6	773.4	779.3	764.7	768.5	746.1	832.8	885.7	951.5
Manufacturing	NA	NA	NA	NA	NA	NA	NA	NA	NA	547.2	606.2	639.9	682.1
Food, kindred, and tobacco products	S	S	S	S	S	9.4	9.8	9.6	10.3	8.9	9.8	10.4	11.0
Textiles and apparel	2.6	2.4	2.4	2.5	2.8	S	2.8	3.1	3.3	2.9	3.8	3.7	3.2
Lumber, wood products, and furniture	S	1.3	1.3	1.4	S	S	1.5	1.6	3.3	2.3	2.9	3.4	3.6
Paper and allied products	6.4	6.0	6.1	6.4	8.5	S	10.7	10.6	10.6	10.6	11.0 S	11.2 S	10.4
Chemicals and allied products	75.8	75.2	75.8	78.3	80.4	81.6	85.6	86.5	93.1	99.6	94.3	89.0	89.5
Industrial chemicals	24.9	S	S	S	S	S	29.9	26.4	28.8	33.7	28.4	31.2	26.7
Drugs and medicines	31.8	32.6	33.0	34.4	34.3	35.4	38.7	42.3	48.5	50.3	49.0	43.1	45.3
Other chemicals	19.1	20.2	20.3	18.8	18.9	17.0	17.0	17.8	15.8	15.6	16.9	14.8	17.5
Petroleum refining and extraction	10.4	9.9	9.5	10.7	11.1	11.4	11.5	11.0	9.7	8.4	8.4	9.0	7.9
Rubber products	S	S	S	S	S	S	14.8	13.0	9.0	9.8	10.2	9.9	11.5
Stone, clay, and glass products	7.5	8.6	8.6	7.6	7.0	6.0	5.3	5.1	4.0	4.3	3.9	3.7	3.9 S
Primary metals	5.7	5.5	5.6	5.5	5.2	4.6	5.1	4.6	5.1	6.5	4.1	5.5 S	6.4 S
Ferrous metals and products	2.5	S	S	2.3	S	S	S	1.6	2.6	1.9	1.6	1.6	4.0 S
Nonferrous metals and products	3.2	3.4	3.3	3.3	3.3	3.0	1.7	1.6	2.5	4.6	2.5	3.9 S	2.5
Fabricated metal products	S	9.9	10.5	9.9	10.1	S	8.7	7.9	10.2	9.2	9.1	10.0	12.8
Machinery	89.7	95.8	98.4	100.4	113.3	109.7	99.3	97.4	70.4	68.6	87.3	88.7	112.1
Office, computing, and accounting machines	71.9	73.4	74.4	75.0	84.7	77.6	67.1	65.8	34.6	31.8	33.7	45.3	66.1
Other machinery, except electrical	17.8	22.4	24.0	25.4	28.6	32.1	32.2	31.6	35.8	36.8	53.6	43.4	46.0
Electrical equipment	117.9	130.4	132.5	122.5	105.2	95.9	91.9	89.2	96.5	103.3	125.8	135.9	171.6
Radio and TV receiving equipment	1.8	1.2	1.3	1.5	0.8	1.0	1.0	1.0	0.8	0.9	1.3	1.6	1.2
Communication equipment	65.0	71.9	73.1	58.0	47.1	35.8	31.2	S	36.0	41.7	34.7 S	37.6 S	69.0
Electronic components	S	43.7	44.3	42.8	S	32.6	28.4	28.9	40.1	40.8	64.6 S	75.3 S	66.8 S
Other electrical equipment	16.5	13.6	S	S	21.3	S	31.2	28.8	19.6	20.0	25.3	21.4	34.6
Transportation equipment	179.2	187.3	188.2	185.4	170.2	149.7	141.1	147.5	129.6	120.8	156.6	161.8 S	144.1 S
Motor vehicles and motor vehicles equipment	33.9	46.5	47.3	45.8	49.4	45.3	44.5	45.1	51.0	51.1	57.0	63.8	64.1
Other transportation equipment	3.1 S	S	S	S	S	S	S	S	5.8	S	6.2	4.0	3.1
Aircraft and missiles	144.8	136.3	136.4	134.8	115.3	100.2	92.9	97.9	72.8	63.5	95.5	94.6 S	77.0 S
Professional and scientific instruments	S	S	S	S	S	S	S	S	100.6	84.4	70.9 S	64.3 S	66.3 S
Scientific and mechanical measuring instruments	S	S	S	S	S	S	S	S	66.4	63.2	48.2 S	42.6 S	43.7 S
Optical, surgical, photographic, and other													

1.6. Number of Full-Time-Equivalent (FTE) R&D Scientists and Engineers in R&D-Performing Companies, by Industry and Size of Company, United States, 1986–98 (in thousands) (continued).

Industry and size of company	1986	1987	1988	1989	1990	1991	1992	1993	1994	1995	1996	1997	Projected 1998
instruments	24.0	24.6	24.9	14.5	8.1	S	S	20.8	34.2	21.2	22.7	21.7	22.6
Other manufacturing industries[1]	S	6.3	6.4	5.4	5.6	S	6.0	5.8	15.5	7.5	8.2	33.4	27.8
Distribution by industry:													
Nonmanufacturing[2]	75.1	96.4	101.9	125.2	S	S	202.6	196.5	197.4	198.9	226.6	245.8	269.5
Transportation and utilities	NA	NA	NA	NA	NA	NA	NA	NA	NA	NA	29.3	29.4	15.2
Communications	NA	NA	NA	NA	NA	NA	NA	NA	NA	NA	25.4	24.7	8.4 S
Telephone communications	NA	NA	NA	NA	NA	NA	NA	NA	NA	NA	24.7	23.9	7.8 S
Other communications	NA	NA	NA	NA	NA	NA	NA	NA	NA	NA	0.8	0.8	0.6
Electric, gas, and sanitary services	NA	NA	NA	NA	NA	NA	NA	NA	NA	NA	3.2 S	4.0 S	1.1
Other transportation and utilities	NA	NA	NA	NA	NA	NA	NA	NA	NA	NA	0.7	0.8	5.7
Trade	NA	NA	NA	NA	NA	NA	NA	NA	NA	NA	47.5	39.6	46.4
Finance, insurance, and real estate	NA	NA	NA	NA	NA	NA	NA	NA	NA	NA	5.6	15.9	18.3
Services	NA	NA	NA	NA	NA	NA	NA	NA	NA	NA	139.6	143.9	168.7
Business services	NA	NA	NA	NA	NA	NA	NA	NA	NA	NA	80.2	87.8	105.1
Computer and data processing services	NA	NA	NA	NA	NA	NA	NA	NA	NA	NA	77.3	85.1	102.6
Other business services	NA	NA	NA	NA	NA	NA	NA	NA	NA	NA	3.0	2.8	2.5
Health services	NA	NA	NA	NA	NA	NA	NA	NA	NA	NA	5.0	6.6	4.7
Offices and clinics of medical doctors, hospitals, medical and dental labs	NA	NA	NA	NA	NA	NA	NA	NA	NA	NA	4.0	6.3	4.0
Other health services	NA	NA	NA	NA	NA	NA	NA	NA	NA	NA	0.6	0.3 S	0.7
Engineering and management services	NA	NA	NA	NA	NA	NA	NA	NA	NA	NA	52.9	45.3	54.5
Engineering, architectural, and surveying	NA	NA	NA	NA	NA	NA	NA	NA	NA	NA	22.8 S	18.0	20.9
Research, development, and testing	NA	NA	NA	NA	NA	NA	NA	NA	NA	NA	28.8	24.9	28.3
Other engineering and management services	NA	NA	NA	NA	NA	NA	NA	NA	NA	NA	1.4	2.3 S	5.3
Other services	NA	NA	NA	NA	NA	NA	NA	NA	NA	NA	1.5	4.2	4.4
Other nonmanufacturing industries[1]	NA	NA	NA	NA	NA	NA	NA	NA	NA	NA	4.6	17.0	20.8
Distribution by size of company: [Number of employees]													
Total	671.0	695.8	708.6	722.5	743.6	773.4	779.3	764.7	768.5	746.1	832.8	885.7	951.5
Fewer than 500	S	105.2	109.0	105.4	S	S	142.1	125.1	148.6	139.8	165.1	177.3	201.8
500 to 999	S	18.4	19.3	18.0	18.6	18.6	46.2	46.1	27.9	29.7	36.6	37.7	41.3
1,000 to 4,999	66.7	76.4	81.9	76.1	75.4	79.3	94.2	99.3	108.9	110.8	124.4	131.6	137.8
5,000 to 9,999	38.9	40.5	40.2	47.3	57.2	55.1	57.6	56.4	60.3	52.8	61.1	82.2	100.6
10,000 to 24,999	88.4	92.0	94.5	87.0	73.9	90.3	99.9	102.6	90.6	91.8	97.0	104.9	116.5
25,000 or more	365.3	363.3	363.7	388.7	404.2	408.4	339.2	335.2	332.1	321.2	348.6	351.9 S	353.5

Notes:

[1]Beginning in 1996 manufacturing companies with 50 or fewer employees and nonmanufacturing companies with 15 or fewer employees were sampled separately without regard to industry classification to minimize year-to-year variation in survey estimates. Estimates for manufacturing companies in this group are combined with those for companies in "Other manufacturing industries." Estimates for nonmanufacturing companies in this group are combined with those for companies in "Other nonmanufacturing industries." As a result, statistics for "Other manufacturing industries" and for "Other nonmanufacturing industries" for 1996 are not comparable with statistics for prior years.

[2]Until 1995, data were not broken down below this level of detail for most nonmanufacturing industries.

S = Indicates imputation of more than 50 percent. For years prior to 1993, data have been withheld. When a company returns an incomplete form, missing information is imputed, i.e., ascribed, based on the information that is provided, prior-year submissions, and information provided by other companies within the industry. When a company fails to return a form, information is imputed from prior years, adjusted to reflect changes noted among responding companies. If no previous year information is available for the company, data are imputed from other companies in the same industry.

NA = Not available.

As a result of a new sample design, statistics for 1988–91 have been revised since originally published. These statistics now better reflect R&D performance among firms in the nonmanufacturing industries and small firms in all industries. As a result of the new sample design, statistics for 1991 and later years are not directly comparable with statistics for 1990 and earlier years.

Source: National Science Foundation, Division of Science Resources Studies. *Research and Development in Industry: 1997 (Early Release Tables).* Arlington, VA: National Science Foundation, 1999. Table A-52. http://www.nsf.gov/sbe/srs/indus/start.htm

EMPLOYMENT IN SCIENCE AND TECHNOLOGY

1.7. Employment Status of Scientists and Engineers, by Broad Occupation and Highest Degree Received, United States, 1995.

Occupation	Total	Employed	%	Unemployed	%
All degree recipients					
All S&E occupations	3,256,200	3,185,600	98%	70,600	2%
Computer and math scientists	966,200	949,500	98%	16,800	2%
Life scientists	311,500	305,300	98%	6,200	2%
Physical scientists	281,800	274,300	97%	7,500	3%
Social scientists	321,400	317,500	99%	3,900	1%
Engineers	1,375,200	1,339,000	97%	36,200	3%
Bachelor's degree recipients					
All S&E occupations	1,883,400	1,844,000	98%	39,400	2%
Computer and math scientists	635,300	625,000	98%	10,400	2%
Life scientists	123,900	121,500	98%	2,400	2%
Physical scientists	131,000	128,100	98%	2,900	2%
Social scientists	61,600	60,600	98%	1,000	2%
Engineers	931,500	908,800	98%	22,800	2%
Master's degree recipients					
All S&E occupations	915,800	892,700	97%	23,000	3%
Computer and math scientists	273,600	268,000	98%	5,600	2%
Life scientists	65,200	64,000	98%	1,200	2%
Physical scientists	69,800	67,200	96%	2,700	4%
Social scientists	138,000	135,800	98%	2,300	2%
Engineers	369,100	357,900	97%	11,300	3%
Ph.D. degree recipients					
All S&E occupations	425,700	418,300	98%	7,500	2%
Computer and math scientists	54,600	53,800	99%	800	1%
Life scientists	104,500	102,400	98%	2,000	2%
Physical scientists	80,600	78,900	98%	1,800	2%
Social scientists	113,900	113,300	99%	700	1%
Engineers	72,100	69,900	97%	2,200	3%
Other professional degree recipients					
All S&E occupations	31,300	30,600	98%	700	2%
Computer and math scientists	2,700	2,700	100%	–	0%
Life scientists	17,900	17,400	97%	600	3%
Physical scientists	300	200	67%	100	33%
Social scientists	7,900	7,900	100%	–	0%
Engineers	2,500	2,500	100%	–	0%

Notes:
— = Weighted value of less than 50.
Totals may be slightly off due to independent rounding.

Source: National Science Foundation, Science Resources Studies Division. *SESTAT (Scientists and Engineers Statistics Data System) Surveys of Science and Engineering College Graduates 1995.* Arlington, VA: National Science Foundation. Cited in National Science Board. *Science and Engineering Indicators 1998.* Arlington, VA: National Science Foundation, 1998. Table 3-3.

1.8. Number of Employed Scientists and Engineers, by Occupation and Highest Degree Received, United States, 1995.

Occupation	Total	Bachelor's degree	Master's degree	Ph.D. degree	Other professional degree
All S&E occupations	3,185,600	1,844,000	892,700	418,300	30,600
Computer and math scientists	949,500	625,000	268,000	53,800	2,700
Computer and information scientists	839,600	595,200	219,800	22,100	2,500
Mathematicians	37,900	16,300	14,000	7,600	–
Postsecondary teachers	72,000	13,500	34,200	24,100	200
Life scientists	305,300	121,500	64,000	102,400	17,400
Agricultural scientists	43,400	24,700	9,300	9,300	100
Biological scientists	168,600	69,700	32,800	57,900	8,200
Environmental life scientists	20,100	13,400	5,800	900	–
Postsecondary teachers	73,200	13,600	16,200	34,400	9,000
Physical scientists	274,300	128,100	67,200	78,900	200
Chemists, except biochemists	111,400	65,300	20,200	25,900	100
Earth scientists	70,700	36,100	24,300	10,400	–
Physicists and astronomers	29,000	7,100	7,700	14,100	–
Other physical scientists	17,000	9,300	5,100	2,600	100
Postsecondary teachers	46,200	10,300	9,900	25,900	–
Social scientists	317,500	60,600	135,800	113,300	7,900
Economists	33,100	10,900	15,100	7,000	100
Political scientists	8,900	5,100	2,600	1,200	–
Psychologists	167,200	25,400	88,700	48,200	4,900
Sociologists and anthropologists	16,000	7,800	5,200	3,000	–
S&T historians and other social scientists	12,600	3,800	4,900	3,100	900
Postsecondary teachers	79,700	7,600	19,300	50,800	2,000
Engineers	1,339,000	908,800	357,900	69,900	2,500
Aerospace engineers	72,800	42,600	25,900	4,200	100
Chemical engineers	71,100	46,100	19,200	5,800	–
Civil and architectural engineers	198,900	143,400	51,000	3,900	600
Electrical and related engineers	357,400	241,000	102,000	13,200	1,200
Industrial engineers	69,600	52,400	16,400	800	–
Mechanical engineers	255,100	191,600	55,000	8,100	300
Other engineers	282,800	186,300	79,900	16,400	200
Postsecondary teachers	31,300	5,300	8,400	17,500	–

Notes:
— = Weighted value of less than 50.
S&E = Science and engineering.
S&T = Science and technology.

Source: National Science Foundation, Science Resources Studies Division. *SESTAT (Scientists and Engineers Statistics Data System) Surveys of Science and Engineering College Graduates 1995.* Arlington, VA: National Science Foundation. Cited in National Science Board. *Science and Engineering Indicators 1998.* Arlington, VA: National Science Foundation, 1998. Table 3-4.

1.9. Cost per R&D Scientist or Engineer in R&D-Performing Companies, by Industry and Size of Company, United States, 1986–97 (cost in dollars).

Industry and size of company	1986	1987	1988	1989	1990	1991	1992	1993	1994	1995	1996	1997
Distribution by industry:												
All Industries	$128,500	$128,800	$132,300	$134,500	$141,300	$148,600	$157,912	$153,336	$157,601	$167,339	$168,362	$171,495
Manufacturing	NA	NA	NA	NA	NA	NA	NA	NA	NA	$173,523	$179,538	$183,097
Food, kindred, and tobacco products	D	S	D	D	D	$130,000	$144,215	$135,261	$153,609	$167,017	$154,853	$166,821
Textiles and apparel	D	D	D	D	D	D	D	D	D	D	D	D
Lumber, wood products, and furniture	S	S	D	S	S	D	D	D	D	D	D	$99,055
Paper and allied products	D	D	D	S	S	D	D	D	D	D	D	D
Chemicals and allied products	$117,100	$125,000	$139,400	$149,000	$159,000	$167,600	$183,508	D	D	$180,971	D	D
Industrial chemicals	$150,200	S	S	S	S	$181,000	$193,871	D	D	D	D	D
Drugs and medicines	$113,600	D	$142,800	D	D	D	$196,631	$202,607	$195,145	$205,637	$212,185	$262,201
Other chemicals	$83,100	D	$105,600	D	D	D	$135,423	D	D	D	$158,355	D
Petroleum refining and extraction	D	$187,400	$182,700	$194,000	$201,800	$217,000	$205,318	$208,360	$209,439	$209,495	$189,981	D
Rubber products	$118,000	$122,500	D	D	D	D	D	D	D	D	D	D
Stone, clay, and glass products	D	D	D	D	D	$142,200	D	$137,990	$141,842	$110,172	$124,685	$160,632 S
Primary metals	$138,800	$131,500	$118,600	D	D	D	$114,057	$138,552	$116,556	$111,743	D	$165,397 S
Ferrous metals and products	S	D	S	D	D	D	D	$138,851	D	D	D	D
Nonferrous metals and products	D	D	S	D	D	D	D	$138,326	S	D	D	$158,211
Fabricated metal products	S	$76,600	$88,100	$130,000	$138,500	$115,500	$123,828	$129,131	$113,168	$112,010	$152,894	$184,263
Machinery	D	D	D	D	$147,800	$147,800	$152,797	$100,631	$118,717	D	D	$230,471
Office, computing, and accounting machines	$119,200	$98,300	$99,400	D	D	D	D	$99,564	$126,476	D	D	$126,651
Other machinery, except electrical	$120,700	$124,300	$124,100	D	D	D	D	$102,213	$111,691	$111,421	D	$159,879
Electrical equipment	S	$97,900	$124,000	$132,400	$142,800	$147,100	$150,118	$144,725	$151,872	$163,626	$171,907	D
Radio and TV receiving equipment	$88,700	D	S	$103,400	$113,200	S	D	D	D	D	D	D
Communication equipment	$141,300	$155,300	$160,300	$170,600	$177,100	S	D	D	D	D	D	D
Electronic components	D	D	S	S	$128,400	D	$127,801	$154,366	$148,761	D	D	$175,531
Other electrical equipment	D	$98,400	$68,500	S	S	D	D	D	D	D	D	
Transportation equipment	$170,700	$183,400	$195,600	$211,700	$215,700	$189,400	$191,274	$196,777	$228,881	$233,968	$205,665 S	$209,164 S
Motor vehicles and motor vehicles equipment	D	D	D	D	D	D	D	D	D	D	D	D
Other transportation equipment.	S	S	D	D	D	D	D	D	D	D	D	D
Aircraft and missiles	$149,800	$180,400	$193,300	$207,300	$213,700	$177,000	$180,552	$176,450	$217,219	$213,328	$170,733 S	$189,972 S
Professional and scientific instruments	S	S	S	S	S	S	S	S	$119,693 S	$154,251	$179,723 S	$206,112 S
Scientific and mechanical measuring instruments							S	$115,180 S	$102,570 S	$128,237	D	$188,560 S
Optical, surgical, photographic, and other instruments			$316,700				S	$94,036 S				$240,293
Other manufacturing industries[1]	S	D	D	D	D	D	S	$161,721	$161,435	$220,383	D	$91,458
Distribution by industry:												
Nonmanufacturing[2]	$86,800	$69,100	$74,900	S	S	$137,400	$152,411	$159,188	$142,125	$150,578	$138,882	$141,730
Transportation and utilities	NA	NA	NA	NA	NA	NA	NA	NA	NA	$370,933	$159,442	$135,133
Communications	NA	NA	NA	NA	NA	NA	NA	NA	NA	D	D	D
Telephone communications	NA	NA	NA	NA	NA	NA	NA	NA	NA	D	D	D
Other communications	NA	NA	NA	NA	NA	NA	NA	NA	NA	$157,435	$94,495	$84,028
Electric, gas, and sanitary services	NA	NA	NA	NA	NA	NA	NA	NA	NA	$274,704 S	$98,427 S	D
Other transportation and utilities	NA	NA	NA	NA	NA	NA	NA	NA	NA	D	D	D
Trade	NA	NA	NA	NA	NA	NA	NA	NA	NA	D	$146,774	D

1.9. Cost per R&D Scientist or Engineer in R&D-Performing Companies, by Industry and Size of Company, United States, 1986–97 (cost in dollars) (continued).

Industry and Size of Company	1986	1987	1988	1989	1990	1991	1992	1993	1994	1995	1996	1997
Finance, insurance, and real estate	NA	NA	NA	NA	NA	NA	NA	NA	NA	D	D	D
Services	NA	NA	NA	NA	NA	NA	NA	NA	NA	$256,085	$134,175	$143,277
Business services	NA	NA	NA	NA	NA	NA	NA	NA	NA	$231,680	$126,625	$123,974
Computer and data processing services	NA	NA	NA	NA	NA	NA	NA	NA	NA	$234,483	D	$124,766
Other business services	NA	NA	NA	NA	NA	NA	NA	NA	NA	$158,453	D	$95,896
Health services	NA	NA	NA	NA	NA	NA	NA	NA	NA	$305,132	D	D
Offices and clinics of medical doctors, hospitals, medical and dental labs	NA	NA	NA	NA	NA	NA	NA	NA	NA	D	$133,871 S	D
Other health services	NA	NA	NA	NA	NA	NA	NA	NA	NA	D	D	$100,372
Engineering and management services	NA	NA	NA	NA	NA	NA	NA	NA	NA	$289,459	$149,040	$186,234
Engineering, architectural, and surveying	NA	NA	NA	NA	NA	NA	NA	NA	NA	$257,067 S	$81,271 S	$104,608
Research, development, and testing	NA	NA	NA	NA	NA	NA	NA	NA	NA	D	$204,445	D
Other engineering and management services	NA	NA	NA	NA	NA	NA	NA	NA	NA	D	$93,929 S	D
Other services	NA	NA	NA	NA	NA	NA	NA	NA	NA	$221,147	D	D
Other nonmanufacturing industries[1]	NA	NA	NA	NA	NA	NA	NA	NA	NA	$179,020	D	$85,625
Distribution by size of company [Number of employees]												
Total	$128,500	$128,800	$132,300	$134,500	$141,300	$148,600	$157,912	$153,336	$157,601	$167,339	$168,362	$171,495
Fewer than 500	$70,600	$66,800	$67,100	$66,200	S	$103,400	$111,358	$106,888	$85,793	$109,319	$118,280	$126,925
500 to 999	$103,400	$92,600	S	$98,200	S	$178,700	$176,585	$87,233	$138,366	$141,660	$124,765	$125,611
1,000 to 4,999	$106,400	$91,900	$100,600	$101,900	$91,400	$119,400	$127,135	$128,565	$139,693	$144,213	$142,747	$145,436
5,000 to 9,999	$108,400	$102,900	$100,400	$102,500	$110,800	$150,800	$148,285	$156,899	$152,227	$167,284	$161,035	$156,090
10,000 to 24,999	$122,300	$132,700	$143,000	$127,300	$135,300	$167,300	$156,190	$160,229	$182,243	$180,849	$199,722	$194,323
25,000 or more	$154,300	$157,700	$160,700	$168,200	$193,700	$180,100	$182,839	$184,804	$195,890	$200,606	$199,286 S	$207,375

Notes:

[1] Beginning in 1996 manufacturing companies with 50 or fewer employees and nonmanufacturing companies with 15 or fewer employees were sampled separately without regard to industry classification to minimize year-to-year variation in survey estimates. Estimates for manufacturing companies in this group are combined with those for companies in "Other manufacturing industries." Estimates for nonmanufacturing companies in this group are combined with those for companies in "Other nonmanufacturing industries." As a result, statistics for "Other manufacturing industries" and for "Other nonmanufacturing industries" for 1996 are not comparable with statistics for prior years.

[2] Until 1995, data were not broken down below this level of detail for most nonmanufacturing industries.

D = Data have been withheld to avoid disclosing operations of individual companies.

S = Indicates imputation of more than 50 percent. For years prior to 1993, data have been withheld. When a company returns an incomplete form, missing information is imputed, i.e., ascribed, based on the information that is provided, prior-year submissions, and information provided by other companies within the industry. When a company fails to return a form, information is imputed from prior years, adjusted to reflect changes noted among responding companies. If no previous year information is available for the company, data are imputed from other companies in the same industry.

NA = Not available.

The number of full-time-equivalent R&D scientists and engineers used to estimate the cost per R&D scientist or engineer is the arithmetic mean of the numbers of R&D scientists and engineers reported for January in two consecutive years. This number is then divided into the total R&D expenditures of the earlier years, and the ratio is attributed to the earlier year. As a result of a new sample design, statistics for 1988–91 have been revised since originally published. These statistics now better reflect R&D performance among firms in the nonmanufacturing industries and small firms in all industries. As a result of the new sample design, statistics for 1991 and later years are not directly comparable with statistics for 1990 and earlier years.

Source: National Science Foundation, Division of Science Resources Studies. *Research and Development in Industry: 1997 (Early Release Tables)*. Arlington, VA: National Science Foundation, 1999. Table A-54. http://www.nsf.gov/sbe/srs/indus/start.htm

PATENTS, HIGH-TECH TRADE, AND HIGH-TECH COMPANY FORMATION

1.10. Patents Granted by Date of Patent Grant (Granted January 1, 1963–December 31, 1998) (number of patents).

	Pre-1984	1984	1985	1986	1987	1988	1989	1990	1991	1992	1993	1994	1995	1996	1997	1998	Total
Total	1,349,397	67,200	71,661	70,860	82,952	77,924	95,537	90,364	96,513	97,444	98,342	101,676	101,419	109,646	111,983	147,520	2,770,438
U.S. Origin	931,868	38,367	39,555	38,126	43,520	40,496	50,186	47,390	51,179	52,253	53,231	56,066	55,739	61,104	61,707	80,294	1,701,087
Foreign Origin	417,529	28,833	32,106	32,734	39,432	37,428	45,351	42,974	45,334	45,191	45,111	45,610	45,680	48,542	50,276	67,226	1,069,351
Japan	94,400	11,110	12,746	13,209	16,557	16,158	20,168	19,525	21,026	21,925	22,293	22,384	21,764	23,053	23,179	30,841	390,338
Germany	102,515	6,323	6,718	6,856	7,884	7,353	8,352	7,614	7,680	7,309	6,893	6,731	6,600	6,818	7,008	9,095	211,758
United Kingdom	55,205	2,269	2,494	2,405	2,775	2,579	3,094	2,789	2,800	2,425	2,295	2,234	2,478	2,453	2,678	3,464	94,440
France	39,117	2,162	2,400	2,369	2,874	2,661	3,140	2,866	3,030	3,029	2,909	2,779	2,821	2,788	2,958	3,674	81,578
Canada	22,241	1,206	1,342	1,314	1,594	1,489	1,959	1,859	2,036	1,964	1,944	2,008	2,104	2,233	2,379	2,974	50,646
Switzerland	23,786	1,174	1,233	1,211	1,374	1,245	1,363	1,284	1,335	1,197	1,127	1,169	1,056	1,112	1,090	1,278	42,034
Italy	13,337	794	919	995	1,183	1,076	1,297	1,259	1,209	1,271	1,285	1,215	1,078	1,200	1,239	1,582	30,939
Sweden	14,678	701	857	883	948	777	837	768	716	626	636	706	806	854	867	1,225	26,885
Netherlands	12,349	726	766	722	922	806	1,060	960	992	855	800	852	799	797	808	1,226	25,440
Taiwan	469	99	174	208	343	457	591	732	906	1,001	1,189	1,443	1,620	1,897	2,057	3,100	16,286
South Korea	142	30	41	46	84	97	159	225	405	538	779	943	1,161	1,493	1,891	3,259	11,293
Australia	4,087	292	340	374	389	416	502	432	463	410	378	467	459	472	478	720	10,679
Belgium	4,888	240	240	243	295	302	359	313	324	325	350	352	397	488	515	693	10,324
Austria	4,580	256	318	357	345	337	401	393	359	370	312	289	337	361	376	387	9,781
USSR	5,560	214	147	116	121	96	161	174	178	66	65	53	12	16	4	6	9,781
Israel	1,540	162	179	189	245	238	325	299	304	335	314	350	384	484	452	595	6,635
Finland	1,571	167	200	210	275	232	230	304	331	361	293	312	358	444	333	391	6,335
Denmark	2,767	150	187	182	204	151	221	158	210	193	197	207	199	241	177	248	5,991
Norway	1,546	87	90	81	135	121	126	112	111	108	117	126	130	139	142	198	3,379
Spain	1,318	69	78	97	115	126	131	130	153	133	158	141	148	157	101	115	3,369
South Africa	1,371	82	96	88	107	103	134	114	105	97	93	101	123	111	101	50	2,941
Hungary	1,119	111	108	131	127	94	129	93	85	97	61	46	50	43	25	9	2,360
Czechoslovakia	1,725	33	54	35	46	33	34	39	27	17	13	19	15	8	9	57	2,116
Mexico	1,143	42	32	37	49	44	39	32	29	39	45	44	40	39	45	114	1,756
New Zealand	534	50	33	52	68	55	58	51	41	44	39	37	44	52	85	160	1,357
Ireland	310	29	30	28	38	43	65	54	56	55	53	50	50	78	81	74	1,174
Hong Kong	275	24	25	31	34	29	48	52	50	60	60	57	86	88	73	74	1,086
Brazil	376	20	30	27	34	41	36	41	62	40	57	60	63	63	62	43	1,074
Argentina	456	20	11	17	18	16	20	17	16	20	24	32	31	30	35	72	806
Poland	497	15	11	14	13	8	14	17	8	5	8	8	8	15	11	15	721
People's Rep. of China	108	2	1	9	23	47	52	47	50	41	53	48	62	46	47	85	667
Luxembourg	251	24	37	31	22	29	29	17	27	26	28	22	24	18	22	20	659
India	249	12	10	18	12	14	14	23	22	24	30	27	37	35	94	120	627
Liechtenstein	305	16	13	17	16	10	11	15	11	16	11	17	17	12	111	189	603
Bulgaria	278	22	21	21	32	23	16	27	10	5	5	4	1	1	11	16	555
Venezuela	150	11	15	21	24	20	23	20	25	22	31	23	29	25	237	303	514
Others (121)	2,286	89	110	90	107	102	153	119	142	151	166	254	289	378			5,216
Ownership:																	
U.S. Corporations	705,979	29,999	31,181	29,490	33,726	31,437	38,664	36,093	39,133	40,308	41,826	44,036	44,035	48,741	50,220	66,062	1,310,929
U.S. Government	31,676	1,235	1,139	1,022	981	733	880	983	1,183	1,161	1,166	1,258	1,028	923	944	1,018	47,330
U.S. Individuals	227,405	8,911	9,265	9,477	10,887	10,122	13,028	12,542	13,207	12,751	12,281	12,805	12,885	13,729	12,914	16,407	408,617
Foreign Corporations	307,416	23,238	25,957	26,545	32,371	30,960	37,506	35,548	37,594	38,239	38,401	38,788	38,688	41,476	42,907	57,668	853,301
Foreign Government	4,456	438	483	479	555	453	441	423	472	463	434	296	245	259	273	256	10,426
Foreign Individuals	72,465	3,379	3,636	3,847	4,432	4,219	5,018	4,775	4,924	4,522	4,234	4,493	4,538	4,518	4,725	6,109	139,835

Notes:
The location of patent origin is determined by the residence of the first-named inventor at the time of grant. The U.S. and Foreign Corporation ownership categories count predominantly corporate patents; however, patents assigned to other organizations such as small businesses, nonprofit organizations, universities, etc., are also included in this category.

Source: U.S. Patent and Trademark Office, Office of Electronic Information Products, Office of Electronic Information Products/TAF Program. *All Technologies Report, January 1963—December 1998.* Springfield, VA: National Technical Information Service, March 1999. Page A1-1. http://www.uspto.gov/web/offices/ac/ido/oeip/taf/all_tech.pdf

1.11. Patents Granted by Date of Patent Grant (Granted January 1, 1963–December 31, 1998) (percent of patents).

	Pre-1984	1984	1985	1986	1987	1988	1989	1990	1991	1992	1993	1994	1995	1996	1997	1998	Total
Total	100	100	100	100	100	100	100	100	100	100	100	100	100	100	100	100	100
U.S. Origin	69	57	55	54	52	52	53	52	53	54	54	55	55	56	55	54	61
Foreign Origin	31	43	45	46	48	48	47	48	47	46	46	45	45	44	45	46	39
Japan	7	17	18	19	20	21	21	22	22	23	23	22	21	21	21	21	14
Germany	8	9	9	10	10	9	9	8	8	8	7	7	7	6	6	6	8
United Kingdom	4	3	3	3	3	3	3	3	3	2	3	2	2	2	2	2	3
France	3	3	3	3	3	3	3	3	3	2	3	3	3	3	3	2	3
Canada	2	2	2	2	2	2	2	2	2	2	2	2	2	2	2	2	2
Switzerland	2	2	2	2	2	2	2	2	2	2	1	1	1	1	1	1	2
Italy	1	1	1	1	1	1	1	1	1	1	1	1	1	1	1	1	1
Sweden	1	1	1	1	1	1	1	1	1	1	1	1	1	1	1	2	1
Netherlands	1	1	1	1	1	1	1	1	1	1	1	1	1	1	1	2	1
Taiwan	—	—	—	—	1	1	1	1	1	1	1	1	2	2	2	2	1
Australia	—	—	—	—	—	—	—	—	—	—	—	—	—	—	—	—	0
Belgium	—	—	—	1	—	—	—	—	—	—	—	—	—	—	—	—	0
Austria	—	—	—	1	—	—	—	—	—	—	—	—	—	—	—	—	0
USSR	—	—	—	—	—	—	—	—	—	—	—	—	—	—	—	—	0
South Korea	—	—	—	—	—	—	—	—	—	—	1	1	1	1	2	2	0
Israel	—	—	—	—	—	—	—	—	—	—	—	1	1	1	1	1	0
Finland	—	—	—	—	—	—	—	—	—	—	—	—	—	—	—	—	0
Denmark	—	—	—	—	—	—	—	—	—	—	—	—	—	—	—	—	0
Norway	—	—	—	—	—	—	—	—	—	—	—	—	—	—	—	—	0
Spain	—	—	—	—	—	—	—	—	—	—	—	—	—	—	—	—	0
South Africa	—	—	—	—	—	—	—	—	—	—	—	—	—	—	—	—	0
Hungary	—	—	—	—	—	—	—	—	—	—	—	—	—	—	—	—	0
Czechoslovakia	—	—	—	—	—	—	—	—	—	—	—	—	—	—	—	—	0
Mexico	—	—	—	—	—	—	—	—	—	—	—	—	—	—	—	—	0
New Zealand	—	—	—	—	—	—	—	—	—	—	—	—	—	—	—	—	0
Ireland	—	—	—	—	—	—	—	—	—	—	—	—	—	—	—	—	0
Hong Kong	—	—	—	—	—	—	—	—	—	—	—	—	—	—	—	—	0
Brazil	—	—	—	—	—	—	—	—	—	—	—	—	—	—	—	—	0
Argentina	—	—	—	—	—	—	—	—	—	—	—	—	—	—	—	—	0
Poland	—	—	—	—	—	—	—	—	—	—	—	—	—	—	—	—	0
People's Rep. of China	—	—	—	—	—	—	—	—	—	—	—	—	—	—	—	—	0
Luxembourg	—	—	—	—	—	—	—	—	—	—	—	—	—	—	—	—	0
India	—	—	—	—	—	—	—	—	—	—	—	—	—	—	—	—	0
Liechtenstein	—	—	—	—	—	—	—	—	—	—	—	—	—	—	—	—	0
Bulgaria	—	—	—	—	—	—	—	—	—	—	—	—	—	—	—	—	0
Venezuela	—	—	—	—	—	—	—	—	—	—	—	—	—	—	—	—	0
Others (121)	—	—	—	—	—	—	—	—	—	—	—	—	—	—	—	—	0
Ownership:																	
U.S. Corporations	52	45	44	42	41	40	40	40	41	41	43	43	43	44	45	45	47
U.S. Government	2	2	2	1	1	1	1	1	1	1	1	1	1	1	1	1	2
U.S. Individuals	17	13	13	13	13	13	14	14	14	13	12	13	13	13	12	11	15
Foreign Corporations	23	35	36	37	39	40	39	39	39	39	39	38	38	38	38	39	31
Foreign Government	—	—	—	—	—	—	—	—	—	—	—	—	—	—	—	—	0
Foreign Individuals	5	5	5	5	5	5	5	5	5	5	4	4	4	4	4	4	5

Notes:
The location of patent origin is determined by the residence of the first-named inventor at the time of grant. The U.S. and Foreign Corporation ownership categories count predominantly corporate patents; however, patents assigned to other organizations such as small businesses, nonprofit organizations, universities, etc., are also included in this category.

Source: U.S. Patent and Trademark Office, Office of Electronic Information Products/TAF Program. *All Technologies Report, January 1963–December 1998.* Springfield, VA: National Technical Information Service, March 1999. Page A1-2. http://www.uspto.gov/web/offices/ac/ido/oeip/taf/all_tech.pdf

1.12. Ranked List of Some Organizations with 1,000 or More Patents, as Distributed by Year of Patent Grant, and as Distributed by Year of Patent Application (Granted January 1, 1969—December 31, 1998).

Organizational Patenting	Pre-1984	1984	1985	1986	1987	1988	1989	1990	1991	1992	1993	1994	1995	1996	1997	1998	Total
GENERAL ELECTRIC COMPANY																	
Patents by Year of Grant:	13,196	789	778	714	779	689	819	787	809	937	932	970	758	819	664	729	25,169
Patents by Year of Application:	14,749	726	696	664	707	781	821	900	883	949	858	855	804	591	181	0	25,169
INTERNATIONAL BUSINESS MACHINES CORP.																	
Patents by Year of Grant:	7,891	609	578	598	591	549	623	609	679	842	1,085	1,298	1,383	1,867	1,724	2,657	23,583
Patents by Year of Application:	9,307	487	509	547	541	637	708	874	1,220	1,468	1,196	1,734	2,258	1,666	427	4	23,583
HITACHI, LTD.																	
Patents by Year of Grant:	4,668	596	693	731	845	908	1,054	908	928	956	913	976	910	963	903	1,094	18,046
Patents by Year of Application:	6,250	682	720	827	924	1,052	1,064	1,020	997	908	892	948	985	611	165	1	18,046
CANON KABUSHIKI KAISHA																	
Patents by Year of Grant:	2,210	430	427	523	846	723	954	870	827	1,109	1,037	1,096	1,087	1,541	1,381	1,928	16,989
Patents by Year of Application:	3,155	585	652	787	710	813	960	989	1,162	1,218	1,247	1,706	1,718	893	394	0	16,989
TOSHIBA CORPORATION																	
Patents by Year of Grant:	2,357	543	698	694	824	751	962	893	1,014	1,023	1,039	968	969	914	862	1,170	15,681
Patents by Year of Application:	3,796	616	661	693	823	961	1,012	1,039	1,181	928	875	1,063	982	802	246	3	15,681
EASTMAN KODAK COMPANY																	
Patents by Year of Grant:	5,302	255	223	229	296	433	589	721	863	775	1,007	888	772	768	795	1,124	15,040
Patents by Year of Application:	5,733	174	289	346	459	696	833	908	927	1,026	762	745	746	1,003	390	3	15,040
AT&T CORPORATION																	
Patents by Year of Grant:	8,179	488	546	437	406	375	387	430	484	440	448	595	638	510	46	150	14,559
Patents by Year of Application:	9,335	386	361	360	308	335	509	492	461	564	690	470	181	83	24	0	14,559
U.S. PHILIPS CORPORATION																	
Patents by Year of Grant:	5,615	438	466	503	687	581	746	637	650	501	441	396	504	477	473	725	13,840
Patents by Year of Application:	6,651	474	534	600	569	654	668	553	492	438	438	546	587	483	153	0	13,840
E.I. DU PONT DE NEMOURS AND COMPANY																	
Patents by Year of Grant:	6,870	348	342	329	419	375	443	481	597	599	568	486	441	395	311	393	13,397
Patents by Year of Application:	7,502	322	380	394	406	434	573	640	551	503	443	413	413	294	128	1	13,397
SIEMENS AKTIENGESELLSCHAFT																	
Patents by Year of Grant:	5,564	406	418	410	539	562	658	508	475	398	371	376	419	418	454	626	12,602
Patents by Year of Application:	6,472	366	453	528	528	544	463	459	379	406	396	450	489	472	195	2	12,602
MOTOROLA, INC.																	
Patents by Year of Grant:	2,763	225	256	334	414	341	384	394	613	660	729	837	1,012	1,064	1,058	1,406	12,490
Patents by Year of Application:	3,328	253	287	370	330	361	581	682	782	833	940	1,180	1,302	975	285	1	12,490
MITSUBISHI DENKI KABUSHIKI KAISHA																	
Patents by Year of Grant:	970	285	364	359	518	543	770	868	940	959	926	972	973	934	892	1,080	12,354
Patents by Year of Application:	1,648	372	386	484	623	897	959	969	1,074	969	920	957	1,038	772	282	3	12,354
WESTINGHOUSE ELECTRIC CORP.																	
Patents by Year of Grant:	7,176	348	372	398	652	434	452	436	354	358	276	248	170	132	72	81	11,959
Patents by Year of Application:	7,983	455	454	390	392	445	403	375	305	228	194	182	86	50	17	0	11,959
BAYER AKTIENGESELLSCHAFT																	
Patents by Year of Grant:	5,819	363	359	389	371	441	470	499	492	472	443	342	327	323	357	381	11,848
Patents by Year of Application:	6,528	349	420	408	385	448	518	482	392	413	345	328	366	333	132	1	11,848
GENERAL MOTORS CORPORATION																	
Patents by Year of Grant:	6,171	324	286	294	370	383	412	379	437	399	438	331	282	297	277	305	11,385
Patents by Year of Application:	6,702	261	273	375	341	395	406	396	451	440	339	323	307	280	96	0	11,385
UNITED STATES OF AMERICA, NAVY																	
Patents by Year of Grant:	7,287	285	248	216	170	103	125	265	381	297	344	378	330	285	288	341	11,343
Patents by Year of Application:	8,172	175	179	82	84	160	195	348	307	346	337	289	304	248	99	2	11,343
XEROX CORPORATION																	
Patents by Year of Grant:	4,654	180	272	219	227	258	283	252	354	473	561	611	551	703	606	769	10,973

1.12. Ranked List of Some Organizations with 1000 or More Patents, as Distributed by Year of Patent Grant, and as Distributed by Year of Patent Application (Granted January 1, 1969 through December 31, 1998) *(continued)*.

Organizational Patenting	Pre-1984	1984	1985	1986	1987	1988	1989	1990	1991	1992	1993	1994	1995	1996	1997	1998	Total
Patents by Year of Application:	5,109	234	202	201	252	267	335	485	558	580	634	663	659	528	257	9	10,973
FUJI PHOTO FILM CO., LTD.																	
Patents by Year of Grant:	2,434	278	380	448	494	589	892	768	733	641	632	545	504	510	467	547	10,862
Patents by Year of Application:	3,039	470	526	577	713	790	749	659	647	567	505	509	509	464	137	0	10,862
MATSUSHITA ELECTRIC INDUSTRIAL CO., LTD.																	
Patents by Year of Grant:	2,711	233	249	224	305	277	365	343	456	608	713	771	854	841	746	1,034	10,730
Patents by Year of Application:	3,227	234	265	286	291	342	451	591	747	777	840	905	870	694	207	3	10,730
NEC CORPORATION																	
Patents by Year of Grant:	1,306	127	168	234	375	353	480	437	428	453	594	897	1,005	1,043	1,095	1,627	10,622
Patents by Year of Application:	1,714	247	307	311	360	461	469	420	609	883	772	1,325	1,150	1,185	404	5	10,622
DOW CHEMICAL COMPANY																	
Patents by Year of Grant:	5,143	328	336	371	469	421	431	400	332	320	328	267	226	196	163	174	9,905
Patents by Year of Application:	5,755	311	505	451	363	400	387	313	287	270	260	223	211	131	37	1	9,905
SONY CORPORATION																	
Patents by Year of Grant:	1,793	204	238	227	332	303	320	238	299	419	565	656	754	855	859	1,316	9,378
Patents by Year of Application:	2,326	210	274	275	260	295	300	385	556	709							9,378
CIBA-GEIGY CORPORATION																	
Patents by Year of Grant:	4,221	290	307	244	286	279	346	409	414	319	332	299	272	271	238	16	8,543
Patents by Year of Application:	4,784	252	274	294	324	384	359	351	317	288	290	308	277	40	1	0	8,543
MINNESOTA MINING AND MANUFACTURING CO.																	
Patents by Year of Grant:	2,762	193	228	232	256	279	381	323	365	388	414	519	564	537	548	554	8,543
Patents by Year of Application:	3,183	203	261	265	320	359	320	402	412	449	527	621	665	423	131	1	8,543
TEXAS INSTRUMENTS, INC.																	
Patents by Year of Grant:	2,655	169	194	178	277	195	400	285	373	394	369	479	527	600	607	611	8,313
Patents by Year of Application:	3,112	238	190	194	261	358	346	398	400	421	508	667	737	364	118	1	8,313
BASF AKTIENGESELLSCHAFT																	
Patents by Year of Grant:	3,092	260	265	250	263	297	354	394	384	396	360	318	301	317	276	399	7,926
Patents by Year of Application:	3,588	251	270	248	328	389	388	383	344	313	322	319	361	263	156	3	7,926
RCA CORPORATION																	
Patents by Year of Grant:	5,885	429	454	484	504	129	2	0	1	0	0	0	0	0	0	0	7,888
Patents by Year of Application:	6,839	379	394	237	38	0	1	0	0	0	0	0	0	0	0	0	7,888
FUJITSU LIMITED																	
Patents by Year of Grant:	765	179	183	204	241	245	283	261	358	412	452	592	724	869	903	1,189	7,860
Patents by Year of Application:	1,238	174	184	194	242	261	356	434	541	592	662	973	1,048	724	236	1	7,860
HOECHST AKTIENGESELLSCHAFT																	
Patents by Year of Grant:	3,334	223	237	231	234	227	260	282	282	331	356	335	311	351	353	327	7,674
Patents by Year of Application:	3,794	209	249	237	233	298	299	282	295	365	328	363	463	178	79	2	7,674
PHILLIPS PETROLEUM COMPANY																	
Patents by Year of Grant:	4,837	279	291	222	206	153	142	131	157	171	193	149	101	91	76	75	7,274
Patents by Year of Application:	5,352	256	240	147	116	126	127	181	169	169	114	92	114	52	19	0	7,274
UNITED STATES OF AMERICA, ARMY																	
Patents by Year of Grant:	4,721	187	205	198	169	104	163	132	171	144	153	186	172	137	168	165	7,175
Patents by Year of Application:	5,181	175	136	112	93	124	155	139	148	178	178	174	183	143	51	2	7,175
MOBIL OIL CORPORATION																	
Patents by Year of Grant:	2,803	373	351	341	292	248	376	355	366	299	284	286	202	167	91	67	6,901
Patents by Year of Application:	3,452	343	303	231	335	347	363	309	323	280	232	204	129	37	13	0	6,901
ROBERT BOSCH GMBH																	
Patents by Year of Grant:	2,588	295	314	241	262	223	200	204	261	278	363	316	312	306	267	348	6,778
Patents by Year of Application:	3,194	228	243	208	186	165	249	285	332	372	298	336	294	286	102	0	6,778
UNITED STATES OF AMERICA, DEPT. OF ENERGY																	
Patents by Year of Grant:	3,435	327	293	234	236	224	215	223	195	242	172	199	114	71	68	61	6,309
Patents by Year of Application:	4,076	253	198	208	168	203	227	210	220	171	164	76	68	58	9	0	6,309

1.12. Ranked List of Some Organizations with 1,000 or More Patents, as Distributed by Year of Patent Grant, and as Distributed by Year of Patent Application (Granted January 1, 1969, through December 31, 1998) *(continued)*.

Organizational Patenting	Pre-1984	1984	1985	1986	1987	1988	1989	1990	1991	1992	1993	1994	1995	1996	1997	1998	Total
UNISYS CORPORATION																	
Patents by Year of Grant:	4,237	278	285	181	136	97	95	82	99	85	51	80	112	133	109	122	6,182
Patents by Year of Application:	4,837	185	122	69	101	90	103	69	63	72	113	126	132	75	24	0	6,182
SHELL OIL COMPANY																	
Patents by Year of Grant:	2,957	181	206	203	158	206	370	263	267	270	223	226	161	125	123	143	6,082
Patents by Year of Application:	3,268	208	196	189	262	292	255	272	244	233	195	166	144	119	36	2	6,082
NISSAN MOTOR COMPANY, LTD																	
Patents by Year of Grant:	2,244	373	330	281	304	238	275	375	334	340	212	139	119	102	138	164	5,968
Patents by Year of Application:	2,996	280	252	254	212	326	348	317	283	159	109	118	149	128	37	0	5,968
ALLIED-SIGNAL, INC.																	
Patents by Year of Grant:	2,270	164	303	249	285	245	274	246	250	319	245	270	196	189	119	140	5,764
Patents by Year of Application:	2,749	254	274	208	231	248	271	277	308	257	215	163	145	119	44	0	5,764
EXXON RESEARCH + ENGINEERING CO.																	
Patents by Year of Grant:	3,477	256	268	223	178	106	125	126	136	111	109	127	89	72	73	105	5,581
Patents by Year of Application:	3,958	247	192	114	116	95	124	129	119	105	107	85	95	58	37	0	5,581
FORD MOTOR COMPANY																	
Patents by Year of Grant:	2,336	157	186	121	138	139	163	155	180	229	286	320	334	381	265	147	5,537
Patents by Year of Application:	2,657	111	135	116	175	154	146	212	246	302	393	360	339	139	50	2	5,537
HEWLETT-PACKARD COMPANY																	
Patents by Year of Grant:	799	63	56	77	130	153	242	218	318	344	338	428	470	501	530	805	5,472
Patents by Year of Application:	933	89	81	156	206	221	332	336	393	411	444	527	584	591	166	2	5,472
SHARP KABUSHKI KAISHA																	
Patents by Year of Grant:	531	111	123	189	284	237	339	350	385	393	373	454	388	352	395	560	5,464
Patents by Year of Application:	867	157	245	243	238	366	387	392	428	442	401	390	455	353	99	1	5,464
MONSANTO COMPANY																	
Patents by Year of Grant:	3,910	138	100	111	91	76	88	84	129	129	103	76	75	82	97	77	5,366
Patents by Year of Application:	4,162	111	94	77	65	93	125	114	93	86	82	81	107	50	26	0	5,366
RICOH COMPANY, LTD.																	
Patents by Year of Grant:	1,336	123	143	199	205	180	274	293	314	360	287	304	273	295	336	406	5,328
Patents by Year of Application:	1,654	154	181	183	184	306	333	382	306	316	290	331	339	284	84	1	5,328
UNITED TECHNOLOGIES CORPORATION																	
Patents by Year of Grant:	2,547	175	168	191	206	189	232	152	164	235	205	203	214	159	123	116	5,279
Patents by Year of Application:	2,964	173	147	171	211	199	169	177	225	217	196	206	127	74	22	1	5,279
HUGHES AIRCRAFT COMPANY																	
Patents by Year of Grant:	1,744	105	89	98	183	216	298	250	349	296	327	411	323	272	130	50	5,144
Patents by Year of Application:	1,977	129	156	220	218	275	287	309	347	411	344	251	166	48	2	0	5,144
IMPERIAL CHEMICAL INDUSTRIES PLC																	
Patents by Year of Grant:	3,130	137	131	110	110	162	157	178	181	189	232	128	93	66	61	63	5,128
Patents by Year of Application:	3,414	116	119	138	151	172	208	158	192	168	89	70	80	42	10	0	5,128
HONDA GIKEN KOGYO KABUSHIKI KAISHA																	
Patents by Year of Grant:	686	193	309	280	395	365	352	365	244	195	243	223	248	293	341	389	5,123
Patents by Year of Application:	1,208	249	329	345	361	340	317	185	217	254	239	297	393	282	106	1	5,123
TOYOTA JIDOSHA K.K.																	
Patents by Year of Grant:	1,800	239	256	228	376	306	274	152	149	149	145	132	156	158	211	387	5,118
Patents by Year of Application:	2,268	224	295	357	238	173	152	160	138	155	153	161	199	308	137	0	5,118
AMERICAN CYANAMID COMPANY																	
Patents by Year of Grant:	3,114	111	115	92	107	101	114	136	106	126	135	148	98	106	107	112	4,828
Patents by Year of Application:	3,338	135	97	111	110	100	86	129	124	133	141	102	151	39	31	1	4,828
ROCKWELL INTERNATIONAL CORP.																	
Patents by Year of Grant:	2,707	196	208	161	158	134	123	115	152	124	136	150	117	119	75	77	4,752
Patents by Year of Application:	3,207	106	128	128	110	118	148	125	146	137	132	109	112	36	9	0	4,752

1.12. Ranked List of Some Organizations with 1000 or More Patents, as Distributed by Year of Patent Grant, and as Distributed by Year of Patent Application (Granted January 1, 1969 through December 31, 1998) (continued).

Organizational Patenting	Pre-1984	1984	1985	1986	1987	1988	1989	1990	1991	1992	1993	1994	1995	1996	1997	1998	Total
UNION CARBIDE CORPORATION																	
Patents by Year of Grant:	3,428	231	242	208	159	141	132	43	8	0	0	0	0	0	0	0	4,592
Patents by Year of Application:	3,870	200	181	138	107	64	32	0	0	0	0	0	0	0	0	0	4,592
CATERPILLAR INC.																	
Patents by Year of Grant:	3,023	54	75	46	54	44	58	52	76	60	107	107	143	206	212	197	4,514
Patents by Year of Application:	3,138	45	49	53	44	44	54	86	83	117	130	200	224	197	50	0	4,514
UNITED STATES OF AMERICA, NASA																	
Patents by Year of Grant:	2,510	136	144	93	122	93	127	128	151	151	166	160	122	104	96	95	4,398
Patents by Year of Application:	2,774	100	105	93	96	145	139	137	170	156	167	120	102	73	21	0	4,398
PROCTER + GAMBLE CO.																	
Patents by Year of Grant:	1,527	100	93	75	127	92	143	127	146	156	155	205	226	346	406	454	4,378
Patents by Year of Application:	1,712	95	101	100	136	115	137	146	148	181	239	331	426	370	137	4	4,378
UOP																	
Patents by Year of Grant:	2,869	132	99	63	73	68	161	120	99	103	115	103	109	97	60	88	4,359
Patents by Year of Application:	3,053	90	67	80	127	137	94	99	98	124	95	127	82	61	25	0	4,359
HONEYWELL, INC.																	
Patents by Year of Grant:	2,079	131	133	154	169	147	213	167	162	137	139	167	173	126	100	123	4,320
Patents by Year of Application:	2,377	157	120	149	144	198	175	150	154	155	183	128	110	93	27	0	4,320
TEXACO, INC.																	
Patents by Year of Grant:	2,618	196	183	140	146	96	126	115	100	107	109	87	107	65	25	34	4,254
Patents by Year of Application:	2,955	155	131	128	111	104	95	113	110	101	94	83	42	24	8	0	4,254
MERCK & CO., INC.																	
Patents by Year of Grant:	2,057	135	105	103	123	125	173	134	152	164	222	163	125	126	161	178	4,246
Patents by Year of Application:	2,307	113	126	108	138	123	134	163	200	223	125	157	150	124	54	1	4,246
UNITED STATES OF AMERICA, AIR FORCE																	
Patents by Year of Grant:	2,289	181	143	180	186	115	137	101	135	139	132	131	80	99	86	79	4,213
Patents by Year of Application:	2,650	152	164	143	107	113	107	144	129	124	105	92	92	68	20	0	4,213
OLYMPUS OPTICAL CO., LTD.																	
Patents by Year of Grant:	1,171	196	199	146	146	180	196	179	161	197	203	228	210	223	213	278	4,126
Patents by Year of Application:	1,594	137	129	132	170	182	176	204	223	221	272	253	230	140	61	2	4,126
SAMSUNG ELECTRONICS CO., LTD.																	
Patents by Year of Grant:	0	1	0	1	7	8	26	61	148	250	347	410	423	485	583	1,304	4,055
Patents by Year of Application:	1	1	8	5	22	48	182	231	376	402	406	477	618	1,022	253	3	4,055
MINOLTA CAMERA CO., LTD.																	
Patents by Year of Grant:	946	90	80	70	99	171	350	376	315	282	237	160	114	164	158	304	3,916
Patents by Year of Application:	1,123	63	100	113	256	399	382	290	266	184	135	163	164	175	99	3	3,916
PPG INDUSTRIES, INC.																	
Patents by Year of Grant:	2,364	137	128	124	118	118	112	100	94	69	83	60	52	51	74	97	3,781
Patents by Year of Application:	2,597	114	124	130	121	103	88	85	81	47	63	66	76	64	22	0	3,781
AMP, INC.																	
Patents by Year of Grant:	1,748	142	132	126	216	204	243	224	267	231	129	2	0	1	0	0	3,665
Patents by Year of Application:	2,006	124	164	196	179	201	246	225	212	111	0	1	0	0	0	0	3,665
BRISTOL-MYERS SQUIBB COMPANY																	
Patents by Year of Grant:	1,798	87	110	105	125	120	116	123	124	129	119	100	96	102	134	137	3,525
Patents by Year of Application:	1,977	121	128	110	116	105	120	145	111	119	126	109	122	82	32	2	3,525
GTE PRODUCTS CORP.																	
Patents by Year of Grant:	2,237	110	106	129	133	142	168	125	148	94	79	35	8	2	0	3	3,519
Patents by Year of Application:	2,441	105	130	137	125	165	136	111	99	61	7	2	0	0	0	0	3,519
HOFFMAN-LA ROCHE, INC.																	
Patents by Year of Grant:	2,187	78	74	84	86	59	93	78	78	72	87	89	103	106	79	72	3,425
Patents by Year of Application:	2,360	67	80	69	72	93	85	78	60	92	89	118	94	47	18	1	3,425

1.12. Ranked List of Some Organizations with 1,000 or More Patents, as Distributed by Year of Patent Grant, and as Distributed by Year of Patent Application (Granted January 1, 1969, through December 31, 1998) *(continued)*.

Organizational Patenting	Pre-1984	1984	1985	1986	1987	1988	1989	1990	1991	1992	1993	1994	1995	1996	1997	1998	Total
NIKON CORPORATION																	
Patents by Year of Grant:	687	88	71	69	119	79	69	73	84	104	132	113	224	423	479	575	3,389
Patents by Year of Application:	848	76	86	77	54	65	90	85	138	121	171	379	602	414	183	0	3,389
EATON CORPORATION																	
Patents by Year of Grant:	1,242	126	152	126	123	103	140	143	114	141	134	148	165	182	145	164	3,348
Patents by Year of Application:	1,531	101	109	89	132	127	121	134	133	153	185	184	140	154	55	0	3,348
SUMITOMO CHEMICAL COMPANY, LTD.																	
Patents by Year of Grant:	1,442	91	103	87	97	123	147	114	146	130	147	164	139	138	123	148	3,339
Patents by Year of Application:	1,639	85	95	114	133	130	140	121	134	156	144	175	144	99	30	0	3,339
BOEING COMPANY																	
Patents by Year of Grant:	1,158	141	121	138	169	168	214	192	197	182	106	80	92	97	114	157	3,327
Patents by Year of Application:	1,514	135	150	138	168	206	202	157	129	86	87	66	181	90	18	0	3,327
FMC CORPORATION																	
Patents by Year of Grant:	2,115	99	102	70	74	68	68	76	78	77	80	71	53	70	71	97	3,269
Patents by Year of Application:	2,299	77	69	77	63	67	79	72	94	58	72	71	72	67	31	1	3,269
NIPPONDENSO CO., LTD.																	
Patents by Year of Grant:	737	156	137	122	184	124	116	76	76	86	123	161	183	311	312	308	3,212
Patents by Year of Application:	1,011	128	173	125	113	82	82	74	107	162	201	263	375	293	23	0	3,212
CHEVRON RESEARCH & TECHNOLOGY CO.																	
Patents by Year of Grant:	2,024	140	146	161	133	92	75	59	77	85	73	64	24	5	5	0	3,163
Patents by Year of Application:	2,330	142	115	108	79	44	62	101	64	71	36	8	2	1	0	0	3,163
ITT CORPORATION																	
Patents by Year of Grant:	2,178	106	84	99	73	33	56	57	59	62	61	79	81	62	34	22	3,146
Patents by Year of Application:	2,384	105	60	22	46	44	58	66	71	69	88	70	49	12	2	0	3,146
KONICA CORPORATION																	
Patents by Year of Grant:	585	146	150	117	105	124	173	211	226	203	222	228	177	155	125	166	3,113
Patents by Year of Application:	877	95	118	111	152	181	197	248	223	196	186	187	154	138	50	0	3,113
UPJOHN COMPANY																	
Patents by Year of Grant:	2,528	109	66	66	26	15	30	26	23	31	30	23	23	33	21	1	3,051
Patents by Year of Application:	2,701	65	32	24	21	21	25	24	30	27	30	26	23	2	0	0	3,051
POLAROID CORPORATION																	
Patents by Year of Grant:	2,071	101	112	83	91	66	87	51	36	39	39	40	55	59	60	60	3,050
Patents by Year of Application:	2,271	113	71	61	74	61	41	32	50	30	45	71	64	50	16	0	3,050
ELI LILLY AND COMPANY																	
Patents by Year of Grant:	1,376	80	107	82	85	68	75	73	66	63	79	77	147	245	222	180	3,025
Patents by Year of Application:	1,555	116	92	62	52	64	60	67	64	98	150	150	342	85	67	1	3,025
DAIMLER-BENZ ARTIENGESELLSCHAFT																	
Patents by Year of Grant:	NA	1,331*	80	77	93	100	134	107	107	97	120	95	104	129	159	241	2,974
Patents by Year of Application:	NA	1,480*	71	97	100	114	116	101	104	111	105	140	171	183	80	1	2,974
GOODYEAR TIRE & RUBBER COMPANY																	
Patents by Year of Grant:	1,636	113	92	67	80	82	83	83	88	64	82	82	64	70	93	108	2,887
Patents by Year of Application:	1,811	88	82	70	73	81	86	82	74	79	60	71	95	84	49	2	2,887
BENDIX CORPORATION (NOW ALLIED-SIGNAL, INC.)																	
Patents by Year of Grant:	2,655	145	25	0	0	0	0	0	0	0	0	0	0	0	0	0	2,825
Patents by Year of Application:	2,821	4	0	0	0	0	0	0	0	0	0	0	0	0	0	0	2,825
TRW, INC.																	
Patents by Year of Grant:	1,644	47	59	68	77	56	74	52	40	50	82	104	96	98	101	144	2,792
Patents by Year of Application:	1,771	58	59	57	61	63	46	43	84	111	87	79	123	114	35	1	2,792
THOMSON-CSF																	
Patents by Year of Grant:	1,217	116	174	121	131	129	104	114	138	107	90	84	90	52	52	66	2,785
Patents by Year of Application:	1,585	122	85	99	92	104	131	120	99	94	73	65	66	37	13	0	2,785

1.12. Ranked List of Some Organizations with 1,000 or More Patents, as Distributed by Year of Patent Grant, and as Distributed by Year of Patent Application (Granted January 1, 1969, through December 31, 1998) (continued).

Organizational Patenting	Pre-1984	1984	1985	1986	1987	1988	1989	1990	1991	1992	1993	1994	1995	1996	1997	1998	Total
W.R. GRACE & CO.-CONN.																	
Patents by Year of Grant:	1,442	57	45	43	44	61	111	105	121	129	122	116	86	85	73	56	2,696
Patents by Year of Application:	1,534	50	42	57	97	117	116	114	135	100	109	69	109	33	14	0	2,696
CORNING, INC.																	
Patents by Year of Grant:	1,530	82	73	47	54	74	55	83	59	82	102	104	72	85	78	72	2,652
Patents by Year of Application:	1,677	42	61	54	61	65	76	72	90	106	94	100	89	34	30	0	2,652
HOECHST CELANESE CORPORATION																	
Patents by Year of Grant:	1,150	94	67	66	71	109	134	188	128	133	113	93	78	72	75	68	2,639
Patents by Year of Application:	1,309	44	92	106	128	141	148	136	120	98	77	79	91	51	18	0	2,639
INTEL CORPORATION																	
Patents by Year of Grant:	NA	162*	16	15	24	33	49	45	56	73	125	203	271	423	405	701	2,601
Patents by Year of Application:	NA	190*	22	25	52	49	50	74	111	206	308	433	517	439	125	0	2,601
OLIN CORPORATION																	
Patents by Year of Grant:	1,480	112	117	81	58	60	80	74	105	114	72	74	47	56	28	21	2,579
Patents by Year of Application:	1,673	105	75	57	61	61	92	119	83	73	64	52	44	12	7	1	2,579

Notes:
* = Pre-1985.

Source: U.S. Patent and Trademark Office, Office of Electronic Information Products/TAF Program. All Technologies Report, January 1963–December 1998. Springfield, VA: National Technical Information Service, March 1999. Pages B-1 to B-22. http://www.uspto.gov/web/offices/ac/ido/oeip/taf/all_tech.pdf

1.13. U.S. Trade in Advanced Technology Products, 1990–96 (millions of dollars).

Product category	1990	1991	1992	1993	1994	1995	1996
			U.S. exports				
Total	$94,727.6	$101,641.5	$107,091.3	$108,356.6	$120,743.3	$138,416.5	$154,909.2
Biotechnology	$661.2	$706.0	$745.8	$892.7	$1,029.2	$1,055.5	$1,197.4
Life science	$4,860.3	$5,492.5	$5,826.0	$6,133.7	$6,821.3	$8,571.5	$9,255.6
Opto-electronics	$524.0	$627.9	$604.0	$701.7	$926.3	$1,164.6	$1,418.6
Computers and telecommunications	$31,375.0	$30,726.3	$32,569.2	$34,198.8	$39,859.3	$47,890.5	$52,780.1
Electronics	$7,535.5	$8,925.6	$9,968.4	$11,987.4	$16,235.6	$31,391.7	$36,548.0
Flexible manufacturing	$3,095.7	$3,251.4	$3,412.6	$4,039.0	$5,191.0	$7,469.6	$8,583.6
Advanced materials	$6,403.0	$6,226.1	$7,153.6	$8,404.2	$10,406.2	$4,519.5	$1,693.6
Aerospace	$36,972.7	$41,904.5	$42,445.5	$37,348.1	$34,955.4	$30,983.1	$38,088.7
Weapons	$687.9	$851.7	$784.1	$745.1	$730.6	$1,040.5	$1,466.9
Nuclear technology	$1,260.5	$1,304.2	$1,502.4	$1,375.6	$1,560.6	$1,272.0	$1,258.9
Software technology	$1,351.8	$1,625.2	$2,079.7	$2,530.2	$3,027.9	$3,057.9	$2,617.7
			U.S. imports				
Total	$59,381.2	$63,252.1	$71,871.5	$81,233.1	$98,116.5	$124,787.0	$130,361.6
Biotechnology	$32.1	$48.7	$48.8	$59.2	$73.3	$444.8	$548.8
Life science	$3,417.6	$4,305.8	$4,821.4	$4,607.5	$4,821.5	$6,607.2	$7,291.6
Opto-electronics	$1,138.0	$2,038.4	$2,570.3	$2,531.0	$2,544.1	$2,816.6	$3,172.8
Computers and telecommunications	$30,110.5	$29,153.4	$33,848.5	$39,790.2	$49,440.0	$58,865.6	$61,346.1
Electronics	$10,955.3	$12,391.7	$14,205.3	$17,824.2	$25,507.3	$38,232.6	$36,756.8
Flexible manufacturing	$1,676.6	$1,789.7	$1,684.5	$2,222.2	$2,899.7	$4,947.5	$5,740.7
Advanced materials	$1,045.6	$1,051.5	$1,548.4	$2,052.9	$1,091.8	$1,527.6	$1,219.8
Aerospace	$10,713.8	$12,106.0	$12,687.2	$11,613.3	$11,135.6	$10,540.5	$12,805.4
Weapons	$129.9	$167.8	$156.9	$164.7	$143.9	$205.0	$265.5
Nuclear technology	$4.5	$3.0	$5.2	$7.9	$22.7	$39.8	$626.1
Software technology	$157.4	$196.0	$295.0	$360.0	$436.5	$559.8	$588.0
			U.S. trade balance				
Total	$35,346.4	$38,389.4	$35,219.8	$27,123.6	$22,626.8	$13,629.5	$24,547.6
Biotechnology	$629.1	$657.3	$697.0	$833.5	$956.0	$610.7	$648.6
Life science	$1,442.8	$1,186.7	$1,004.6	$1,526.2	$1,999.7	$1,964.3	$1,964.1
Opto-electronics	– $613.9	– $1,410.5	– $1,966.3	– $1,829.3	– $1,617.8	– $1,652.0	– $1,754.2
Computers and telecommunications	$1,264.5	$1,572.9	– $1,279.3	– $5,591.4	– $9,580.7	– $10,975.1	– $8,566.0
Electronics	– $3,419.9	– $3,466.1	– $4,236.9	– $5,836.8	– $9,271.7	– $6,840.9	– $208.8
Flexible manufacturing	$1,419.1	$1,461.7	$1,728.1	$1,816.8	$2,291.3	$2,522.1	$2,842.9
Advanced materials	$5,357.5	$5,174.6	$5,605.3	$6,351.3	$9,314.4	$2,991.9	$473.8
Aerospace	$26,258.9	$29,798.5	$29,758.3	$25,734.8	$23,819.7	$20,442.7	$25,283.3
Weapons	$558.0	$683.9	$627.2	$580.4	$586.7	$835.6	$1,201.4
Nuclear technology	$1,256.0	$1,301.1	$1,497.2	$1,367.7	$1,537.9	$1,232.2	$632.9
Software technology	$1,194.5	$1,429.2	$1,784.7	$2,170.3	$2,591.4	$2,498.1	$2,029.7

Source: Cited in National Science Board. *Science and Engineering Indicators 1998.* Arlington, VA: National Science Foundation, 1998. Table 6-6. http://www.nsf.gov/sbe/srs/seind98/start.htm

1.14. High-tech Companies Formed in the United States, 1980–94.

Period formed	All high-tech fields	Automation	Biotechnology	Computer hardware	Advanced materials	Photonics and optics	Software	Electronic components	Telecom-munications	Other fields
	Number of companies									
Total, (since 1960)	29,358	1,939	735	2,845	1,045	977	7,661	2,923	1,556	9,677
1980–94	16,660	917	546	1,907	487	507	5,196	1,293	933	4,874
1980–84	7,727	483	213	842	212	221	2,467	629	408	2,252
1985–89	6,510	331	225	756	194	191	1,962	508	370	1,973
1990–94	2,423	103	108	309	81	95	767	156	155	649
	Percentage of all high-tech companies formed during each period									
Total, all years	100%	6.6%	2.5%	9.7%	3.6%	3.3%	26.1%	10.0%	5.3%	33.0%
1980–94	100%	5.5%	3.3%	11.4%	2.9%	3.0%	31.2%	7.8%	5.6%	29.3%
1980–84	100%	6.3%	2.8%	10.9%	2.7%	2.9%	31.9%	8.1%	5.3%	29.1%
1985–89	100%	5.1%	3.5%	11.6%	3.0%	2.9%	30.1%	7.8%	5.7%	30.3%
1990–94	100%	4.3%	4.5%	12.8%	3.3%	3.9%	31.7%	6.4%	6.4%	26.8%
	Percentage of all U.S. high-tech companies									
Total, all years	100 %	100 %	100 %	100 %	100 %	100 %	100 %	100 %	100 %	100 %
1980–94	56.7%	47.3%	74.3%	67.0%	46.6%	51.9%	67.8%	44.2%	60.0%	50.4%
1980–84	26.3%	24.9%	29.0%	29.6%	20.3%	22.6%	32.3%	21.5%	26.2%	23.3%
1985–89	22.2%	17.1%	30.6%	26.6%	18.6%	19.5%	25.6%	17.4%	23.8%	20.4%
1990–94	8.3%	5.3%	14.7%	10.9%	7.8%	9.7%	10.0%	5.3%	10.0%	6.7%

Notes:
Other fields are chemicals, defense-related, energy, environmental, manufacturing equipment, medical, pharmaceuticals, test and measurement, and transportation.

Source: National Science Board. *Science and Engineering Indicators 1996.* Springfield, VA: National Science Foundation, 1996. Table 6-16. http://www.nsf.gov/sbe/srs/seind96/startse.htm

Chapter 2. Agriculture

If it weren't for agriculture, we wouldn't be able to pop out for a Big Mac, an ice cream cone, or Chinese food. Nor could we live in a city, drive a car, use a computer, or watch TV. Agriculture doesn't just fill our stomachs. It enables humans to control our food supply, which means that we no longer have to chase mastodons with spears or wait for berries to emerge on bushes. Agriculture allows us to settle in one place because the food supply is relatively secure, and that stability lets us divide labor, develop specialties and expertises, and invent new technologies.

When we think of agriculture, most of us envision fields full of crops, but agriculture involves a lot more than corn stalks undulating in the breeze. The object is to raise large amounts of high-quality food at a good profit while preserving natural resources and food safety. Farmers and scientists need to consider irrigation techniques, erosion prevention, insects and weeds, plant and animal diseases, soil composition issues, genetics and breeding, fertilizers, mechanized processing, weather, biodiversity issues, transportation and storage, energy, manpower, and economic matters. Agriculture can involve raising water-based life (aquaculture), animals (for meat, milk, fiber, transportation, and labor), ornamental plants, food plants, and insects (for honey and as friendly pest-eradicating weapons).

Hot topics in farming technology today include

- Use of "green" technologies. Environmentally friendly methods that increase yields and lower costs should be in great demand, but their adoption depends on
- How profitable they are for their producers.
- Whether they're directly beneficial to the farmer, or only to society at large.
- Whether they're easy to use and/or integrate into existing methods.
- How much improvement they facilitate.
- Whether they work well in the soil, water, and climate types of the particular farm.
- How much they cost to acquire.

- Use of high-tech devices for communications and management.

- Intellectual property rights for new plant varieties and biological inventions. Until this century, the Patent Act of 1790 did not apply to these categories. Technological developments (hybrid seed technologies and biotechnology applications such as gene transfer), laws, and Patent and Trademark Office decisions have changed that, making private-sector agricultural R&D profitable.

AGRICULTURAL METHODS

As population growth and the global economy have increased agricultural demand and lowered profits, farmers have devised ways of increasing crop yields, such as adding important nutrients to soils, controlling pests, and using water efficiently. At the same time, environmental concerns revolving around the use of toxic chemicals, polluting runoff, and soil erosion have influenced farming methods. Many new nutrient and pest-controlling products were developed in the 1940s and 1950s, and their use as well as the amount of planted acreage soared until the early 1980s. After that, commodity prices fell and large amounts of land were taken out of production by federal programs. Table 2.1 shows the dramatic increase in fertilizer use during the 1960s and 1970s, from 24.9 million tons in 1960 to a high of 54.0 million tons in 1981. You can also see that the proportion of primary nutrients in the fertilizers has changed over time. Nitrogen use in particular has grown rapidly, from 2.7 million tons in 1960 to a high of 12.6 million tons in 1994, as nitrogenous fertilizers have produced dazzling results.

Table 2.5 shows that pesticide use has increased dramatically as well, more than doubling between 1964, when 215,008,000 pounds of active ingredients were used, and 1995, when 565,639,000 pounds were used, with a peak of 572,448,000 in 1982. (Fertilizer use also peaked in the early 1980s.) However, the types of pesticides used over that time have changed substantially. Herbicide use has skyrocketed from 48,158,000 pounds of active ingredients in 1964 to 323,791,000 in 1995, with a peak of 430,345,000 pounds in 1982, but insecticide use has fallen from 123,304,000 pounds of active ingredients in the 1964 to 69,599,000 in 1995 because of new compounds that require lower rates of application. "Other" pesticides, including soil fumigants, growth regulators, desiccants, and harvest aids, many of which do not destroy pests but affect the growth or functions of the plants, have grown more than any other class. However, such numbers can be misleading because some of them require extremely heavy application, though over tiny areas representing a small percentage of areas under pesticide use.

While pesticide use has increased yields, its adverse effects are well known. As alternatives to pesticides, farmers have adopted "green" pest management practices like the use of beneficial insects, resistant plants, pheromone traps, and other techniques (Table 2.4). As production has intensified, many farmers have also instituted crop residue management systems to protect the soil surface from erosion. Table 2.6 shows that the percentage of planted areas under conservation tillage has increased from 25.6% in 1989 to 35.8% in 1996, an increase of about 32 million acres.

As competition for water supplies among agricultural and other users has increased and as growing use of agricultural chemicals has affected water quality, farmers have looked to better ways of managing water use. The two main irrigation system types are gravity flow, relying on gravity to distribute water across the field, and pressurized, where water is forced through the system by pressure. Table 2.2 shows that pressurized systems, which include sprinkler and low-flow irrigation systems, have become more popular since 1979, with sprinkler systems in use in over 17% more acres by 1994, and center pivot sprinkler systems in particular—a self-propelled mechanism with reduced labor and energy requirements—showing a 72% increase in acreage. Sprinkler irrigation helps conserve water. Low-flow irrigation grew a whopping 445% during that time, although it still accounts for a small share of irrigation systems (less than 4%). Even though the use of gravity flow systems fell in relation to the whole, gravity systems still account for about 54% of all irrigation systems.

FARM EQUIPMENT

Many city dwellers fail to realize how completely farmers have embraced the new communications and computer technologies. Table 2.7 shows that they have—enthusiastically. In 1998, over 76% of farmers used computers, and over 71% had cell phones. About a third used online services, the Internet, and e-mail. Since the mid-1980s, farmers' use of most types of traditional farm equipment grew or held steady, despite increases in complexity, price, and in many cases, size, but also efficiency (Tables 2.8 and 2.9).

GENETICS AND FARMING

One of the most dramatic changes in technology affecting not only agriculture, but also medicine and environmental remediation, has been the introduction of biotechnology techniques. These methods have allowed farmers and scientists to breed plants that are pest resistant, contain high nutrient contents, and can grow quickly. The application of biotechnology to agriculture is not without controversy, however, as people question what will happen to the world's seed stock if natural varieties disappear. Tables 2.11 and 2.12 show the types of plant technologies for which patents have been issued and the types of field testing being performed on genetically modified plants.

AGRICULTURAL METHODS

2.1. Commercial Fertilizer Use, United States, 1960–95 (million tons).

Year Ending June 30th	Total Fertilizer Materials [1]	Primary Nutrient Use			
		Nitrogen (N)	Phosphate (P_2O_5)	Potash (K_2O)	Total
1960	24.9	2.7	2.6	2.2	7.5
1961	25.6	3.0	2.6	2.2	7.8
1962	26.6	3.4	2.8	2.3	8.4
1963	28.8	3.9	3.1	2.5	9.5
1964	30.7	4.4	3.4	2.7	10.5
1965	31.8	4.6	3.5	2.8	10.9
1966	34.5	5.3	3.9	3.2	12.4
1967	37.1	6.0	4.3	3.6	14.0
1968	38.7	6.8	4.4	3.8	15.0
1969	38.9	6.9	4.7	3.9	15.5
1970	39.6	7.5	4.6	4.0	16.1
1971	41.1	8.1	4.8	4.2	17.2
1972	41.2	8.0	4.9	4.3	17.2
1973	43.3	8.3	5.1	4.6	18.0
1974	47.1	9.2	5.1	5.1	19.3
1975	42.5	8.6	4.5	4.4	17.6
1976	49.2	10.4	5.2	5.2	20.8
1977	51.6	10.6	5.6	5.8	22.1
1978	47.5	10.0	5.1	5.5	20.6
1979	51.5	10.7	5.6	6.2	22.6
1980	52.8	11.4	5.4	6.2	23.1
1981	54.0	11.9	5.4	6.3	23.7
1982	48.7	11.0	4.8	5.6	21.4
1983	41.8	9.1	4.1	4.8	18.1
1984	50.1	11.1	4.9	5.8	21.8
1985	49.1	11.5	4.7	5.6	21.7
1986	44.1	10.4	4.2	5.1	19.7
1987	43.0	10.2	4.0	4.8	19.1
1988	44.5	10.5	4.1	5.0	19.6
1989	44.9	10.6	4.1	4.8	19.6
1990	47.7	11.1	4.3	5.2	20.6
1991	47.3	11.3	4.2	5.0	20.5
1992	48.8	11.5	4.2	5.0	20.7
1993	49.2	11.4	4.4	5.1	20.9
1994	52.3	12.6	4.5	5.3	22.4
1995	50.7	11.7	4.4	5.1	21.3

Notes:

Includes Puerto Rico. Fertilizer statistics include commercial fertilizers and natural processed and dried organic materials. Purchased natural-processed and dried organic materials historically have represented about 1% of total nutrient use.

Fertilizer use estimates for 1960–84 are based on USDA data; those for 1985–94 are Tennessee Valley Authority (TVA) estimates; those for 1995 are the Association of Plant Food Control Officials estimates.

Totals may not add due to rounding.

[1] Includes secondary and micronutrients. Most of the difference between primary nutrient tons and total fertilizer materials is filler material.

Source: Based on Tennessee Valley Authority, *Commercial Fertilizers,* 1994 and earlier issues; Association of American Plant Food Control Officials, *Commercial Fertilizers, 1995.* Cited in U.S. Department of Agriculture, Economic Research Service, Natural Resources and Environment Division. *Agricultural Resources and Environmental Indicators, 1996–97.* Section 3.1, Nutrients. Table 3.1.1, Page 100. Agricultural Handbook No. 712. U.S. Department of Agriculture. Washington, DC, 1997. http://www.econ.ag.gov/epubs/pdf/ah712/

2.2. Changes in Irrigation System Acreage, United States, 1979–94.

System	1979	1994	Change 1979–94
	Million acres		Percent
All systems	50.1	46.4	-7%
Gravity-flow systems	31.2	25.1	-20%
Sprinkler systems	18.4	21.5	17%
Center pivot	8.6	14.8	72%
Mechanical move	5.1	3.7	-27%
Hand move	3.7	1.9	-48%
Solid set and permanent	1.0	1.0	2%
Low-flow irrigation (drip/trickle)	0.3	1.8	445%
Subirrigation	0.2	0.4	49%

Source: Based on U.S. Department of Commerce data from 1982 and 1986 by the U.S. Department of Agriculture, Economic Research Service. Cited in U.S. Department of Agriculture, Economic Research Service, Natural Resources and Environment Division. *Agricultural Resources and Environmental Indicators, 1996.* Section 4.6, Irrigation Water Management. Agricultural Handbook No. 712. U.S. Department of Agriculture. Springfield, VA: National Technical Information Service, 1997. Page 228. http://www.econ.ag.gov/epubs/pdf/ah712/

2.3. Irrigation Application Systems, by Type, United States, 1994.

System	Share of Acres Million	All Systems Percent
All systems	46.4	100%
Gravity flow systems	25.1	54%
Row/furrow application	14.2	31%
Open ditches	5.0	11%
Above-ground pipe	7.4	16%
Underground pipe	1.8	4%
Border/basin application	7.5	16%
Open ditches	5.1	11%
Above-ground pipe	0.9	2%
Underground pipe	1.5	3%
Uncontrolled flooding application	2.3	5%
Open ditches	2.3	5%
Above-ground pipe	0.0	0%
Underground pipe	0.0	0%
Sprinkler systems	21.5	46%
Center pivot	14.8	32%
High pressure	3.2	7%
Medium pressure	5.9	13%
Low pressure	5.7	12%
Mechanical move	3.7	8%
Linear and wheel-move	3.0	7%
All other	0.6	1%
Hand move	1.9	4%
Solid set & permanent	1.0	2%
Low-flow irrigation (drip/trickle)	1.8	4%
Subirrigation	0.4	1%

Notes:
Percents may not add to totals due to multiple systems on some irrigated acres, and rounding.
Center pivot sprinkler systems. The primary irrigation technology involving pressurization. Center pivot sprinklers are self-propelled systems involving a single pipeline supported by a row of mobile towers suspended above the ground. They may be a quarter of a mile long and can irrigate up to 132-acre circular fields.
Gravity-flow systems. Irrigation in which the water is propelled by gravity. Types of gravity-flow systems include the release of water through ditches or pipes, runoff control and return, and management of land and furrow shape and length.
Hand-move sprinkler systems. Portable sprinkler systems used for small, irregular fields with low crops. Hand-move systems are labor-intensive.
Mechanical-move sprinkler systems. Sprinklers mounted on carts, trailers, or wheels.
Solid-set and permanent sprinkler systems. Stationary sprinkler systems in which water supply pipelines are fixed—usually below the soil—with nozzles above the surface. Solid-set systems are used in orchards and vineyards, and for turf production and landscaping.
Subirrigation. A method of irrigation involving regulating water tables through the use of drainage systems.

Source: Based on U.S. Department of Commerce data by the U.S. Department of Agriculture, Economic Research Service. Cited in U.S. Department of Agriculture, Economic Research Service, Natural Resources and Environment Division. *Agricultural Resources and Environmental Indicators, 1996–97.* Section 4.6, Irrigation Water Management. Agricultural Handbook No. 712. Washington, DC: National Technical Information Service, 1997. Page 228. http://www.econ.ag.gov/epubs/pdf/ah712/

2.4. Use of Selected Biological and Cultural Pest Management Practices on Fruit, Vegetable, and Nut Crops, Major Producing States, United States, 1990s.

Crop	Acres planted, in thousands	Scouting — Cultural Methods[2]					Biological Methods[2]						
	In surveyed states[1]	Consultants	Grower/ family member	Chemical dealer	Other	Total	Beneficial insects	Habitat provision	Pheromone traps[3]	Resistant varieties	Water management	Field sanitation	Adjust planting dates
Fruit:					Percent of acres								
Grapes, all	730	68%	NA	NA	NA	NA	18%	NA	14%	31%	41%	64%	NA
Oranges	613	75%	NA	NA	NA	NA	22%	NA	28%	21%	27%	48%	NA
Apples	381	54%	NA	NA	NA	NA	2%	NA	66%	16%	22%	73%	NA
All fruits & nuts	3,251	65%	NA	NA	NA	NA	19%	NA	37%	22%	31%	60%	NA
Vegetables[4]:													
Sweet corn	640	33%	22%	2%	27%	84%	*	NA	17%	NA	7%	NA	8%
Tomatoes	357	5%	15%	47%	1%	68%	5%	NA	6%	NA	21%	NA	47%
Lettuce, head	259	32%	26%	26%	9%	93%	3%	NA	1%	NA	4%	NA	26%
All vegetables	2,914	21%	19%	19%	15%	74%	3%	NA	7%	NA	11%	NA	15%
	Number of growers surveyed				Percent of surveyed growers								
Certified organic vegetables:													
Sweet corn	64	**	91%	0%	3%	94%	46%	67%	NA	80%	33%	NA	56%
Tomatoes	55	**	94%	0%	1%	95%	48%	57%	NA	71%	46%	NA	41%
Lettuce, head	33	**	97%	0%	3%	100%	60%	60%	NA	73%	80%	NA	50%
All vegetables	303	**	91%	0%	6%	97%	46%	58%	NA	75%	44%	NA	54%

Notes:

* = Used on less than .5 percent.

** = Included in other.

NA = Not available.

Scouting is a method of pest monitoring that helps farmers determine when to apply pesticides.

Field sanitation is the practice of removing or destroying crops that are diseased, provide pest habitats, or encourage pest problems.

[1] Data is from the 1991 U.S. Department of Agriculture Chemical Use Survey for fruits and nuts (13 states, major producers), the 1992 Survey for vegetables (14 states, major producers), and the 1994 Survey for certified organic vegetables (same states as 1992 survey).

[2] Used for any type of pest in 1991 and 1992, and for three specific types (insects, disease, or weeds) in 1994 (highest use for a specific type is shown).

[3] Reported for all uses (pest control and monitoring) in 1991 and 1994, and for control only in 1992.

[4] Includes fresh and processing crops.

Source: Based on Chemical Use survey data, by the U.S. Department of Agriculture, Economic Research Service, and National Agricultural Statistics Service. Cited in U.S. Department of Agriculture, Economic Research Service, Natural Resources and Environment Division. *Agricultural Resources and Environmental Indicators, 1996-97.* Section 4.4, Pest Management. Agricultural Handbook No. 712. Springfield, VA: National Technical Information Service, 1997. Page 190. http://www.econ.ag.gov/epubs/pdf/ah712/

2.5. Overall Pesticide Use on Selected U.S. Crops by Pesticide Type, 1964–95.

Commodities	1964	1966	1971	1976	1982	1990	1991	1992	1993	1994	1995
	(1,000 pounds of active ingredients)										
Herbicides	48,158	79,384	175,668	341,390	430,345	344,638	335,177	350,534	323,510	350,449	323,791
Insecticides	123,304	119,240	127,709	131,730	82,651	57,392	52,828	60,047	58,096	67,896	69,599
Fungicides	22,167	23,237	29,308	26,632	25,219	27,762	29,439	34,922	36,583	43,059	44,804
Other pesticides	21,379	18,747	31,710	30,741	34,232	67,900	79,451	90,019	97,810	129,639	127,445
Total on selected crops	215,008	240,608	364,395	530,493	572,448	497,693	496,895	535,522	515,999	591,044	565,639
	(1,000 cropland acres)										
Area represented	174,552	175,040	190,638	233,221	255,866	228,508	226,021	231,531	226,586	232,804	227,855
Total cropland used for crops	335,000	332,000	340,000	340,800	383,000	341,000	337,000	338,000	330,000	338,500	338,000
	(Pounds of active ingredient per planted acre)										
Herbicides	0.276	0.454	0.921	1.464	1.682	1.508	1.483	1.514	1.426	1.505	1.421
Insecticides	0.706	0.681	0.670	0.565	0.323	0.251	0.234	0.259	0.256	0.292	0.305
Fungicides	0.127	0.133	0.154	0.113	0.099	0.121	0.130	0.151	0.161	0.185	0.197
Other pesticides	0.122	0.107	0.166	0.127	0.134	0.297	0.352	0.389	0.432	0.557	0.559
Total on selected crops	1.232	1.375	1.911	2.275	2.237	2.178	2.198	2.313	2.277	2.539	2.482
Percent of crop area represented	*52%	53%	56%	68%	67%	67%	67%	69%	69%	69%	67%

Notes:
* = Share of total for the selected crops to total cropland used for crops.
Estimates include corn, soybeans, wheat, cotton, potatoes, other vegetables, citrus fruit, apples, and other fruit.

Source: For data prior to 1993: B. Lin, M. Padgitt, L. Bull, H. Delvo, D. Shank, and H. Taylor. *Pesticide and Fertilizer Use and Trends in U.S. Agriculture.* Springfield, VA: National Technical Information Service 1995. AER-717. For data after 1993: Unpublished U.S. Department of Agriculture Survey data. Cited in U.S. Department of Agriculture, Economic Research Service, Natural Resources and Environment Division. *Agricultural Resources and Environmental Indicators, 1996–97.* Section 3.2, Pesticides. Page 117. Agricultural Handbook No. 712. Springfield, VA: National Technical Information Service, 1997. http:// www.econ.ag.gov/epubs/pdf/ah712/

2.6. National Use of Crop Residue Management Practices, United States, 1989–96.

Item	1989	1990	1991	1992	1993	1994	1995	1996
			Million acres					
Total area planted [1]	279.6	280.9	281.2	282.9	278.1	283.9	278.7	290.2
Area planted with:								
No-till	14.1	16.9	20.6	28.1	34.8	39.0	40.9	42.9
Ridge-till	2.7	3.0	3.2	3.4	3.5	3.6	3.4	3.4
Mulch-till	54.9	53.3	55.3	57.3	58.9	56.8	54.6	57.5
Total conservation tillage	71.7	73.2	79.1	88.7	97.1	99.3	98.9	103.8
Other tillage types:								
Reduced tillage (15–30% residue)	70.6	71.0	72.3	73.4	73.2	73.1	70.1	74.8
Conventional tillage (< 15% residue)	137.3	136.7	129.8	120.8	107.9	111.4	109.7	111.6
Total other tillage types	207.9	207.7	202.1	194.2	181.0	184.6	179.7	186.4
Percentage of area with:			**Percent**					
No-till	5.1%	6.0%	7.3%	9.9%	12.5%	13.7%	14.7%	14.8%
Ridge-till	1.0%	1.1%	1.1%	1.2%	1.2%	1.3%	1.2%	1.2%
Mulch-till	19.6%	19.0%	19.7%	20.2%	21.2%	20.0%	19.6%	19.8%
Total conservation tillage	25.6%	26.1%	28.1%	31.4%	34.9%	35.0%	35.5%	35.8%
Other tillage types:								
Reduced tillage (15–30% residue)	25.3%	25.3%	25.7%	25.9%	26.3%	25.8%	25.2%	25.8%
Conventional tillage (< 15% residue)	49.1%	48.7%	46.1%	42.7%	38.8%	39.3%	39.3%	38.4%
Total other tillage types	74.4%	73.9%	71.9%	68.6%	65.1%	65.0%	64.5%	64.2%

Notes:
[1]Total area planted does not include newly established permanent pastures, fallow, annual conservation use, and Conservation Reserve Program (CRP) acres. However, it does include newly seeded alfalfa and other rotational forage crops in the year they are planted.
Crop Residue Management and Tillage Definitions
Conservation tillage. A tillage or planting system that covers 30% or more of the soil with crop residue to reduce erosion by water, or that maintains at least 1,000 pounds of residue per acre to reduce wind erosion.
Conventional tillage. Tillage that leaves less than 30% residue cover.
Crop residue management. A method of soil and water preservation that relies on reduced or special tillage practices.
Mulch till. A method of tillage in which the soil is disturbed prior to planting (as opposed to no-till and ridge-till, in which the soil is left undisturbed). Herbicides and/or cultivation are used to control weeds.
No-till. An agricultural technique in which the soil is left undisturbed between harvest and planting, except for the injection of nutrients. Seeds are applied in narrow beds or slots created by a variety of tools. Cultivation is used only for emergency weed control.
Reduced tillage. A type of tillage that leaves 15 to 30% residue on the soil after planting (to protect against water erosion), or 500 to 1,000 pounds of residue per acre (to guard against wind erosion).
Ridge till. An agricultural technique in which the soil is left undisturbed between harvest and planting except for the injection of nutrients. Seeds are deposited in beds on ridges, which are built at planting time and reconstructed during weeding. Residue is left on the surface between ridges.

Source: Based on Conservation Technology Information Center (CTIC) data from Crop Residue Management Surveys, by the U.S. Department of Agriculture, Economic Research Service. Cited in U.S. Department of Agriculture, Economic Research Service, Natural Resources and Environment Division. *Agricultural Resources and Environmental Indicators, 1996–97.* Section 4.2, Crop Residue Management. Page 158. Agricultural Handbook No. 712. Springfield, VA: National Technical Information Service. http://www.econ.ag.gov/epubs/pdf/ah712/Farm Equipment

FARM EQUIPMENT

2.7. Farmers' Use of Communications Tools, United States, 1995–98.

	1995	1996	1997	1998
Computer	71.0%	68.5%	83.78%	76.69%
Online services	11.6%	8.6%	17.11%	31.42%
Fax machine	27.0%	31.4%	41.89%	40.2%
Cell phone	59.0%	62.6%	73.16%	71.28%
Pager	—	—	—	17.57%
Internet	—	10.5%	32.15%	36.15%
E-Mail	7.44%	9.3%	22.71%	31.76%
Home page	—	—	0.88%	2.03%
Home satellite	21.0%	19.5%	28.91%	39.86%
Big dish	—	—	40.22%	17.8%
Small dish	—	—	59.78%	82.2%

Source: American Farm Bureau. Young Farmer and Rancher Survey. Park Ridge, Illinois. February 1998. Unpublished.

2.8. Domestic Farm Machinery Unit Sales, United States, 1986–96.

Machine Category	1986	1987	1988	1989	1990	1991	1992	1993	1994	1995	1996
Tractors:											
Two-wheel drive											
40–99 horsepower	30,800	30,700	33,100	35,000	38,400	33,900	34,500	35,500	39,100	39,700	41,200
100 horsepower and over	14,300	15,900	16,100	20,600	22,800	20,100	15,600	19,000	20,400	20,500	21,400
Four-wheel drive	2,000	1,700	2,700	4,100	5,100	4,100	2,700	3,300	3,700	4,400	4,400
All farm wheel tractors	47,100	48,400	51,700	59,700	66,300	58,100	52,800	57,800	63,200	64,600	67,000
Self-propelled combines	7,700	7,200	6,000	9,100	10,400	9,700	7,700	7,850	8,500	9,200	9,000

Totals may be slightly off due to independent rounding.

Source: U.S. Department of Agriculture, Economic Research Service. Based on Equipment Manufacturers Institute. Various years. Cited in U.S. Department of Agriculture, Economic Research Service. Natural Resources and Environment Division. *Agricultural Resources and Environmental Indicators, 1996–97.* Section 3.4, Farm Machinery, page 143. Agricultural Handbook No. 712. Springfield, VA: National Technical Information Service, 1997. http://www.econ.ag.gov/epubs/pdf/ah712/

2.9. Farm Machines, Selected Types: Units Shipped in the United States, 1987–96.

Year	Moldboard plows	Subsoilers, deep tillage	Chisel plows or tillers (chisel sweep type)	Field cultivators	Cultivators[1]	Power sprayers and dusters[2]	Crop dryers[3]
1987	867	1,202	3,644	4,198	6,720	NR	6,587
1988	1,574	2,102	4,596	4,207	10,154	43,832	3,923
1989	1,952	2,661	4,446	6,403	13,106	46,224	5,401
1990	2,665	2,516	4,225	7,255	15,449	49,992	4,533
1991	1,382	1,999	3,413	3,829	10,580	48,845	10,201
1992	NA	2,968	NA	3,186	9,274	19,138	14,906
1993	NA	3,350	NA	2,885	8,741	19,708	24,561
1994	NA	4,173	NA	5,074	8,378	NA	22,891
1995	NA	3,486	NA	5,903	7,319	NA	23,546
1996	NA	4,148	NA	5,124	8,023	NA	NA

Year	Side delivery rokes	Hay balers[4]	Field forage harvesters	Combines (harvester threshers)	Small grain headers for combines	Corn picking units (combine attachments)	Cotton strippers and pickers[5]
1987	4,230	18,002	2,209	9,484	9,212	4,485	2,978
1988	7,127	25,974	3,508	7,573	14,302	NA	4,242
1989	10,926	33,373	3,958	11,229	20,316	NA	3,879
1990	12,251	36,422	4,631	14,629	24,573	NA	3,872
1991	8,738	25,336	3,192	11,555	16,969	NA	NA
1992	7,288	20,607	1,719	9,236	13,014	NA	NA
1993	7,931	21,705	1,976	NA	NA	NA	NA
1994	9,933	26,547	2,310	NA	NA	NA	NA
1995	10,204	24,098	1,788	NA	NA	NA	NA
1996	9,055	21,241	1,598	NA	NA	NA	NA

Year	Peanut combines, diggers, shakers, and windrowers	Potato harvesters[6]	Manure spreaders	Feed grinders and crushers	Silo and grain bin unloaders	Milking machines[7]	Mower-conditioners (combination)[8]
1987	1,031	260	11,815	2,537	7,273	37,517	11,738
1988	1,422	289	14,383	2,322	11,596	39,539	15,856
1989	3,621	393	12,935	2,301	14,446	38,709	21,035
1990	5,431	466	12,978	2,211	18,490	42,870	29,306
1991	NA	NA	9,562	2,000	31,294	25,928	21,176
1992	NA	NA	9,940	2,432	33,768	31,214	14,842
1993	NA	NA	9,914	2,293	34,822	33,211	17,240
1994	NA	NA	11,141	2,259	17,292	NA	22,938
1995	NA	NA	8,034	1,864	36,847	NA	19,894
1996	NA	NA	5,956	NA	39,822	NA	NA

Notes:
NA = Not available.
NR = Figure not reliable.
[1] Row cultivators, tractor-drawn or mounted; corn and cotton type; rotary cultivators not included.
[2] Does not include foggers and mist sprayers. Starting in 1993, includes only self-propelled, tractor-mounted and other power take-off (a drive shaft that delivers power from the tractor to another machine) and engine-driven sprayers.
[3] Beginning in 1991, includes heated and cold air crop dryers.
[4] Beginning in 1991, includes stackers and loaders.
[5] Beginning in 1986, includes potato diggers, corn harvesting equipment, picker-shellers, and field shelling attachments for corn pickers.
[6] Beginning in 1987, beet harvesters are included.
[7] Includes all mechanical milking machines, vacuum pumping outfits, and complete pipeline milking units. Beginning in 1991, excludes vacuum pumping outfits.
[8] Includes self-propelled windrowers.

Source: U.S. Department of Agriculture, National Agricultural Statistics Service. USDA-NASS Agricultural Statistics 1998. Chapter 9, Farm Resources, Income, and Expenses.

2.10. Farm Machinery and Equipment: Number of Certain Kinds on Farms, and Tractor Horsepower, United States, Census Years, 1954–92.

| Year[1] | Tractors (exclusive of garden) | | | Horsepower (millions) | Motor-trucks (thousands) | Grain combines[3] (thousands) | Cornheads[4] (thousands) | Pickup balers[5] (thousands) | Field forage harvesters[6] (thousands) |
	Total[2] (thousands)	Wheel (thousands)	Crawler (thousands)						
1954	4,345	4,185	160	126	2,702	979	688	448	202
1959	4,688	4,489	199	153	2,834	1,042	792	680	291
1964	4,786	4,601	186	176	3,030	910	690	751	316
1969	4,622	4,419	203	203	2,985	790	635	708	304
1974	4,467	4,312	155	222	3,038	524	615	666	255
1978	NA	4,626[7]	NA	301	3,358	655	694	744	295
1982	NA	4,524	NA	309	3,435	644	684	800	285
1987	NA	4,609	NA	NA	3,437	667	NA	823	NA
1992	NA	4,305	NA	NA	3,295	569	NA	790	NA

Notes:

Data are from the Census of Agriculture, U.S. Bureau of the Census.

[1] Data as of December 31.

[2] Includes wheel and crawler-type tractors.

[3] Data for 1974 and after are for self-propelled combines only.

[4] Includes corn pickers and picker shellers.

[5] Does not include balers producing bales weighing more than 200 pounds.

[6] Data for 1978 and after do not include flail-type forage harvesters.

[7] U.S. totals for 1978 are not directly comparable with totals for 1974 or earlier Census years because they include state-level data from farm operators represented on the Census mailing list, plus estimates from the direct enumeration sample for farms not represented on the mailing list. As a result, figures for nearly all categories are somewhat higher than they would be using the earlier base.

NA = Not available.

Source: U.S. Department of Agriculture, National Agricultural Statistics Service. *USDA-NASS Agricultural Statistics 1998.* Chapter 9, Farm Resources, Income, and Expenses. Washington, DC: U.S. Government Printing Office. 1998. Table 9-16. http://www.usda.gov/nass/pubs/agr98/acro98.htm

GENETICS AND FARMING

2.11. Utility Patents Issued for Multicellular Organisms through 1994, United States.

Item	Patents Issued (number)
Technology[1]:	
Animal	38
Plant:	286
Plant, seedling, or plant part	154
Recombinant plant	103
Somatic cell fusion-derived plant	10
Mutant plant	25
Grafted plant	3
Total Technology	324
Plant Commodity[2]:	
Corn	83
Tomato	24
Tobacco	23
Soybean	17
Rice	15
Sunflower	10
Potato	9
Wheat	8
Canola	8
Cotton	8
Mushrooms	8

Notes:
[1] A single patent may involve more than one technology or commodity.
[2] Only commodities with eight or more patents are listed.

Source: U.S. Department of Agriculture, Economic Research Service. Adapted from CASSIS database, Patent and Trademark Office, U.S. Department of Commerce. Cited in Keith Fuglie et al. U.S. Department of Agriculture, Economic Research Service, Natural Resources and Environment Division. *Agricultural Research and Development: Public and Private Investments under Alternative Markets and Institutions.* Agricultural Economic Report No. 735. Washington, DC: U.S. Department of Agriculture, n.d. Table 12, Page 39. http://www.econ.ag.gov/epubs/pdf/aer735/

2.12. Field Test Permits Issued for Genetically Modified Plants through June 1993, United States.

Crop	Herbicide tolerance	Insect resistance	Virus resistance	Product quality	Research	Total
Corn	31	22	12	5	6	76
Tomato	11	15	13	27	8	74
Potato	2	7	39	10	6	64
Soybean	48	0	1	4	4	57
Cotton	25	14	0	0	0	39
Tobacco	6	11	9	3	6	35
Rapeseed	4	1	0	11	0	16
Alfalfa	3	0	8	1	0	12
Melon	0	0	10	0	0	10
Cantaloupe	0	0	10	0	0	10
Rice	1	2	1	1	2	7
Other	1	6	12	5	8	32
Total	132	78	115	67	40	432
Percent	31%	18%	27%	16%	9%	100%

Source: U.S. Department of Agriculture, Economic Research Service. Compiled from Michael Ollinger and Leslie Pope. *Plant Biotechnology: Out of the Laboratory and into the Field.* AER-697. U.S. Department of Agriculture, Economic Research Service. Washington, DC, April 1995. Cited in Keith Fuglie et al. U.S. Department of Agriculture, Economic Research Service, Natural Resources and Environment Division. *Agricultural Research and Development: Public and Private Investments under Alternative Markets and Institutions.* Agricultural Economic Report No. 735. Washington, DC: U.S. Department of Agriculture, n.d. Table 14, Page 40. http://www.econ.ag.gov/epubs/pdf/aer/735/

Chapter 3. Business, Manufacturing, and Materials

BUSINESS

The abacus and stylus were our original business machines. Millennia later the typewriter and adding machine burst onto the scene to transform offices from Scrooge's counting house to Clark Kent's *Daily Planet* newsroom, courtesy of IBM, Royal, and their ilk. Starting in the 1960s, hitherto unaffordable miracle inventions became available at reasonable prices—photocopiers and desktop and handheld calculators, then fax machines, personal computers, printers, and scanners—and began to work their magic at a dizzying pace, changing the speed and nature of our work while placing astonishingly powerful tools in each individual's hands.

But the office isn't the only venue to install new business technologies. The ATM (automatic teller machine), the point-of-sale terminal, and security devices of all types are as ubiquitous as the personal computer. Everyone knows what a bar code is, and we carry a larger and larger variety of plastic cards with us.

In Table 3.1 you will find much-in-demand numbers relating to the use of computers, faxes, scanners, computer printers, and digital cameras. As Table 3.2 shows, the Europeans are way ahead of the rest of the world when it comes to using smart cards, which can contain digital information relating to our medical history, financial transactions, and consumer incentive programs. But people in the United States do like their automatic teller machines. Table 3.4 shows that the number of monthly transactions per machine didn't changed dramatically between 1988 and 1997, but Table 3.3 indicates that the *number* of machines doubled from 1990 to 1997.

Hot topics emerging with respect to business technologies include

- Privacy.
- Biometric devices that identify us by our faces, voices, retinas, and other bodily characteristics.

MANUFACTURING

Of course, the big news in manufacturing was the Industrial Revolution of the nineteenth century. However, today's advances make pretty big news themselves: computer-assisted just-about-everything, including design, machine control, inventory control, robotic materials handling, automated inspection, and more. The goals of new manufacturing technologies are efficiency, productivity, and competitiveness, with an eye on safety. Experts classify these technologies as either hard—hardware- or software-based—or soft—referring to techniques.

Important manufacturing technologies include

- Hard technologies, found in Tables 3.5, 3.6 and 3.8.
  - Computer-aided design (CAD) (an application of computer graphics that allows engineers to create and manipulate designs quickly).
  - Computer-aided manufacturing (CAM) (the use of computers to control and monitor manufacturing hardware).
  - Computer numerical control (CNC) (machines that can be programmed locally).
  - Robots. Tables 3.5, 3.6, and 3.8.

- Soft technologies, found in Table 3.7, with manufacturing cells also covered in Table 3.6
  - Just-in-time (JIT) manufacturing (a practice that involves materials arriving when needed rather than being stored as inventory).
  - Manufacturing cells (small physically proximate groups of workers and machines that produce similar items).

MATERIALS

Materials are like songs: unique but limitless compositions made from a small possible number of building blocks. Think of how many different songs can be written using only a few notes! Then imagine the multitude and diversity of materials you could make from the 92 stable elements found in nature.

But unlike songs, materials take physical form. Not only can we touch and see them, we can harness them to do work for us. Advances in materials allow us to keep ourselves more comfortable (both houses and clothing), fly through the atmosphere without burning up, build more sophisticated and hardier structures and machines, use less energy, and keep production costs down. Glass windows changed the world by allowing us to keep the elements out and the heat in. By providing lightweight, shatter-proof packagings, casings, and structural elements, plastics let us build gadgets, tools, vehicles, and even buildings more cheaply and quickly than ever before. New materials have brought us the record, lightweight sub-zero clothing, space flight, and that new car smell we love.

Some types of materials are

- Metals and alloys. Metals comprise ferrous (iron-based) and nonferrous types. Iron and iron alloys such as steel are ferrous metals. Nonferrous metals include aluminum, copper, and titanium. Alloys are metals bonded together in the liquid state. Table 3.9 shows the production of major metals and alloys in 1994 through 1997. Look at the difference in the values of lead and gold, where amounts produced are similar but values light-years apart.

- Polymers. A polymer is a material composed of smaller units called monomers. Most polymers contain carbon molecules in long chains or networks. Different types of polymers, covered in Tables 3.13 through 3.18, include
  - Thermoplastics. Thermoplastics will repeatedly soften when heated and harden when cooled. They are formed by applying heat and pressure to create polymers in unconnected chains. The longer the chain, the tougher and more resistant to cracking the thermoplastic is. Examples of thermoplastics include polystyrene, polyethylene, and nylon. Applications are computer housings, foam cups, and automotive components. Thermoplastics are used in greater volume than thermosets, because they are usually less expensive, can be processed more rapidly, and satisfy most product requirements. Tables 3.13 and 3.15 illustrate this trend.
  - Thermosets. Thermosets are more temperature-resistant and provide better dimensional stability than thermoplastics, but they can't be remelted and reformed as thermoplastics can.

Epoxy and unsaturated polyester represent types of thermosets. Applications include circuit-breaker components and glass-reinforced boat hulls. Thermosets are widely used in nonmolded applications such as paints, coatings, and adhesives.
  - Elastomers and rubbers. Unlike most plastics, elastomers and rubbers are stretchy: They return to their original shape after being pulled. Rubbers are elastomers that recover particularly quickly when deformed. Typical uses for elastomers and rubbers are tires and hoses.

- Composites. Composites combine two or more disparate materials. Each material is different from the others in form and chemical composition, and none of the materials will dissolve in the others. Composites may be used in aircraft (especially the noses) and sports equipment. Information about composites appears in Table 3.18.

- Ceramics. There are three types of ceramics: traditional, technical, and glass. Traditional ceramics include brick, porcelain, and silica. Technical ceramics are highly processed, much stronger than the traditional types, and can operate at far higher temperatures. Glass ceramics are fine-grained, nonporous, crystalline materials with electrical insulating properties, but they are not useful at temperatures above 2000 degrees Fahrenheit. Ceramics can include metallic and nonmetallic elements that are chemically bonded. They may be crystalline or noncrystalline, or both. Most are very hard and strong at high temperatures, but they can be brittle. They are used for valves, transistor bases, and engine components.

- Refractories. Refractories are extremely hardy materials with a melting point above 2876 degrees Fahrenheit. They can be clay, metal, or other minerals, or artificial composites such as silicon carbide. Refractories are used in furnaces and as mechanical components for very high-temperature applications. Information about refractories appears in Tables 3.10 and 3.11.

- Carbon. Because carbon is self-lubricating and corrosion-resistant, it is used in sliding parts of mechanical devices and can be found in seals, motors, pumps, and bearings.

Hot topics in materials include

- Nanotechnology. The ability to move and arrange atoms and molecules one-by-one heralds a sea change not only in materials science, but also in medicine, environmental engineering, and other fields. Scientists predict that they will be able to make materials 50 times stronger than steel of the same weight and custom, and design shatter-proof diamonds for use in electronics and spacecraft. (Natural diamonds can't be coaxed into many shapes, and they break.)

- Smart materials. These high-tech constructions rely on chemical switches and mechanical sensors to tell them when and how to respond to conditions around them.

- Phase-change materials. These chameleons shift between solid and liquid form, giving off and absorbing heat as they change. Such materials are useful for insulation and temperature regulation.
- Superconductivity. If electricity can flow without resistance, energy transfer is extremely efficient. Superconducting materials will facilitate huge advances in computer chip and electric power efficiencies.

BUSINESS TECHNOLOGY

3.1. Manufacturers' Shipments of Business Appliances, United States and Canada, 1995–96 (units).

Product	1995	1996
Calculators, desktop	13,550,000	11,640,000
Calculators, handheld	45,975,000	44,770,000
Computers, mainframe	11,400	11,700
Computers, micro [1]	17,099,000	21,373,000
Computers, mini	255,800	252,300
Computers, notebook	3,693,000	4,590,000
Computers, palmtop	420,000	440,000
Computers, sub-notebook	575,000	600,000
Copiers, plain paper	1,645,800	1,686,250
Dictation equipment, desk	259,000	259,200
Dictation equipment, portable	387,000	391,200
Digital cameras [2]	94,500	151,220
Facsimile equipment	2,778,000	4,345,000
Multifunction devices	575,000	1,951,000
Printers, ink jet	8,690,000	8,765,000
Printers, laser (desktop) [3]	1,013,000	908,370
Printers, laser (personal) [4]	1,680,000	1,450,000
Printers, serial, dot matrix	1,660,000	997,200
Scanners [5]	887,730	1,736,800
Typewriters, electronic	920,000	NA

Notes:
[1] Includes text-processing workstations and low-end units.
[2] Includes low-, mid-, and high-end still devices and image database units.
[3] 8–12 pages per minute.
[4] Under 8 pages per minute.
[5] Includes handheld, sheet-fed, and low- and mid-end flatbed units, film scanners, high-speed drum scanners, and high-resolution and large-format units.

Source: "Business Appliances." *Appliance Magazine* 54 (1997): 40. Reprinted by permission of *Appliance Magazine*. Copyright Dana Chase Publications, Inc., April, 1997.

3.2. World Smart Card* Usage by Global Region, 1995–2002.

Number of Cards (millions)	1997	1998 Projected	1999 Projected	2000 Projected	2001 Projected	2002 Projected
Europe	900	1,020	1,200	1,400	1,600	1,800
Asia/Pacific	270	380	510	680	920	1,300
U.S.	10	50	100	180	280	380
Other	20	30	50	100	200	370
Total	1,200	1,480	1,860	2,360	3,000	3,850

Notes:
* = A smart card is a computer the size of a credit card. Smart cards can contain databases, usually composed of personal information and often password-protected. Applications include credit, debit, and cash cards; personal medical histories; consumer loyalty rewards and other consumer information, etc.

Source: Datamonitor. *Global Smart Card Opportunities, 1997–2002.* New York, NY: Datamonitor, 1997. Page 96.

3.3. Number of Automated Teller Machines Deployed in the United States, 1990–97.

Year	Number of ATMs
1990	80,156
1991	83,545
1992	87,330
1993	94,822
1994	109,080
1995	122,706
1996	139,134
1997	160,000*

Note:
* = *Estimated.*

Source: "Non-Bank ATM Deployers: Quick Moves, Big Ideas." *Bank Technology News* 10 (December 1997): 1+.

3.4. Number of Monthly Transactions per Automated Teller Machine in the United States, 1988–97.

Year	Monthly Transactions per ATM
1988	5,151
1989	5,638
1990	5,980
1991	6,403
1992	6,876
1993	6,772
1994	6,459
1995	6,580
1996	6,399
1997	5,545

Note:
June 1997 and August 1996 data. All other years are September data.

Source: "Fast Times in ATM Land." *Financial Services Online* (November 1997): 55–58.

MANUFACTURING TECHNOLOGY

3.5. U.S. Robot Population, 1982–97.

Year	Units	Year	Units
1982	6,300	1990	40,000
1983	9,000	1991	42,000
1984	14,500	1992	46,000
1985	20,000	1993	50,000
1986	25,000	1994	55,000
1987	29,000	1995	65,000
1988	33,000	1996	72,000
1989	36,000	1997	82,000

Note:
These are year-end figures.

Source: Robotic Industries Association. Ann Arbor, Michigan. Unpublished.

3.6. Manufacturers' Use and Planned Use of Certain Advanced Technologies in the United States, 1988 and 1993.

Technology	1988		1993	
	In use	**Plan to use within 5 years**	**In use**	**Plan to use within 5 years**
At least one of the 17 advanced technologies	68.4%	59.7%	75.0%	52.3%
Design and engineering				
Computer-aided design (CAD) or computer-aided engineering	39.0%	19.6%	58.8%	9.5%
CAD output used to control manufacturing machines	16.9%	21.1%	25.6%	15.3%
Digital representation of CAD output used in procurement	9.9%	17.5%	11.3%	11.8%
Fabrication/machining and assembly				
Flexible manufacturing cells or systems	10.7%	11.5%	12.7%	9.7%
Numerically controlled or computer numerically controlled machines	41.4%	7.9%	46.9%	6.3%
Materials working lasers	4.3%	9.1%	5.0%	7.1%
Pick and place robots	7.7%	12.2%	8.6%	8.5%
Other robots	5.7%	11.2%	4.8%	7.5%
Automated material handling				
Automatic storage and retrieval systems	3.2%	5.8%	2.6%	4.5%
Automatic guided vehicle systems	1.5%	3.8%	1.1%	2.1%
Automated sensor-based inspection or testing				
Performed on incoming or in-process materials	10.0%	11.9%	9.9%	9.8%
Performed on final product	12.5%	12.4%	12.5%	10.9%
Communication and control				
Local area network (LAN) for technical data	18.9%	17.2%	29.3%	15.1%
LAN for factory use	16.2%	19.1%	22.1%	18.6%
Intercompany computer network linking plant to subcontractors, suppliers, or customers	14.8%	20.3%	17.9%	18.8%
Programmable controllers	32.1%	10.7%	30.4%	8.6%
Computers used for control on factory floor	27.3%	22.0%	26.9%	18.1%

Notes:
The U.S. survey included establishments with 20 or more employees selected to represent a universe of almost 40,000 manufacturing establishments classified in Standard Industrial Classification codes 34–38.

Sources: U.S. Bureau of the Census. *Manufacturing Technology 1988.* SMT(88)-1. Washington, DC: U.S. Bureau of the Census, 1989; U.S. Bureau of the Census. *Manufacturing Technology: Prevalence and Plans for Use 1993.* SMT(93)-3. Washington, DC: U.S. Bureau of the Census, 1994.

3.7. Soft Technology Use among Small and Large Plants, United States, 1997.

	Large Plants	Small Plants
Just-in-Time Production (JIT)/Variation of JIT	76.6%	51.9%
Manufacturing Cells	72.9%	38.4%
Total Quality Management (TQM)	72.7%	42.1%
Material Requirements Planning (MRP)	68.9%	32.5%
Bar Codes	67.1%	36.0%
Statistical Quality Control (SQC/SPC)	66.1%	37.7%
Concurrent Engineering	63.6%	32.6%
Manufacturing Resource Planning	49.5%	15.4%
Simulation/Modeling	39.0%	15.6%

Notes:
Based on a survey of 1,000 manufacturing plants in 1997.
"Hard" technology is based on hardware and software.
"Soft" technology refers to techniques as opposed to hardware and software.
A small plant employs fewer than 100 people.
A large plant employs 100 or more people.

Source: Paul M. Swamidass. *Technology on the Factory Floor III: Technology Use and Training in U.S. Manufacturing Firms.* Washington, DC: The Manufacturing Institute, August, 1998. Figure 2, page 3.

3.8. Hard Technology Use among Small and Large Plants, United States, 1997.

	Large Plants	Small Plants
CAD (Computer-aided design)	94.9%	76.3%
Local Area Networks (LAN)	79.2%	51.9%
Machines with Computerized Numerical Control	75.7%	61.9%
CAM (Computer-aided manufacturing) [1]	72.0%	47.7%
Robots	39.5%	12.8%
Automated Inspection	34.1%	13.5%
Computer-integrated Manufacturing (CIM)	32.9%	13.9%
Flexible Manufacturing Systems (FMS) [2]	18.9%	5.3%

Notes:
Based on a survey of 1,000 manufacturing plants in 1997.
"Hard" technology is based on hardware and software.
"Soft" technology refers to techniques as opposed to tangibles like hardware and software.
A small plant employs fewer than 100 people.
A large plant employs 100 or more people.
[1] CAM includes programmable automation of single or multimachine systems.
[2] FMS comprises automated multimachine systems linked by an automated material-handling system.

Source: Paul M. Swamidass. *Technology on the Factory Floor III: Technology Use and Training in U.S. Manufacturing Firms.* Washington, DC: The Manufacturing Institute, August 1998. Figure 3, page 3.

MATERIALS TECHNOLOGY

3.9. Nonfuel Mineral Production in the United States, 1994–97.

Mineral	1994 Quantity	1994 Value	1995 Quantity	1995 Value	1996 Quantity	1996 Value	1997p Quantity	1997p Value
Metals:								
Antimony[4] metric tons	0	0	262	W	242	W	W	W
Beryllium concentrates metric tons	4,330	$ 5	5,040	$ 6	5,260	$ 6	5,300	6
Copper[3]	1,810	$ 4,430,000	1,850	$ 5,640,000	1,920	$ 4,610,000	1,920	$ 4,570,000
Gold[3] kilograms	327,000	$ 4,050,000	317,000	$ 3,950,000	321,000	$ 4,030,000	341,000	$ 3,720,000
Iron ore (usable)	57,600	$ 1,580,000	61,100	$ 1,730,000	62,200	$ 1,770,000	61,600	$ 1,750,000
Iron oxide pigments (crude) metric tons	46,400	$ 6,010	51,700	$ 6,720	44,700	$ 6,990	46,000	$ 7,240
Lead[3] metric tons	363,000	$ 298,000	386,000	$ 359,000	426,000	$ 459,000	440,000	$ 456,000
Magnesium metal metric tons	128,000	$ 389,000	142,000	$ 476,000	133,000	$ 455,000	120,000	$ 378,000
Molybdenum[4] metric tons	46,000	$ 284,000	W	W	57,900	$ 456,000	58,500	$ 467,000
Nickel ore metric tons	0	0	1,560	W	1,330	W	0	0
Palladium kilograms	6,440	$ 29,400	5,260	$ 22,000	6,100	$ 25,500	8,340	$ 46,700
Platinum kilograms	1,960	$ 25,300	1,590	$ 20,800	1,840	$ 23,500	2,500	$ 31,700
Rare-earth metal concentrates metric tons	W	W	22,200	W	20,400	W	20,000	W
Silver[3] metric tons	1,490	$ 253,000	1,560	$ 259,000	1,570	$ 263,000	1,600	$ 224,000
Zinc[3] metric tons	570,000	$ 619,000	614,000	$ 756,000	600,000	$ 676,000	607,000	$ 1,080,000
Combined value of bauxite, manganiferous ore, mercury, titanium concentrates, tungsten, vanadium, zircon concentrates, and values indicated by symbol W	0	0	XX	$ 812,000	XX	$ 190,000	XX	$ 188,000
Total metals	XX	$12,000,000	XX	$14,000,000	XX	$13,000,000	XX	$12,900,000
Industrial minerals (excluding fuels):								
Asbestos metric tons	10,100	$ 5,120	10,200	W	9,550	W	9,070	W
Barite	583	$ 19,100	543	$ 10,400	662	$ 14,700	700	$ 17,500
Boron minerals (B_2O_3) metric tons	1,110,000	$ 443,000	1,190,000	$ 560,000	1,150,000	$ 519,000	622,000[8]	503,000
Bromine metric tons	195,000	$ 155,000	218,000	$ 186,000	227,000	$ 150,000	250,000	$ 247,000
Cement:								
Masonry	3,610	$ 286,000	3,600	$ 307,000	3,470	$ 321,000 /e	3,540	$ 334,000 /e
Portland	74,300	$ 4,460,000	73,300	$ 4,920,000	75,800	$ 5,310,000 /e	77,300	$ 5,520,000 /e
Clays:								
Ball	0	0	993	$ 45,500	973	$ 43,100	1,030	$ 50,700
Bentonite	0	0	3,820	$ 138,000	3,740	$ 134,000	3,780	$ 133,000
Common	0	0	25,600	$ 151,000	26,200	$ 144,000	26,400	$ 164,000
Fire	0	0	583	$ 12,800	505	$ 10,700	403	$ 9,990
Fuller's earth	0	0	2,640	$ 269,000	2,600	$ 278,000	2,570	$ 249,000
Kaolin	0	0	9,480	$ 1,110,000	9,120	$ 1,100,000	8,990	$ 1,110,000
Diatomite	613	$ 152,000	687	$ 171,000	698	$ 176,000	705	$ 178,000
Feldspar metric tons	765,000	$ 31,200	882,000	$ 37,400	890,000	$ 39,400	930,000	$ 41,900
Garnet (industrial) metric tons	51,000	$ 6,100	53,000	$ 9,690	68,200	$ 10,200	70,800	$ 7,580
Gemstones	NA	$ 50,500	NA	$ 48,700	NA	$ 43,600	NA	$ 38,200

3.9. Nonfuel Mineral Production in the United States, 1994–97 (continued).

Mineral	1994 Quantity	1994 Value	1995 Quantity	1995 Value	1996 Quantity	1996 Value	1997[p] Quantity	1997[p] Value
Gypsum (crude)	17,200	$115,000	16,600	$121,000	17,500	$124,000	17,000	$121,000
Helium:								
Crude million cubic meters	39	$38,500	36	$32,100	37	$33,100	38	$33,900
Grade A million cubic meters	100	$199,000	99	$196,000	97	$193,000	100	$198,000
Iodine metric tons	1,630	$12,800	1,220	$12,500	1,270	$14,600	1,330	$24,000
Lime	17,400	$1,020,000	18,500	$1,100,000	19,100	$1,140,000	19,300	$1,150,000
Mica (scrap)	110	$5,780	108	$5,630	97	$7,820	92	$8,430
Peat	552	$15,300	660[5]	17,000[5]	640[5]	$18,500[5]	617[5]	$15,900[5]
Perlite metric tons	644,000	$19,400	700,000	$21,600	684,000	$21,300	703,000	$22,400
Phosphate rock	41,100	$869,000	43,500	$947,000	45,400	$1,060,000	46,300	$1,090,000
Potash (K_2O)	2,970	$284,000	2,880	$284,000	2,960	$299,000	2,990	$318,000
Pumice and pumicite metric tons	490,000	$11,800	529,000	$13,200	612,000	$14,800	538,000	$14,200
Salt	39,700	$990,000	40,800	$1,000,000	42,900	$1,060,000	41,400	$958,000
Sand and gravel:								
Construction	891,000	$3,740,000	910,000	$3,910,000	914,000	$4,000,000	961,000	$4,290,000
Industrial	27,300	$488,000	28,200	$502,000	27,800	$497,000	28,300	$512,000
Silica stone[6] metric tons	514	$3,990	374	W	410	3,810	NA	NA
Sodium compounds:								
Soda ash	9,320	$724,000	10,100	$829,000	10,200	$926,000	10,400	$822,000
Sodium sulfate (natural)	298	$24,200	327	$27,700	306	$27,200	320	$28,800
Stone (crushed)[7]	1,230,000	$6,620,000	1,260,000	$6,750,000	1,330,000	$7,180,000	1,390,000	$7,720,000
Sulfur (Frasch)	3,010	$162,000	3,070	$207,000	W	W	W	W
Tripoli metric tons	82,300	$10,900	80,100	$10,500	79,600	$18,400	NA	NA
Vermiculite metric tons	177,000	$14,200	171,000	W	W	W	W	W
Zeolites metric tons	57,600	NA	46,800	NA	39,300	NA	NA	NA
Combined value of brucite, emery, greensand marl, kyanite, lithium minerals, magnesite, magnesium compounds, olivine, staurolite, stone (dimension) sulfur (Frasch), talc and pyrophyllite, wollastonite, and values indicated by symbol W	XX	0	0	XX	$626,000	XX	$818,000	XX
Total industrial minerals	XX	$21,000,000	XX	$24,600,000	XX	$25,800,000	XX	$26,700,000
Grand total	XX	$33,000,000	XX	$38,600,000	XX	$38,800,000	XX	$39,600,000

(At top of the 1997 Value column, a floating entry reads: XX $743,000.)

Notes:
p = Preliminary
NA = Not available
W = Withheld to avoid disclosing company proprietary data; value included with "Combined value."
XX = Not applicable
e = Estimated
[1] Production as measured by mine shipments, sales, or marketable production (including consumption by producers)
[2] Data are rounded to three significant digits; may not add to totals shown
[3] Recoverable content of ores, etc.
[4] Content of ore and concentrate
[5] Data series changed to production beginning in 1995; prior years shipment data may not be comparable
[6] Includes grindstones, pulpstones, and sharpening stones; excludes mill liners and gringing pebbles
[7] Excludes abrasive stone and bituminous limestone and sandstone; all included elsewhere in table
[8] Weight reported as B_2O_3 and is not comparable to prior years.

Source: Stephen D. Smith, *Preliminary Statistical Summary of Nonfuel Minerals* 1996 and 1997 editions. Washington, DC: U.S. Geological Survey. Table 1 in both editions. http://minerals.er.usgs.gov/minerals/pubs/state/

3.10. Refractories, Value of Shipments, United States, 1986–97 (thousands of dollars).

Year	Total	Clay	Nonclay
1986	$1,521,323	$670,002	$851,321
1987	$1,682,338	$739,770	$942,568
1988	$1,950,032	$813,739	$1,136,293
1989	$2,010,947	$823,266	$1,187,681
1990	$2,002,833	$771,315	$1,231,518
1991	$1,946,842	$784,331	$1,162,511
1992	$1,956,083	$786,078	$1,170,005
1993	$1,930,233	$772,986	$1,157,247
1994	$2,046,979	$906,165	$1,140,814
1995	$2,222,384	$940,835	$1,281,549
1996	$2,295,688	$944,926	$1,350,762
1997	$2,403,518	$967,640	$1,435,878

Note:
Refactories are very hardy materials with a melting point above 2876 degrees Farenheit. They can be clay, metal, other minerals, or artificial composites.

Source: U.S. Bureau of the Census. *Current Industrial Reports, Series MA32C: Refractories.*Washington, DC: U.S. Bureau of the Census, 1997. Table 1. http://www.census.gov/cir/www/ma32c.html

3.11. Changes in Use of Refractory Raw Materials, United States, 1970–94 (in thousands of tons, except synthetic materials, which is percent of total).

	1970	1975	1980	1985	1990	1994
Quartzite	356.4	211.7	116.4	55.8	31.5	24.3
Fireclay	1499.3	830.6	657.3	352.5	297.0	192.3
Bauxite	120.1	118.2	157.2	171.4	188.9	170.0
Natural Magnesite	15.7	29.6	78.0	122.2	142.7	164.3
Synthetic Materials (percent of total)	23.9%	31.6%	33.5%	35.2%	38.5%	41.6%

Source: Charles E. Semler. "The Refractories Industry Never Sleeps." *Ceramic Industry* 147 (August 1997): 32.

3.12. U.S. Chemical Specialties Market, 1997, and Percent Change in 1998 (billions of dollars).

	1997 Sales	% Change in 1998
Industrial coatings	$ 8.3	2%
Agricultural chemicals	$ 7.7	2%
Adhesives and sealants	$ 6.4	5%
I&I cleaners**	$ 6.1	2%
Electronic chemicals	$ 5.5	8%
Specialty polymers	$ 4.6	8%
Food ingredients	$ 3.3	6%
Plastics additives	$ 3.1	4%
Construction chemicals	$ 2.8	3%
Water management	$ 2.6	3%
Specialty surfactants	$ 2.0	3%
Flavors and fragrances	$ 1.9	6%
Lubricants and additives	$ 1.8	2%
Dyes and pigments	$ 2.8	4%
Catalysts	$ 1.4	4%
Printing inks	$ 1.1	4%
Oil field chemicals	$ 1.0	4%
Paper additives	$ 0.8	3%
Flame retardants	$ 0.7	5%
Rubber chemicals	$ 0.5	3%
Others	$ 17.5	4%

Notes:
* = Total market is $83 billion. Column total varies due to rounding.
** = Industrial and institutional cleaners.

Source: Strategic Analysis. Wyomissing, PA. Cited in Chemical Week (January 7, 1998): 32.

3.13. Plastics: Production, Sales, and Captive Use, United States, 1996 and 1997 (millions of pounds, dry weight basis)[1].

Resin	U.S. Production			Total Sales and Captive Use		
	1997	1996	% Change 97/96	1997	1996	% Change 97/96
Epoxy [2]	654	662	-1.20%	650	643	1.1%
Polyester, Unsaturated [7]	1,621	1,557	4.10%	1,657	1,565	5.9%
Urea [3][6]	2,335	2,147	8.8%	2,334	2,139	9.1%
Melamine [3][6]	303	287	5.6%	326	293	11.3%
Phenolic [3][6]	3,734	3,476	7.4%	3,729	3,496	6.5%
Total Thermosets	8,647	8,129	6.4%	8,696	8,136	6.9%
LDPE [2]	7,691	7,784	-1.2%	7,911	7,874	0.5%
LLDPE [2]	6,888	6,361	8.3%	8,373	7,799	7.4%
HDPE [2]	12,557	12,373	1.5%	13,482	13,211	2.1%
PP [2][3]	13,320	11,991	11.1%	13,125	12,121	8.3%
ABS [2][3]	1,374	1,477	-7%	1,337	1,433	-6.7%
SAN [2][3]	96	121	-20.7%	109	124	-12.1%
Other Styrenics [2][3]	1,623	1,492	8.8%	1,604	1,518	5.7%
PS [2]	6,380	6,065	5.2%	6,489	6,071	6.9%
Nylon [2][4]	1,222	1,120	9.1%	1,231	1,105	11.4%
Polyvinyl Chloride [3]	14,084	13,220	6.5%	14,049	13,299	5.6%
Thermoplastic Polyester [2][3]	4,260	4,031	5.7%	4,063	3,962	2.5%
Total Thermoplastics	69,495	66,035	5.2%	71,773	68,517	4.8%
Subtotal	78,142	74,164	5.4%	80,469	76,653	5.0%
Engineering Resins [3][5]	2,619	2,409	8.7%	2,246	2,094	7.3%
All Other	8,045	7,722	4.2%	8,001	7,679	4.2%
Total Engineering and Other	10,664	10,131	5.3%	10,247	9,773	4.9%
Grand Total	88,806	84,295	5.4%	90,716	86,426	5%

Notes:

[1] Except phenolic resins, which are reported on a gross weight basis.

[2] Sales & captive use data include imports.

[3] Canadian production and sales & captive use data included in U.S. domestic data.

[4] Nylon sales to Canada and Mexico included in U.S. sales & captive use.

[5] Includes Acetal, Fluoropolymers, Polyamide-imide, Polycarbonate, Thermoplastic Polyester, Polyimide, Modified Polyphenylene Oxide, Polyphenylene Sulfide, Polysulfone, Polyetherimide, and Liquid Crystal Polymers.

[6] Canadian participation incomplete in 1996.

[7] Includes material produced in Canada and sold in the United States.

Source: Data obtained from Society of the Plastics Industry's Web site at http://www.socplas.org. Attributed to Society of the Plastics Industry Committee on Resin Statistics as compiled by Association Services Group, LLC, April 1998. Similar table can be found in Society of the Plastics Industry. *Facts & Figures of the U.S. Plastics Industry, 1998 Edition.* Washington, DC: The Society of the Plastics Industry, Inc., 1998. Page 37.

3.14. Total Resin Sales and Captive Use by Major Market, United States, 1992–97 (in millions of pounds, dry weight basis).

Major Market	1992	1993	1994	1995	1996	1997
Transportation	2,817	3,221	3,795	3,916	3,964	4,102
Packaging	18,284	19,569	19,551	19,334	21,271	21,767
Building and Construction	11,876	12,885	14,715	14,321	16,199	17,117
Electrical/Electronic	2,766	2,981	3,325	2,966	3,137	3,150
Furniture and Furnishing	2,559	2,759	3,118	3,198	3,477	3,721
Consumer and Institutional	6,093	6,015	9,266	9,054	9,804	10,649
Industrial/Machinery	617	768	836	818	980	950
Adhesives/Inks/Coatings	1,723	1,572	1,789	1,795	1,833	2,019
All Other	6,877	7,234	7,515	8,050	9,361	9,573
Exports	6,950	6,632	6,889	7,742	8,722	9,667
Total	60,562	63,636	70,799	71,194	78,748	82,715

Notes:

Major market volumes are derived from plastic resins sales and captive use data as compiled by Association Services Group. This study reflects data collected for the following individual resin categories:

  Thermosets: Epoxy; Polyester, Unsaturated; Urea and Melamine; Phenolic

  Thermoplastics: Low density Polyethylene; Linear Low Density Polyethylene;

    High Density Polyethylene; Polypropylene; Acrylonitrile-Butadiene-Styrene (ABS);

    Styrene-Acrylonitrile (SAN) and other styrene-based polymers (SBP); Polystyrene;

    Styrene Butadiene Latexes (SBL); Nylon; Polyvinyl Chloride; Thermoplastic Polyester;

    Engineering Resins

Source: Data from 1993 to 1997: Society of the Plastics Industry. *Facts & Figures of the U.S. Plastics Industry, 1998 Edition.* Washington, DC: The Society of the Plastics Industry, Inc., 1998. Page 25.

Data from 1992: Society of the Plastics Industry Committee on Resin Statistics. Available at the Society's Web site at http://www.socplas.org

3.15. Production of Thermosets and Thermoplastics, United States, 1987–97 (millions of pounds on dry-weight basis unless otherwise specified).

	1987	1988	1989	1990	1991	1992	1993	1994	1995	1996	1997
Thermosetting Resins											
Epoxy (unmodified)	433	486	509	499	197	457	512	601	632	662	681
Melamine [1]	212	207	222	202	196	232	270	300	290	287	309
Phenolic [1][2][3]	2,869	3,066	2,879	2,946	2,658	2,923	3,078	3,229	3,204	3,476	3,618
Polyester (unsaturated)	1,367	1,404	1,319	1,221	1,075	1,175	1,264	1,468	1,577	1,557	1,625
Urea [1]	1,382	1,425	1,477	1,496	1,483	1,548	1,744	1,914	1,816	2,147	2,340
Total Thermosetting Resins	6,263	6,588	6,408	6,364	5,909	6,335	6,868	7,512	7,519	8,129	8,573
Thermoplastic Resins											
Polyethylene											
Low density [4]	9,599 [5]	10,397 [5]	6,575	7,255	7,236	7,273	7,226	7,578	7,643	7,784	7,691
Linear low density [4]		—	3,120	3,893	4,346	4,644	4,841	5,022	5,243	6,361	6,888
High density [6]	7,995	8,400	8,102	8,337	9,213	9,808	9,941	11,117	11,211	12,373	12,545
Polypropylene [7]	6,647	7,274	7,238	8,310	8,330	8,421	8,629	9,539	10,890	11,991	13,309
Styrene Polymers											
Polystyrene	4,780	5,187	5,104	5,021	4,954	5,096	5,383	5,848	5,656	6,065	6,572
Styrene-acrylonitrile [8]	126	148	113	135	109	113	105	138	130	121	96
Acrylonitrile-butadiene-styrene and other styrene polymers [8][9]	2,254	2,418	2,394	2,351	2,287	2,610	2,924	3,230	2,908	2,969	3,000
Polyamide, nylon type	507	566	569	558	576	668	768	943	1,020	1,103	1,204
Polyvinyl chloride & copolymers [8]	7,971	8,350	8,478	9,096	9,164	9,989	10,257	11,712	12,295	13,220	14,084
Thermoplastic polyester [8]	1,394	1,652	1,630	1,879	2,115	2,413	2,549	3,196	3,785	4,031	4,303
Total Thermoplastic Resins	41,273	44,392	43,323	46,835	48,330	51,035	52,623	58,323	60,781	66,018	69,692
Grand Total	47,536	50,980	49,731	53,199	54,239	57,370	59,491	65,835	68,300	74,147	78,265

Notes:
[1] 1997 includes data from Canada.
[2] Includes other tar-acid resins.
[3] Gross-weight basis.
[4] Density 0.940 and below.
[5] Low density and linear low density.
[6] Density above 0.940.
[7] 1995 includes data from Canada.
[8] 1994 includes data from Canada.
[9] Includes styrene-butadiene copolymers and other styrene-based polymers.

Source: Society of the Plastics Industry. Cited in "PLASTICS: Production of Thermosets and Thermoplastic Up Strongly Last Year." *Chemical & Engineering News 76* (June 29, 1998): 45.

3.16. World's Most Widely Used Commodity Plastics, 1998.

Plastic	Percent of total	Million Metric Tons
Polyethylene	40.3%	44.0
ABS/SAN*	3.8%	4.2
Polystyrene	10.6%	11.6
Polypropylene	23.3%	25.5
Polyvinyl chloride	22.0%	24.0
Total	100%	109.3

Notes:
*= Acrylonitrile-butadiene-styrene/styrene-acrylonitrile.
Pounds do not equal total due to rounding.

Source: SRI Consulting. World Petro Chemicals Program, 1999 edition. Menlo Park, CA.

3.17. Plastics Consumption by World Region, 1998.

	Percentage
Asia	34%
Japan	9%
Rest of Asia	25%
Western Europe	25%
U.S.	24%
Latin America	5%
Canada	2%
Mexico	2%
Other	8%

Note:
Includes polyethylene, polyvinyl chloride, polypropylene, polystyrene, acrylonitrile- butadiene-styrene/styrene-acrylonitrile.

Source: SRI Consulting. World Petrochemicals Program, 1999 edition. Menlo Park, CA.

3.18. Composites Shipments Comparison, United States, 1987–97 (millions of pounds).

Markets	1987	1988	1989	1990	1991	1992	1993	1994	1995	1996	1997	1998 (Projected)
Aircraft/Aerospace/Defense	36	39	41	39	38.7	32.3	25.4	24.2	23.7	23.7	23.9	24.2
Appliance/Business Equipment	141	150	151	153	135.2	143.2	147.5	180.7	188.8	176.9	185.0	190.6
Construction	508	495	470	468	420	483	530	696.9	629.3	655.1	699.6	729.3
Consumer Products	167	169	158	185	148.7	152.2	185.7	174.8	184	194.2	210.0	218
Corrosion-resistant Equipment	329	349	335	350	355	332.3	352	376.3	394.4	381.1	396.0	410
Electrical/Electronic	214	230	229	241	231.1	280	274.9	299.3	314.8	318.8	348.2	365.8
Marine	413	452	405	375	275	304.4	319.3	363.5	372	367.9	353.0	350
Transportation	656	695	677	705	682.2	750	622.1	945.8	881.8	988.5	1,095.2	1,156.6
Other	75	80	78	79	73.5	83.4	89.3	101.8	108.7	107.3	110.8	116

Note:
Composites combine two or more disparate materials, each of which is different from others in form and chemical composition.

Source: Society of the Plastics Industry, The Composites Institute. Harrison, NY. Unpublished.

3.19. Consumption of Oils and Fats for Industrial Use, United States, 1988–95.

Coconut Oil Consumption

Year	Total consumption	Total edible	Total inedible	Soap	Paint or varnish	Feed	Resins and plastics	Lubricants[1]	Fatty acids	Other products
				Million pounds						
1988/89	688.8	211.2	477.6	130.6	1.4	D	14.6	D	121.9	206.6
1989/90	525.2	160.6	364.6	156.9	2.1	0	9.7	4	134.6	57.3
1991	815.6	153.0	662.6	158.0	D	D	2.4	D	426.7	72.8
1992	875.4	176.3	699.1	121.7	D	0	3.2	D	D	D
1993	936.3	218.0	718.3	132.0	D	0	3.1	D	D	D
1994	969.2	227.1	742.1	146.1	D	0	2.3	D	D	D
1995	676.1	252.2	625.9	92.3	D	0	2.3	0	D	D

Notes:
D = Data withheld to avoid disclosing figures for individual companies.
[1] Includes similar oils.

Inedible Tallow Consumption

Year	Total consumption	Total edible	Total inedible	Soap	Paint or varnish	Feed	Resins and plastics	Lubricants[1]	Fatty acids	Other products
				Million pounds						
1988/89	3,086.7	0	3,086.7	374.9	0	1,925.4	0	70.3	680.0	36.1
1989/90	3219.0	0	3219.0	398.4	0	1,982.9	0	109.0	684.0	44.7
1991	2,949.3	0	2,949.3	391.5	0	1,748.4	0	59.6	700.9	48.9
1992	3,050.1	0	3,050.1	334.4	0	1,954.4	0	63.2	659.0	39.1
1993	3,018.2	0	3,018.2	299.6	0	1,994.7	0	71.5	615.1	37.3
1994	3,189.9	0	3,189.9	300.8	0	2,101.9	0	81.8	634.0	71.4
1995	3,222.8	0	3,222.8	263.9	0	2,166.5	0	89.7	656.9	45.8

Notes:
[1] Includes similar oils.

3.19. Consumption of Oils and Fats for Industrial Use, United States, 1988–95 *(continued)*.

Lard Consumption

Year	Total consumption	Total edible	Total inedible	Soap	Paint or varnish	Feed	Resins and plastics	Lubricants[1]	Fatty acids	Other products
				Million pounds						
1988/89	389.9	324.5	65.4	0	0	D	0	D	D	D
1989/90	369.3	303.8	66.5	D	0	D	0	9.1	D	D
1991	393.1	313.8	79.3	0	0	D	0	5.7	D	4.1
1992	479.7	345.0	134.6	0	0	D	0	10.9	D	13.5
1993	473.3	324.6	149.7	0	0	D	0	8.6	D	28.7
1994	451.9	324.7	127.2	0	0	0	0	8.9	0	118.3
1995	488.7	364.3	124.4	0	0	0	0	27.2	0	97.2

Notes:
D = Data withheld to avoid disclosing figures for individual companies.
[1] Includes similar oils.

Linseed Oil Consumption

Year	Total consumption	Total edible	Total inedible	Soap	Paint or varnish	Feed	Resins and plastics	Lubricants[1]	Fatty acids	Other products
				Million pounds						
1988/89	154.9	0	154.9	0	101.6	0	23.1	D	D	28.2
1989/90	110.5	0	110.5	0	30.3	D	52.5	D	D	23.8
1991	95.8	0	95.8	0	40.7	0	41.6	D	D	12.7
1992	154.4	0	154.4	0	69.0	0	31.3	D	D	D
1993	125.8	0	125.8	0	66.9	0	25.4	D	D	D
1994	124.3	0	124.3	0	33	0	50.9	D	D	40.4
1995	112.8	0	112.8	0	30.2	0	51.4	0	0	31.2

Notes:
D = Data withheld to avoid disclosing figures for individual companies.
[1] Includes similar oils.

3.19. Consumption of Oils and Fats for Industrial Use, United States, 1988–95 *(continued)*.

Rapeseed Oil Consumption[1]

Year	Total consumption	Total edible	Total inedible	Soap	Paint or varnish	Feed	Resins and plastics	Lubricants[2]	Fatty acids	Other products
					Million pounds					
1989/90	D	265.0	D	0	D	D	D	D	D	D
1991	D	285.1	D	0	0	D	0	D	D	D
1992	D	360.5	D	0	0	D	0	D	D	D
1993	D	362.5	D	0	0	0	0	D	D	D
1994	D	446.3	D	0	0	0	0	0	0	D
1995	D	315.8	D	0	0	0	0	0	D	D

Notes:

D = Data withheld to avoid disclosing figures for individual companies.
[1] Includes both canola and industrial rapeseed.
[2] Includes similar oils.

Soybean Oil Consumption

Year	Total consumption	Total edible	Total inedible	Soap	Paint or varnish	Feed	Resins and plastics	Lubricants[1]	Fatty acids	Other products
					Million pounds					
1988/89	9,917.6	9,635.8	281.8	1.5	34.9	D	123.7	D	D	68.2
1989/90	10,808.3	10,536.7	271.6	D	38.2	D	112.4	D	D	52.4
1991	11,267.7	10,966.7	301.0	D	49.2	D	104.7	D	D	40.4
1992	11,471.6	11,168.7	302.8	D	43.5	22.3	94.0	5.9	D	69.8
1993	12,495.6	12,200.9	294.7	D	38.7	23.7	98.1	5.8	D	65.8
1994	12,474.1	12,157.8	316.3	D	47.6	D	119.6	D	D	91.9
1995	12,354.0	12,049.3	304.7	D	47.0	D	122.4	0	D	99.6

D = Data withheld to avoid disclosing figures for individual companies.
[1] Includes similar oils.

3.19. Consumption of Oils and Fats for Industrial Use, United States, 1988–95 (continued).

Tall Oil Consumption

Year	Total consumption	Total edible	Total inedible	Soap	Paint or varnish	Feed	Resins and plastics	Lubricants[1]	Fatty acids	Other products
	Million pounds									
1988/89	1,234.3	0	1,234.3	8.3	31.8	0	18.0	8.1	1,157.3	10.8
1989/90	1,024.7	0	1,024.7	8.4	7.4	0	21.7	7.1	969.9	10.2
1991	940.0	0	940.0	3.5	5.4	0	11.6	4.0	906.5	9.0
1992	883.5	0	883.5	D	D	0	19.4	7.0	841.8	11.4
1993	891.8	0	891.8	D	D	0	23.0	6.3	806.9	D
1994	1,362.5	0	1,362.5	D	D	0	48.4	6.1	1,025.0	D
1995	1,357.7	0	1,357.7	D	D	0	16.0	7.9	908.5	D

Notes:
D = Data withheld to avoid disclosing figures for individual companies.
[1] Includes similar oils.

Tung Oil Consumption

Year	Total consumption	Total edible	Total inedible	Soap	Paint or varnish	Feed	Resins and plastics	Lubricants[1]	Fatty acids	Other products
	Million pounds									
1988/89	7.7	0	7.7	0	3.5	0	1.8	0	0	2.4
1989/90	8.9	0	8.9	0	2.7	0	3.8	0	0	2.4
1991	6.4	0	6.4	0	D	d	2.9	0	0	1.6
1992	7.3	0	7.3	D	D	0	3.3	3.5	0	28.4
1993	11.2	0	11.2	D	1	0	8.6	0	0	1.6
1994	9.3	0	9.3	D	1.2	0	6.6	2.4	0	1.5
1995	20.2	0	20.2	0	D	0	d	0	0	12.1

Notes:
D = Data withheld to avoid disclosing figures for individual companies.
[1] Includes similar oils.

3.19. Consumption of Oils and Fats for Industrial Use, United States, 1988–95 *(continued)*.

Year	Total consumption	Total edible	Total inedible	Total Fats and Oils Consumption						
				Soap	Paint or varnish	Feed	Resins and plastics	Lubricants[1]	Fatty acids	Other products
				Million pounds						
1988/89	19,426.7	13,542.0	5,884.7	744.5	180.3	2,079.3	202.3	115.8	2,074.1	488.4
1989/90	20,036.0	14,382.7	5,653.3	792.0	89.5	2,143.5	222.4	157.1	1,944.7	304.1
1991	20,332.1	14,613.0	5,719.1	832.9	106.8	1,974.0	182.6	101.7	2,234.7	286.4
1992	20,751.7	14,847.3	5,904.4	738.8	123.8	2,176.5	165.5	109.4	2,041.2	549.3
1993	21,590.4	15,744.7	5,845.7	748.5	125.2	2,199.5	170.2	116.0	1,897.6	588.7
1994	22,058.7	15,373.8	6,684.9	770.0	115.1	2,272.5	240.7	219.3	2,306.2	761.1
1995	21,157.4	15,056.3	6,101.1	593.8	102.8	2,340.9	210.7	141.9	1,963.6	747.4

Notes:
[1] Includes similar oils.

Source: Bureau of the Census. Cited in U.S. Department of Agriculture, Economic Research Service, Commercial Agriculture Division. *Industrial Uses of Agricultural Materials.* IUS-6. Washington, DC: U.S. Department of Agriculture, Economic Research Service, Commercial Agriculture Division, August 1996. Tables section IUS-6: Tables 21, 23, 24, 25, 26, 27, 28, 29, and 30. Pages 44–46. http://www.econ.ag.gov/epubs/pdf/ius6/index.htm.

3.20. Industry Shipments and Number of Producers of Adhesives, United States, 1992.

Type	Number of companies	Value (million dollars)
Natural base glues and adhesives (except rubber adhesives):		
Animal glue	12	$47
Protein adhesives: casein, soybean, fish, etc.	8	$29
Starch, dextrin	17	$102
Other except rubber (gums, silicates, asphalt, etc.)	—	$30
Subtotal, natural	—	$208
Synthetic resin and rubber adhesives, including laminating, bonding:		
Epoxy	62	$301
Phenolic and resorcinol type	14	$247
Urea and modified urea-formaldehyde	11	$175
Polyvinyl acetate	71	$624
Polyvinyl chloride and other vinyl types	37	$96
Acrylic (acrylate)	49	$221
Cyanoacrylate	11	$122
Urethane	29	$80
Polyester	9	$84
Styrenic	14	$153
Hot melt adhesives including vinyl, polyamide, polyolefin	52	$423
Adhesive films/tapes, pressure-sensitive adhesives	9	$313
Rubber cement	37	$272
Rubber-synthetic resin combinations	36	$558
Cellulosic, polyamide, anaerobic, etc.	43	$250
Subtotal, synthetic resin and rubber	—	$3,919
Grand total		$4,127

Source: Partially estimated by staff of the U.S. International Trade Commission from 1992 Census of Manufactures. Cited in U.S. International Trade Commission, Office of Industries. *Industry Trade Summary: Adhesives, Glues, and Gelatin.* USITC Publication 3093. Washington, DC: United States International Trade Commission, 1998. Table 3, page 19. ftp://ftp.usitc.gov/pub/reports/studies/PUB3093.PDF

3.21. Adhesives Production by Major Class, United States, 1995.

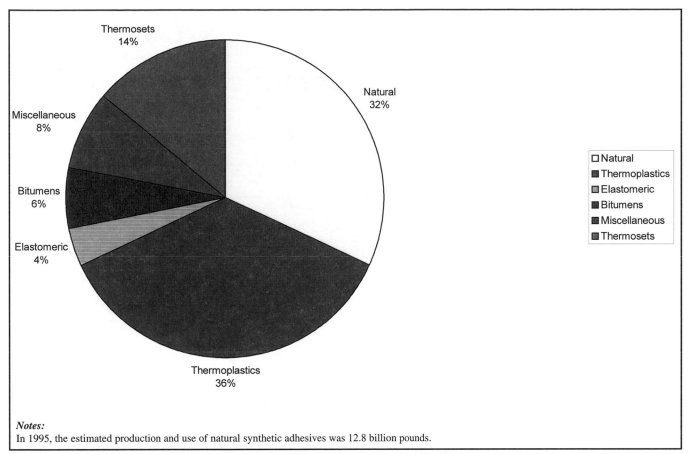

Thermosets
14%

Natural
32%

Miscellaneous
8%

Bitumens
6%

Elastomeric
4%

Thermoplastics
36%

☐ Natural
■ Thermoplastics
▨ Elastomeric
■ Bitumens
■ Miscellaneous
▨ Thermosets

Notes:
In 1995, the estimated production and use of natural synthetic adhesives was 12.8 billion pounds.

Source: Irshad Ahmed, Booz-Allen and Hamilton, Inc., McLean, Virginia, July 1996. Cited in U.S. Department of Agriculture, Economic Research Service, Starches and Sugars Section. Commercial Agriculture Division. *Industrial Uses of Agricultural Materials.* IUS-6. Washington, DC: U.S. Department of Agriculture, Economic Research Service, Commercial Agriculture Division, August 1996. Starches and Sugars section. Figure 2, page 15. http://www.econ.ag.gov/epubs/pdf/ius6/index.htm.

Chapter 4. Communications

In 1799, it took 24 days for the news of George Washington's death to travel from Mount Vernon, Virginia, to western Ohio. In 1830, it took two days for President Andrew Jackson's State of the Union address to go approximately the same distance. Today, we can see and hear astronauts thousands of miles away in space in real time, and we can't imagine what it would have been like to live without telephones.

Communications have changed in two vital ways in the twentieth century: they have sped up, and they have enabled us to reach masses of people. We have so many choices now! We can communicate one-to-one and one-to-multiple, where the multiple can be as low as two or as high as the whole planet. We can choose voice devices (corded phone, cordless phone, cell phone, voice mail, mobile radio), fax, e-mail, pager, broadcast TV, satellite TV, cable TV, AM radio, FM radio, ham radio, Web, local- and wide-area networks, telegraph, and, oh yes, paper mail.

Communications are facilitated by devices we construct—satellites that hug the earth and relay signals around its curves— by coax, fiber optic, copper, and other cables—and by natural phenomena we harness, like radio waves. But communications can also be jammed, slowed, garbled, and invaded by weather, errant digging, wire taps and other snooping devices, scramblers, encryption devices, hackers, electrical outages, and too much traffic.

As we adopt more high-tech communications systems, the world shrinks, and we can meet and interact with people we might never have met otherwise. At the same time, we become highly dependent on that technology for our survival, and vulnerable as a result, as those of us who lost phone service during the 1994 Northridge earthquake will attest.

Hot topics in communications include

- The integration of telephones, televisions, and computers
- Allocation of various frequencies for particular uses and users

The tables that follow offer a smorgasbord covering many familiar communications technologies and the infrastructure that supports them. Tables 4.1 through 4.5 illustrate the growth in communications cables, with Table 4.5 showing economies of scale as more and more transatlantic cable has been laid. Table 4.6, which depicts telephone lines by state (a.k.a. loops), shows that a state doesn't have to be large to support a healthy number of phone companies (see Iowa in particular). Table 4.7 shows how the number of calls we make has skyrocketed, as one would suspect in an age where people communicate more and more by e-mail and fax. Tables 4.14 through 4.17 show how much we have been spending for our phone service over the years: interestingly, as overall rates and expenditures have increased, the daytime cost of long-distance calls has fallen, though it is still higher than calling during off hours. (See Table 4.16.) Tables 4.18, 4.19, and 4.20 show us that we're using up area codes and 800 numbers like mad (no surprise there). Tables 4.26 and 4.27 show that as more and more people have acquired cellular phones, the cost for that service has dropped. Or is it the other way around? And finally, table 4.32 shows how much providers are paying for the privilege of offering various wireless services and how much bidding they are willing to endure to gain it (the people who bid for C block and D, E, & F block Broadband PCS have to be admired for their stamina, but then look at how much money was involved).

COMMUNICATION CABLES

4.1. Fiber Miles of Fiber Deployed by Local Operating Companies, United States, 1985–97. (in thousands)

Company	1985	1986	1987	1988	1989	1990	1991	1992	1993	1994	1995	1996	1997
Ameritech	77.7	111.1	147.1	177.5	228.4	285.5	400.7	585.6	802.1	918.9	1,095.9	1,339.3	1,556.0
Bell Atlantic	83.1	150.8	227.5	311.0	373.4	523.0	809.7	1,026.6	1,568.7	1,919.4	2,169.2	2,403.5	4,374.0
BellSouth	50.8	170.1	218.5	319.2	445.5	609.2	768.6	938.7	1,121.0	1,380.9	1,685.0	2,012.0	2,293.0
NYNEX	83.4	129.7	207.1	290.6	357.8	473.3	637.0	806.7	964.4	1,112.0	1,264.7	1,422.50	NA
Pacific Telesis	84.3	97.8	101.1	110.3	126.9	185.2	246.4	311.7	374.9	424.2	481.5	540.4	605.0
SBC	70.5	151.0	182.9	214.9	270.3	352.3	477.7	576.4	775.0	970.8	1,234.8	1,504.4	1,724.0
US West	47.3	70.1	107.8	164.0	234.9	351.6	542.3	797.6	1,042.5	1,238.8	1,483.3	1,615.3	1,668.0
Bell Totals:	497.1	880.7	1,192.0	1,587.6	2,037.1	2,780.0	3,882.4	5,043.3	6,648.6	7,965.0	9,414.4	10,837.3	12,219.0
GTE Companies	NA	NA	NA	134.7	163.4	317.5	390.5	513.7	672.4	795.2	930.4	1,064.9	1,262.0
Sprint	NA	NA	NA	32.3	54.6	83.5	115.6	139.7	187.0	257.4	353.0	440.9	536.0
Rural Companies	NA	2.0	14.2	28.7	42.3	68.2	NA	NA	NA	NA	NA	NA	NA
Total Reported:	497.1	882.7	1,206.2	1,783.2	2,297.3	3,249.3	4,388.5	5,696.8	7,508.0	9,017.6	10,697.8	12,343.1	14,017.0

Notes:
A fiber mile is a measure of capacity—the number of fiber cables times the number of fibers times the length of the route.

Source: Jonathan M. Kraushaar. *Fiber Deployment Update End of Year 1996 and Fiber Deployment Update End of Year 1997.* Washington, DC: Industry Analysis Division, Common Carrier Division, Federal Communications Commission. Table 6 in both editions. http://www.fcc.gov/Bureaus/Common_Carrier/Reports/FCC-State_Link/Fiber/

4.2. Competitive Access Fiber Systems, United States, 1990–97.

Company Name	Route Miles								Thousands of Fiber Miles							
	1990	1991	1992	1993	1994	1995	1996	1997	1990	1991	1992	1993	1994	1995	1996	1997
American (ACS)	NA	NA	NA	NA	NA	NA	697	1,061	NA	NA	NA	NA	NA	NA	48.8	92.5
Brooks Fiber	109	141	193	264	264	480	1,059	2,494	2.6	3.8	4.3	6.2	18	24.3	71.3	215.2
Eastern Telelogic	140	140	140	194	233	395	438	NA	3.7	3.7	3.7	4.4	4.4	13.8	18.8	NA
Electric Lightwave	NA	6	104	126	225	466	516	952	NA	0.5	6.8	11.7	20.5	NA	61.5	108.4
GST Telecom	NA	NA	NA	NA	NA	NA	305	415	NA	NA	NA	NA	NA	NA	21.5	38.4
Hyperion	NA	NA	NA	NA	NA	NA	2,887	4,761	NA	NA	NA	NA	NA	NA	138.6	220
IntelCom Group	NA	105	132	151	424	637	2,073	2,872	NA	4.8	6.5	8.6	19.0	28.8	69.6	108.1
Intermedia (ICI)	159	165	213	335	372	561	654	605	2.9	3.0	5.2	10.2	11.3	20.5	24.1	35.0
Kansas City Fiber Net	91	94	97	200	200	200	NA	NA	2.5	2.6	2.9	0	3.7	3.8	NA	NA
MCImetro	NA	NA	NA	NA	NA	2,338	2,948	2,948	NA	NA	NA	NA	NA	NA	NA	NA
MFS (WorldCom)	309	546	1,133	1,530	2,387	3,112	3,523	3,858	17.2	29.8	41.4	67.0	106.9	188.0	229.9	283.7
McLeod USA, Inc.	65	75	95	121	116	NA	2,352	NA	1.6	1.8	3.7	5.0	3.0	NA	123.9	NA
Phoenix FiberLink, Inc.	NA	NA	NA	NA	NA	32	76	NA	NA	NA	NA	NA	NA	3.1	7.2	NA
Teleport (TCG)	328	507	1,018	2,082	3,902	5,428	6,744	9,474	18.5	24.7	40.0	96.1	167.3	253.3	346.0	491.1
Time Warner	59	86	88	96	348	3,312	4,232	5,911	0.5	1.2	1.2	1.4	10.4	107.9	151.7	233.5
US Signal	67	115	144	367	554	NA	NA	NA	5.6	6.3	7.3	20.2	31.6	NA	NA	NA
Total Reported:	1,326	1,980	3,357	5,466	9,025	16,961	28,503	35,351	55.1	82.2	122.9	230.7	396.2	643.4	1,312.9	1,826.0

Notes:
A fiber mile is a measure of capacity—the number of fiber cables times the number of fiber times the length of the route.
A route mile is the length of the cable. It does not take into account how many fibers or how many parallel cables there are.

Source: Jonathan M. Kraushaar. *Fiber Deployment Update End of Year 1996 and Fiber Deployment Update End of Year 1997.* Washington, DC: Industry Analysis Division, Common Carrier Division, Federal Communications Commission. Table 14 in both editions. http://www.fcc.gov/Bureaus/Common_Carrier/Reports/FCC-State_Link/Fiber/

4.3. Fiber Miles Installed—Interexchange Carriers, United States, 1985–97 (thousands of fiber miles).

Calendar Year:	1985	1986	1987	1988	1989	1990	1991	1992	1993	1994	1995	1996	1997
AT & T	136.2	261.4	432.0	704.7	838.4	935.7	1,010.9	1,018.5	1,055.6	1,141.6	1,179.1	1,259.0	1,282.2
Consolidated	3.5	3.5	3.7	3.7	3.7	3.7	3.7	3.7	3.7	6.5	NA	15.6	15.6
Elec. Lightwave	NA	NA	NA	NA	NA	NA	NA	NA	NA	NA	14.9	37.7	30.6
Frontier (RCI)	7.0	7.0	7.2	2.6	2.7	2.7	2.7	2.7	2.7	2.6	3.3	3.3	71.1
GST Telecom.	NA	NA	NA	NA	NA	NA	NA	NA	NA	NA	NA	3.6	36.9
IXC Comm.	10.0	10.0	14.0	14.0	14.0	14.2	14.2	14.6	20.8	22.2	22.2	70.5	125.1
LCI	13.7	17.3	22.3	22.3	22.3	22.3	24.7	24.7	24.7	24.7	24.7	24.7	41.4
MCI	83.9	179.1	259.3	278.8	304.2	388.0	413.7	430.0	450.0	525.0	597.4	655.4	663.0
MRC	NA	NA	8.0	8.0	10.1	10.1	10.1	10.2	10.2	10.3	10.2	19.2	19.2
Qwest Comm.	NA	NA	NA	NA	NA	NA	NA	NA	NA	NA	NA	113.3	192.4
Sprint	122.4	249.3	343.2	449.5	450.8	453.4	466.7	466.7	467.2	467.2	467.2	468.7	471.5
TCG	NA	NA	NA	1.7	1.7	1.7	1.7	1.7	1.7	1.7	1.7	1.7	NA
Valley Net	NA	NA	NA	NA	6.1	6.8	7.2	7.2	7.2	NA	NA	NA	NA
WorldCom	79.0	190.8	203.5	237.9	245.5	254.6	255.9	255.9	256.2	256.2	266.2	276.9	470.9
Total Reported:	455.7	918.4	1,293.3	1,723.3	1,899.5	2,093.2	2,211.5	2,235.9	2,300.0	2,458.1	2,587.0	2,949.6	3,420.0

Notes:
A fiber mile is a measure of capacity—the number of fiber cables times the number of fibers times the length of the route. A route mile is the length of the cable. It does not take into account how many fibers or how many parallel cables there are.

Source: Jonathan M. Kraushaar M. *Fiber Deployment Update End of Year 1996 and Fiber Deployment Update End of Year 1997.* Washington, DC: Common Carrier Bureau, Industry Analysis Division. Federal Communications Commission. Table 2 in both editions. http://www.fcc.gov/Bureaus/Common_Carrier/Reports/FCC-State_Link/Fiber/

4.4. Fiber Route Miles—Interexchange Carriers, United States, 1985–97 (fiber system route miles).

Calendar Year:	1985	1986	1987	1988	1989	1990	1991	1992	1993	1994	1995	1996	1997
AT&T	5,677	10,893	18,000	23,324	28,900	32,398	32,500	33,500	35,000	36,022	37,419	39,057	38,704
Consolidated	310	310	332	332	332	332	332	332	332	519	NA	621	621
Electric Lightwave	NA	NA	NA	NA	NA	NA	NA	NA	NA	NA	298	733	1,054
Frontier (RCI)	580	580	796	413	414	415	417	417	417	414	516	516	3,341
GST Telecom	NA	NA	NA	NA	NA	NA	NA	NA	NA	NA	NA	106	769
IXC Communications	382	382	803	803	803	914	914	914	1,257	1,357	1,365	2,025	4,647
LCI	881	950	1,210	1,210	1,210	1,210	1,406	1,406	1,406	1,408	1,408	1,408	2,743
MCI	3,025	6,752	10,267	12,467	13,839	16,000	16,700	17,040	19,793	21,460	21,049	23,096	25,234
MRC	NA	NA	670	670	844	844	844	850	850	850	850	1,100	1,100
Qwest Comm.	NA	NA	NA	NA	NA	NA	NA	NA	NA	NA	NA	2,569	4,358
Sprint	5,300	11,915	17,476	21,938	22,002	22,093	22,725	22,799	22,996	22,996	22,996	23,432	23,574
TCG	NA	NA	NA	84	84	84	84	84	84	84	84	84	NA
Valley Net	NA	NA	NA	NA	520	570	581	581	581	NA	NA	NA	NA
WorldCom	3,884	8,886	9,169	10,262	10,888	11,056	11,093	11,093	11,104	11,104	11,127	12,060	19,619
Total Reported:	20,039	40,668	58,723	71,503	79,836	85,916	87,596	89,016	93,821	96,214	97,112	106,807	125,765

Note:
A route mile is the length of the cable. It does not take into account how many fibers or how many parallel cables there are. This table shows how many miles are covered, point to point.

Source: Jonathan M. Kraushaar. *Fiber Deployment Update End of Year 1996* and *Fiber Deployment Update End of Year 1997.* Washington, DC: Federal Communications Commission, Common Carrier Bureau, Industry Analysis Division. Table 1 in both editions. http://www.fcc.gov/Bureaus/Common_Carrier/Reports/FCC-State_Link/Fiber/

4.5. Transatlantic Cable Systems, 1956–98.

System	Year	Technology	Cost ($ million)	Total 64 kbps Circuits	Usable 64 kbps Circuits[1]	Annual Investment Cost per Usable Circuit[2]	Investment Cost per Minute[3]
TAT-1	1956	Coax Cable	$49.6	44.5	40.1	$213,996	$2.443
TAT-2	1959	Coax Cable	$42.7	49.0	44.1	$167,308	$1.910
TAT-3	1963	Coax Cable	$50.6	87.5.0	78.8	$111,027	$1.267
TAT-4	1965	Coax Cable	$50.4	69.0	62.1	$140,238	$1.601
TAT-5	1970	Coax Cable	$70.4	720.0	648.0	$18,773	$0.214
TAT-6	1976	Coax Cable	$197.0	4,000.0	3,200.0	$10,638	$0.121
TAT-7	1983	Coax Cable	$180.0	4,246.0	3,821.0	$8,139	$0.093
TAT-8	1988	Fiber Optic	$360.0	7,560.0	6,048.0	$10,285	$0.117
TAT-9	1992	Fiber Optic	$406.0	15,120.0	10,584.0	$6,628	$0.076
TAT-10	1992	Fiber Optic	$300.0	22,680.0	18,144.0	$2,857	$0.033
TAT-11	1993	Fiber Optic	$280.0	22,680.0	18,144.0	$2,667	$0.030
TAT-12	1996	Fiber Optic	$378.0	60,480.0	60,480.0	$1,080	$0.012
TAT-13	1996	Fiber Optic	$378.0	60,480.0	60,480.0	$1,080	$0.012
Gemini	1998	Fiber Optic	$520.0	241,920.0	241,920.0	$371	$0.004
AC-1	1998	Fiber Optic	$850.0	483,840.0	483,840.0	$304	$0.003

Notes:

[1]With allowance for redundance, restoration, etc.

[2]The annual investment cost per usable circuit is the annual payment rate for the life of the asset that produces a present value equal to the initial investment cost. This calculation assumes a 25-year cable life and a discount rate equal to the average cost of capital for the firm. For purposes of comparison, the discount rate is based on 40% debt at an embedded cost of debt of 9% and 60% equity at a 14% rate of return, with the latter increased to reflect a 37% income tax rate. These assumptions translate to a discount rate of 16.9333%.

[3]The investment cost per minute assumes that average activated circuits are used 8 hours per day for 365 days each year and that 50% of circuits are idle (not activated). These assumptions probably overstate the utilization rates for early cable systems. The Gemini and AC-1 systems are designed primarily to handle private line circuits.

Source: Linda Blake and Jim Lande. *Trends in the U.S. International Telecommunications Industry.* Washington, DC: Industry Analysis Division, Common Carrier Bureau, U.S. Federal Communications Commission, August, 1998. Table 12. http://www.fcc.gov/Bureaus/Common_carrier/Reports/FCC-State–Link/intl/itltrd98.pdf. Note: File name changes.

4.6. Telephone Loops by State as of December 31, 1997.

State Name	Number of Telephone Companies	Total Loops
Alabama	30	2,404,691
Alaska	25	397,536
Arizona	16	2,732,159
Arkansas	28	1,368,534
California	22	21,482,732
Colorado	28	2,643,505
Connecticut	2	2,152,439
Delaware	1	532,170
District of Columbia	1	919,999
Florida	12	10,490,934
Georgia	36	4,770,210
Hawaii	2	707,995
Idaho	21	680,840
Illinois	56	7,981,248
Indiana	42	3,470,657
Iowa	154	1,588,955
Kansas	39	1,584,824
Kentucky	19	2,064,056
Louisiana	20	2,435,338
Maine	20	808,423
Maryland	2	3,493,609
Massachusetts	3	4,463,949
Michigan	39	6,257,999
Minnesota	88	2,877,876
Mississippi	19	1,321,246
Missouri	44	3,324,016
Montana	18	508,060
Nebraska	41	995,434
Nevada	14	1,206,882
New Hampshire	12	818,122
New Jersey	3	6,200,950
New Mexico	15	901,359
New York	44	12,714,808
North Carolina	26	4,694,604
North Dakota	24	401,815
Ohio	42	6,728,822
Oklahoma	39	1,954,375
Oregon	33	2,022,395
Pennsylvania	36	7,951,437
Rhode Island	1	653,123
South Carolina	27	2,146,610
South Dakota	31	406,294
Tennessee	25	3,270,885
Texas	57	12,006,252
Utah	13	1,099,794
Vermont	10	394,242
Virginia	21	4,381,487
Washington	23	3,499,719
West Virginia	10	958,992
Wisconsin	88	3,295,851
Wyoming	10	284,245
United States	1,432	172,452,497
Northern Marianas	1	20,639
Puerto Rico	2	1,256,646
Virgin Islands	1	60,902
Guam	1	73,185
Grand Total	1,437	173,863,869

Notes:

A loop is one measure of total number of telephone lines. It is used to determine the amount of Universal Service Fund payments to local exchange carriers. Another measure is the number of presubscribed lines, which was used before 1998 to determine the amount of payments by interexchange carriers to support the Universal Service Fund and the Lifeline and Link-Up programs. The last measure, access lines, is published by the United States Telephone Association. The Universal Service Fund is one mechanism for helping companies in high-cost areas recover some costs and hold down local rates.

Source: Cited in Federal Communications Commission, Common Carrier Bureau, Industry Analysis Division. Trends in Telephone Service, Washington, DC: Federal Communications Commission, Common Carrier Bureau, Industry Analys Division, February 1999. Table 20.2. http://www.fcc.gov/Bureaus/ Common_Carrier/Reports/FCC-State_Link/IAD/trend199.pdf

4.7. Statistics of Communications Common Carriers, Selected Operating Statistics, United States, 1984–97.

Year	Number of Carriers	Business			Switched Access Lines					Total Switched
		Total Business	Analog	Digital	Residential					
					Total Residential	Analog	Digital	Payphone	Mobile	
1984	75	27,603,330	NA	NA	77,499,711	NA	NA	1,716,904	NA	106,819,945
1985	55	29,223,623	NA	NA	79,220,307	NA	NA	1,709,914	NA	110,153,844
1986	57	30,366,907	NA	NA	81,210,464	NA	NA	1,715,737	NA	113,293,108
1987	52	31,844,171	NA	NA	83,384,081	NA	NA	1,720,378	NA	116,948,630
1988	52	32,422,985	NA	NA	85,014,956	NA	NA	1,605,476	39,075	119,082,492
1989	51	34,371,315	NA	NA	86,766,613	NA	NA	1,623,946	37,048	122,798,922
1990	51	35,592,942	NA	NA	89,057,341	NA	NA	1,694,575	44,103	126,388,961
1991	52	38,433,693	37,363,953	1,069,740	90,836,057	90,836,051	6	1,680,993	62,252	131,012,995
1992	54	38,735,502	37,500,787	1,234,715	93,236,150	93,235,927	223	1,543,776	82,062	133,597,490
1993	53	40,731,495	39,138,053	1,593,442	95,599,391	95,599,309	82	1,528,723	116,140	137,975,749
1994	52	42,406,369	40,789,515	1,616,854	98,225,105	98,224,590	515	1,524,615	51,773	142,207,862
1995	53	45,589,658	43,432,221	2,157,437	101,333,305	101,329,103	4,202	1,432,843	54,483	148,410,289
1996	51	49,247,530	45,940,727	3,306,803	104,314,789	104,259,359	55,430	1,540,813	61,438	155,164,570
1997	51	52,927,781	48,295,891	4,631,890	108,188,436	108,085,358	103,078	1,748,022	71,542	162,935,781

Year	Special Access Lines			Access Lines	Total Cable	Km of Fiber in Cable[1]		
	Total	Analog	Digital	Total (Switched and Special)	Sheath Km of Copper	Sheath Km of Fiber	Fiber Km Equipped	Total Fiber Km Deployed
1984	1,222,082	NA	NA	108,042,027	NA	NA	NA	NA
1985	1,390,896	NA	NA	111,544,740	NA	NA	NA	NA
1986	1,920,731	NA	NA	115,213,839	NA	NA	NA	NA
1987	2,148,098	NA	NA	119,096,728	NA	NA	NA	NA
1988	3,192,682	NA	NA	122,275,174	NA	NA	719,838	2,539,308
1989	3,037,268	NA	NA	125,836,190	NA	NA	1,059,381	3,979,350
1990	4,035,297	NA	NA	130,424,258	NA	NA	1,648,540	5,328,029
1991	5,725,345	1,174,766	4,550,579	136,738,340	5,570,128	255,541	2,477,451	7,249,442
1992	6,708,337	1,237,007	5,471,330	140,305,827	5,653,859	307,503	3,855,726	10,349,921
1993	11,037,217	2,135,307	8,901,910	149,012,966	5,712,100	363,108	4,015,898	13,328,168
1994	14,964,943	1,974,813	12,990,130	157,172,805	5,763,421	408,210	5,713,076	16,121,035
1995	17,603,651	2,137,734	15,465,917	166,013,940	5,801,044	458,904	5,551,510	17,572,084
1996	22,719,925	1,429,853	21,290,072	177,884,495	5,832,635	567,521	6,466,950	20,357,287
1997	30,998,515	1,159,630	29,838,885	193,934,296	5,840,297	544,246	7,550,624	23,104,987

4.7. Statistics of Communications Common Carriers, Selected Operating Statistics, United States, 1984–97 *(continued)*.

| Year | Number of Carriers | Number of Telephone Calls (thousands) | | Toll Calls Completed (originating) | | | | Interlata Billed Access Minutes (originating and terminating) (thousands) | | |
		Local Calls	Total	Intralata	Total Intralata	Interlata Interstate	Interlata Intrastate	Total	Interstate	Intrastate
1984	75	350,391,981	NA	NA	NA	NA	NA	NA	NA	NA
1985	55	365,304,830	NA	NA	NA	NA	NA	NA	NA	NA
1986	57	372,296,473	NA	NA	NA	NA	NA	NA	NA	NA
1987	52	379,864,264	NA	NA	NA	NA	NA	NA	NA	NA
1988	52	379,035,883	67,547,342	18,983,768	48,563,574	36,752,925	11,810,649	NA	NA	NA
1989	51	389,383,322	68,547,451	19,406,222	49,141,229	37,593,867	11,547,362	NA	NA	NA
1990	51	402,492,293	63,359,346	20,263,554	43,095,792	31,888,748	11,207,044	NA	NA	NA
1991	52	416,213,954	67,333,207	23,337,553	43,995,654	32,126,555	11,869,099	405,456,048	305,745,611	99,710,437
1992	54	434,175,743	71,502,090	22,612,572	48,889,518	36,036,032	12,853,486	432,356,515	327,821,281	104,535,234
1993	53	447,473,714	78,077,246	23,757,662	54,319,584	38,746,788	15,572,796	465,270,369	351,022,599	114,247,770
1994	52	465,207,539	83,441,709	23,796,633	59,645,076	43,244,593	16,400,483	500,297,267	374,996,101	125,301,166
1995	53	484,195,345	94,051,667	23,327,801	70,723,866	50,618,771	20,105,095	549,982,263	405,579,546	144,402,717
1996	51	504,131,507	94,905,927	21,376,847	73,529,080	52,677,037	20,852,043	598,563,946	438,772,880	159,791,066
1997	51	522,025,261	101,112,405	21,844,925	79,267,480	55,927,824	23,339,655	647,813,708	469,638,292	178,175,416

Notes:

Interlata refers to calls directed to and carried by interexchange carriers.

Intralata refers to calls carried by the local operating company within a given local access transport area (lata).

[1] For the years 1988-1990, miles have been converted to kilometers by multiplying the number of miles by 1.6093 and rounding to the nearest whole number.

Note: Totals may be understated because certain data pertaining to the carriers included in this table are unavailable. Between 1987 and 1988, there were significant changes in the definitions of many of the items in this table, with the implementation of a new uniform system of accounts (USOA) in 1988. New categories for reporting of physical plant and network usage were created and defined. Some of these definitions were further refined when the reporting mechanism of the carriers was changed in 1992 for the filing of 1991 data. For these reasons, there may be inconsistencies in the data reported for 1984-1987 compared to what was reported for 1988. There may also be inconsistencies between 1988 and subsequent years when carriers were adapting to this new USOA and the automated reporting requirements.

Source: Compiled from various editions of "Statistics of Communications Common Carriers." Cited in Federal Communications Commission, Common Carrier Bureau, Industry Analysis Division. *Statistics of Common Carriers, 1997, Part 6, Historic Financial and Economic Data.* Washington, DC: Federal Communications Commission, Common Carrier Bureau, Industry Analysis Division, 1998. Table 6–10. Also Statistics of Common Carriers, 1997, Part 6. http://www.fcc.gov/Bureaus/Common_Carrier/Reports/FCC-State_Link/socc.html

COMMUNICATION EQUIPMENT

4.8. Value of Shipments of Communication Equipment by Class of Product, United States, 1988–97. (millions of dollars)

Product Class	Product Description	1988	1989	1990	1991	1992	1993	1994	1995	1996	1997
	Communication equipment, including telephone, telegraph, and other electronic systems and equipment	$35,702.4	$35,415.7	$36,990.3	$37,821.0	$39,501.9	$42,113.8	$49,598.1	$56,362.0	$65,608.6	$74,482.8
36611	Telephone switching and switchboard equipment	$7,399.7	$6,975.5	$7,537.1	$6,956.6	$7,291.3	$7,240.9	$8,067.6	$8,178.4	$9,617.7	$10,618.1
36613	Carrier line equipment and modems	$5,589.0	$4,705.7	$5,013.8	$4,298.4	$4,810.6	$4,993.3	$5,114.0	$5,868.6	$7,544.6	$7,686.8
36614	Other telephone and telegraph equipment and components	$3,066.5	$3,061.8	$3,180.5	$4,909.5	$5,507.3	$7,026.4	$8,479.5	$10,510.0	$14,030.9	$17,846.9
36631	Communication systems and equipment (except broadcast)	$12,213.0	$14,016.3	$14,768.0	$14,987.9	$15,162.8	$16,196.7	$19,977.5	$23,031.9	$25,332.5	$28,550.9
36632	Broadcast, studio, and related electronic equipment	$1,804.2	$1,809.6	$1,856.4	$1,836.3	$1,948.2	$2,077.5	$2,469.5	r/ 2,844.7	r/ $3,000.2	$3,350.5
36693	Intercommunications systems, including inductive paging systems (selective calling)	$392.6	$367.8	$346.1	$303.8	$288.5	$241.7	$283.7	$296.0	$277.8	$252.7
36691	Alarm systems	$1,164.7	$1,132.5	$1,027.2	$1,145.0	$1,370.4	$1,532.4	$1,550.5	$1,662.1	$1,802.8	$1,921.1
36692	Vehicular and pedestrian traffic control equipment and electrical railway signals and attachments	$387.0	$451.1	$470.6	$491.1	$597.3	$573.4	$669.9	$710.7	$762.0	$805.5
36991	Electronic teaching machines, teaching aids, trainers and simulators	$1,243.7	$1,359.6	$1,208.8	$1,411.0	$1,310.8	$847.7	$838.5	$913.3	$865.7	$809.9
36992	Laser sources[1]	NA	NA	NA	NA	NA	NA	$726.4	$787.5	$843.8	$1,020.9
36995	Ultrasonic equipment	$101.3	$137.7	$108.8	$113.7	$119.2	$127.3	$137.5	$172.3	$184.3	$187.8
36999	Other electronic systems and equipment, n.e.c.[2]	$1,446.0	$1,398.1	$1,473.0	$720.8	$1,095.5	$1,256.5	$1,283.5	$1,386.5	$1,346.3	$1,431.8

Notes:

[1] Beginning in 1994, data for laser equipment, instrumentation, and components have been eliminated from this survey. Only laser sources are being collected.
[2] Product class 36997 changed to product class 36999 for 1992. Product classes 36998, 39992, and 39447 are no longer collected on this survey.

NA = Not available.

n.e.c. = Not elsewhere classified.

r/ = Revised by 5 percent or more from previously published data.

Source: Bureau of the Census. Current Industrial Reports, Series MA36P: *Communication Equipment, Including Telephone, Telegraph, and Other Electronic Systems and Equipment.* Washington, DC: U.S. Bureau of the Census, 1996, 1997, and 1998. Table 1. http://www.census.gov/cir/www/ma36p.html

4.9. Quantity and Value of Shipments of Communication Equipment: Modems, Telephone Sets, Other Data Communications Equipment, Video Teleconferencing Equipment, Voice/Call Message Processing Equipment, Facsimile Communication Equipment, United States, 1996–97 (quantity in number of units; value in thousands of dollars).

Product Code	Product Description	1996			1997		
		Number of Companies	Quantity	Value	Number of Companies	Quantity	Value
	Modems:						
36613 61	Up through 10000 BPS	29	[b]1,526,649	[b]$382,433	X	X	X
36613 65	Over 10000 BPS	30	[a]4,810,480	[a]$1,180,076	X	X	X
	Up through 10000 BPS:						
36613 62	Consumer	X	X	X	19	1,985,506	$179,796
36613 63	Nonconsumer[5]	X	X	X	4	D	D
	Over 10000 BPS:						
36613 66	Consumer	X	X	X	18	[b]3,113,424	$702,830
36613 67	Nonconsumer[5]	X	X	X	6	D	D
	Telephone sets:						
36614 21	Single line and ISDN sets[1]	10	D	D	10	2,387,546	$145,621
36614 25	Key sets[1]	7	3,552,560	$218,480	7	477,352	92,431
36614 29	Public pay telephone equipment	8	[r]394,668	$94,580	8	[a]491,806	[b]$ 114,296
36614 36	Cordless handsets[2]	4	D	[br]$ 273,099	3	D	[b]$ 286,160
36614 38	Other telephone sets and accessories (e.g., handsets)[2]	13	X	[c]$229,499	14	X	D
	Other data communications equipment:						
36614 41	Routers and gateways[6]	18	X	$5,803,890	21	X	D
36614 43	Bridges	15	X	[r]$ 114,744	11	X	[a]$ 108,047
36614 45	Terminal servers[3 6]	12	X	D	9	X	D
36614 47	Concentrators[3 6]	13	X	[r]$ 2,411,927	9	X	D
36614 49	Other data communications equipment (front-end processors, translators, etc.)[6]	37	X	[r]$ 869,021	36	X	$12,344,171
36614 59	Video teleconferencing equipment (video CODECS and other systems)	16	X	$522,120	15	X	D
	Voice/call message processing equipment, including answering devices:						
36614 61	Voice mail equipment	12	X	[r]$ 372,475	11	X	[b]$ 463,080
36614 63	Voice response equipment[4]	15	X	D	13	X	D
36614 65	Automated attendance equipment[4]	5	X	D	5	X	D
36614 67	Automated call distributors[4]	3	X	$701,542	3	X	$799,419
36614 69	Other voice or call processing equipment	23	X	$217,517	25	X	[b]$ 235,395
36614 91	Facsimile communication equipment (complete)	8	X	$63,792	6	X	D

Notes:
Product codes are based on SIC (Standard Industrial Classification) codes.
[1]Product code 36614 21 is included with product code 36614 25 to avoid disclosing data for individual companies.
[2]Product code 36614 36 is included with product code 36614 38 to avoid disclosing data for individual companies.
[3]Product code 36614 45 is included with product code 36614 47 to avoid disclosing data for individual companies.
[4]Product codes 36614 63 and 36614 65 are included with product code 36614 67 to avoid disclosing data for individual companies.
[5]Product codes 36613 63 and 36613 67 have a combined total quantity of 966,841 and combined total value of 232,671 for 1997.
[6]Product codes 36614 41, 36614 45, 36614 47, and 36614 49 have been combined for 1997.
D = Withheld to avoid disclosing data for individual companies.
X = Not applicable.
[r]Revised by 5 percent or more from previously published data.
[a]10 25 percent of this item has been estimated.
[b]25 to 50 percent of this item has been estimated.
[c]Over 50 percent of this item has been estimated.

Source: Bureau of the Census. Current Industrial Reports, Series MA36P: *Communication Equipment, Including Telephone, Telegraph, and Other Electronic Systems and Equipment.* Washington, DC: U.S. Bureau of the Census, 1998. Table 2. http://www.census.gov/cir/www/ma36p.html

4.10. Wireless Telephones—U.S. Sales to Dealers, 1993–96, and Estimated for 1984–92 and 1997–98.

Year	Units (thousands)	Dollars (millions)	Average Price	Household Penetration
1984 est.	25	$50	$2,000	0%
1985 est.	75	$116	$1,550	0%
1986 est.	285	$271	$950	0%
1987 est.	565	$503	$890	1%
1988 est.	890	$765	$860	2%
1989 est.	1,500	$1,125	$750	4%
1990 est.	1,830	$1,098	$600	5%
1991 est.	2,390	$825	$345	7%
1992 est.	3,825	$1,301	$340	10%
1993	5,590	$1,666	$298	14%
1994	8,825	$2,365	$268	21%
1995	10,295	$2,574	$250	25%
1996	11,565	$2,660	$230	32%
1997 est.	12,500	$2,750	$220	36%
1998 est.	13,750	$2,750	$200	34%

Notes:
Sales through consumer channels. Includes PCS digital wireless phones.

Source: Consumer Electronics Manufacturers Association (CEMA). *CEMA Consumer Electronics U.S. Sales & Forecasts.* Arlington, VA: CEMA, 1998.

4.11. Corded Telephones—U.S. Sales to Dealers, 1982–96, Estimated for 1997, and Projected for 1998.

Year	Units (thousands)	Dollars (millions)	Household Penetration	Average Unit Price
1982	3,700	$200	NA	NA
1983	15,000	$525	NA	NA
1984	24,000	$790	NA	NA
1985	21,000	$630	NA	NA
1986	18,100	$561	NA	NA
1987	15,900	$461	NA	NA
1988	15,200	$532	NA	NA
1989	19,000	$532	NA	NA
1990	22,003	$638	NA	NA
1991	20,872	$605	NA	NA
1992	23,964	$575	NA	NA
1993	27,080	$617	NA	$23
1994	23,664	$610	NA	$26
1995	25,836	$557	96%	$22
1996	26,013	$553	96%	$21
1997 est.	27,800	$528	96%	$19
1998 proj.	27,000	$513	96%	$19

Notes:
NA = Not available.

Source: Consumer Electronics Manufacturers Association (CEMA). *CEMA Consumer Electronics U.S. Sales & Forecasts.* Arlington, VA: CEMA, 1998.

4.12. Cordless Telephones—U.S. Sales to Dealers, 1980–96, Estimated for 1997, and Projected for 1998.

Year	Units (thousands)	Dollars (millions)	Household Penetration	Average Unit Price
1980	500	$54	0%	NA
1981	1,150	$125	2%	NA
1982	2,200	$200	4%	NA
1983	4,700	$400	7%	NA
1984	6,300	$460	9%	NA
1985	4,000	$280	11%	NA
1986	4,100	$295	14%	NA
1987	6,400	$435	16%	NA
1988	8,200	$681	18%	NA
1989	10,000	$830	23%	NA
1990	10,148	$842	28%	NA
1991	13,232	$1,125	36%	NA
1992	14,944	$1,091	45%	NA
1993	16,183	$1,046	46%	$65
1994	16,772	$1,106	51%	$66
1995	19,510	$1,141	55%	$58
1996	20,555	$1,176	64%	$57
1997 est.	26,900	$1,587	68%	$59
1998 proj.	27,900	$1,618	70%	$58

Source: Consumer Electronics Manufacturers Association (CEMA). *CEMA Consumer Electronics U.S. Sales & Forecasts.* Arlington, VA: CEMA, 1998.

4.13. Fax Machines—U.S. Sales to Dealers, 1990–96, Estimated for 1997, and Projected for 1998.

Year	Unit Sales (thousands)	Dollar Sales (millions)	Household Penetration	Average Unit Price
1990	1,483	$920	NA	$620
1991	1,498	$869	NA	$580
1992	1,731	$826	2%	$477
1993	2,100	$888	2%	$423
1994	2,536	$964	4%	$380
1995	2,827	$919	8%	$325
1996	2,761	$839	9%	$304
1997 est.	2,850	$741	10%	$260
1998 proj.	2,900	$725	10%	$250

Notes:
Sales through consumer channels.
NA = Not available.

Source: Consumer Electronics Manufacturers Association (CEMA). *CEMA Consumer Electronics U.S. Sales & Forecasts.* Arlington, VA: CEMA, 1998.

COMMUNICATION SERVICES, COSTS, AND USAGE

4.14. Average Monthly Residential Rates (in October of each year), United States, 1983–97.

	1983	1984	1985	1986	1987	1988	1989	1990	1991	1992	1993	1994	1995	1996	1997
Representative monthly charge*	$10.50	$12.10	$12.17	$12.58	$12.44	$12.32	$12.30	$12.36	$13.03	$13.05	$13.16	$13.19	$13.62	$13.71	$13.82
Subscriber line charges	$0.00	$0.00	$1.01	$2.04	$2.66	$2.67	$3.53	$3.55	$3.56	$3.55	$3.55	$3.55	$3.54	$3.54	$3.53
Taxes and 911 charges	$1.08	$1.25	$1.36	$1.51	$1.56	$1.58	$1.70	$2.00	$2.12	$2.15	$2.29	$2.31	$2.41	$2.40	$2.44
Touch-tone service	NA	NA	NA	$1.57	$1.52	$1.54	$1.52	$1.33	$1.06	$0.97	$0.94	$0.77	$0.44	$0.30	$0.12
Total monthly charge	NA	NA	NA	$17.70	$18.18	$18.11	$19.05	$19.24	$19.77	$19.72	$19.95	$19.81	$20.01	$19.95	$19.92

Notes:
Average monthly local rates are based on surveys by FCC staff using the same sampling areas and weights used by the Bureau of Labor Statistics in constructing the Consumer Price Index.
* = Rate is based on flat-rate service where available, and on measured/message service with 100 five-minute same-zone business-day calls elsewhere.

Source: Compiled from various editions of *Statistics of Communications Common Carriers*. Cited in Federal Communications Commission, Common Carrier Bureau, Industry Analysis Division. *Statistics of Common Carriers, 1996, Part 8, Trends and Monitoring Tables*. Washington, DC: Federal Communications Commission, Common Carrier Bureau, Industry Analysis Division. Also *Statistics of Common Carriers, 1997, Part 8*. Table 8.4. http://www.fcc.gov/Bureaus/Common_Carrier/Reports/FCC-State_Link/socc.html

4.15. Average Local Rates for Businesses with a Single Line in Urban Areas, United States (in October of each year), 1983–97.

	1983	1984	1985	1986	1987	1988	1989	1990	1991	1992	1993	1994	1995	1996	1997
Representative Rate*	$29.16	$32.74	$33.42	$34.26	$33.71	$31.03	$31.06	$30.97	$32.29	$32.45	$32.70	$32.25	$32.48	$32.58	$32.69
Touch-tone service	**	**	**	**	**	$2.45	$2.43	$2.35	$1.84	$1.71	$1.67	$1.21	$0.97	$0.82	$0.44
Subscriber line charges	$0.00	$0.00	$1.01	$2.04	$2.68	$2.69	$3.55	$3.57	$3.57	$3.56	$3.57	$3.57	$3.57	$3.54	$3.54
Taxes and 911 charges	$3.35	$3.77	$3.96	$4.17	$4.18	$3.95	$4.21	$4.32	$4.42	$4.57	$4.63	$4.61	$4.79	$4.87	$4.99
Total	$32.51	$36.51	$38.39	$40.47	$40.57	$40.12	$41.25	$41.21	$42.12	$42.29	$42.57	$41.64	$41.80	$41.81	$41.65
Average charge for 5 minute same zone daytime business call	$0.085	$0.090	$0.090	$0.092	$0.092	$0.091	$0.093	$0.093	$0.091	$0.093	$0.094	$0.092	$0.093	$0.092	$0.092
Minimum connection charge**	$56.04	$68.84	$70.82	$72.94	$72.15	$70.48	$71.05	$71.36	$72.75	$72.55	$71.41	$69.88	$67.87	$68.47	$68.58
Touch-tone service	**	**	**	**	**	$2.03	$1.70	$1.89	$1.13	$1.19	$1.17	$0.92	$0.27	$0.17	$0.17
Taxes	$3.08	$3.79	$3.90	$4.01	$3.97	$3.92	$4.06	$4.15	$4.32	$4.33	$4.25	$4.13	$4.17	$4.20	$4.42
Total	$59.12	$72.63	$74.72	$76.95	$76.12	$76.43	$76.81	$77.40	$78.20	$78.07	$76.83	$74.93	$72.31	$72.85	$73.18
5 minute payphone call	$0.168	$0.212	$0.222	$0.223	$0.226	$0.228	$0.228	$0.228	$0.228	$0.228	$0.235	$0.24	$0.248	$0.253	NA

Notes:
Average monthly local rates are based on surveys by FCC staff using the same sampling areas and weights used by the Bureau of Labor Statistics in constructing the Consumer Price Index.
* = The representative rate is the monthly single-line rate for touch-tone service with unlimited local calls (where offered) or the measured service rate plus additional charges for the first 200 messages in other cities. The representative business rate includes the additional monthly cost for touch-tone service for 1983 through 1987. The additional charge is shown separately thereafter.
** = Connection charges do not include drop line and block charges. Business connection charges for 1983 through 1987 include the additional connection charge for installing touch-tone service. The charge is shown separately thereafter.

Source: Compiled from various editions of *Statistics of Communications Common Carriers.* Cited in Federal Communications Commission, Common Carrier Bureau, Industry Analysis Division. *Statistics of Common Carriers, 1996, Part 8, Trends and Monitoring Tables.* Washington, DC: Federal Communications Commission, Common Carrier Bureau, Industry Analysis Division. Table 8.5. Also *Statistics of Common Carriers, 1997, Part 8.* Table 8.5. http://www.fcc.gov/Bureaus/Common_Carrier/Reports/FCC-State_Link/socc.html

4.16. Changes in the Price of Directly Dialed Five-minute Long Distance Calls, United States, 1984 and 1998 (AT & T basic rate schedules).

Calling Distance (in airline miles, rate center to rate center)		Residential *			Business **		
		January 1984	June 1998	Percentage Change	January 1984	June 1998	Percentage Change
1–10	Day	$0.96	$1.40	45.8%	$0.96	$1.82	89.3%
	Evening	$0.57	$0.80	40.4%	$0.57	$1.82	218.9%
	Night & Weekend	$0.38	$0.65	71.1%	$0.38	$1.82	378.3%
11–22	Day	$1.28	$1.40	9.4%	$1.28	$1.82	42.0%
	Evening	$0.76	$0.80	5.3%	$0.76	$1.82	139.1%
	Night & Weekend	$0.51	$0.65	27.5%	$0.51	$1.82	256.4%
23–55	Day	$1.60	$1.40	-12.5%	$1.60	$1.82	13.6%
	Evening	$0.96	$0.80	-16.7%	$0.96	$1.82	13.6%
	Night & Weekend	$0.64	$0.65	1.6%	$0.64	$1.82	13.6%
56–124	Day	$2.05	$1.40	-31.7%	$2.05	$1.82	-11.3%
	Evening	$1.22	$0.80	-34.4%	$1.22	$1.82	49.0%
	Night & Weekend	$0.82	$0.65	-20.7%	$0.82	$1.82	121.6%
125–292	Day	$2.14	$1.40	-34.6%	$2.14	$1.82	-15.1%
	Evening	$1.28	$0.80	-37.5%	$1.28	$1.82	42.0%
	Night & Weekend	$0.85	$0.65	-23.5%	$0.85	$1.82	113.8%
293–430	Day	$2.27	$1.40	-38.3%	$2.27	$1.82	-19.9%
	Evening	$1.36	$0.80	-41.2%	$1.36	$1.82	33.6%
	Night & Weekend	$0.90	$0.65	-27.8%	$0.90	$1.82	101.9%
431–925	Day	$2.34	$1.40	-40.2%	$2.34	$1.82	-22.3%
	Evening	$1.40	$0.80	-42.9%	$1.40	$1.82	29.8%
	Night & Weekend	$0.93	$0.65	-30.1%	$0.93	$1.82	95.4%
926–1910	Day	$2.40	$1.40	-41.7%	$2.40	$1.82	-24.3%
	Evening	$1.44	$0.80	-44.4%	$1.44	$1.82	26.2%
	Night & Weekend	$0.96	$0.65	-32.3%	$0.96	$1.82	89.3%
1911–3000	Day	$2.70	$1.40	-48.1%	$2.70	$1.82	-32.7%
	Evening	$1.62	$0.80	-50.6%	$1.62	$1.82	12.2%
	Night & Weekend	$1.08	$0.65	-39.8%	$1.08	$1.82	68.3%
3001–4250	Day	$2.80	$1.40	-50.0%	$2.80	$1.82	-35.1%
	Evening	$1.68	$0.80	-52.4%	$1.68	$1.82	8.2%
	Night & Weekend	$1.12	$0.65	-42.0%	$1.12	$1.82	62.3%
4251–5750	Day	$2.91	$1.40	-51.9%	$2.91	$1.82	-37.5%
	Evening	$1.74	$0.80	-54.0%	$1.74	$1.82	4.50%
	Night & Weekend	$1.16	$0.65	-44.0%	$1.16	$1.82	56.7%

Notes:

* = AT&T initiated a new rate structure for residential customers on November 8, 1997. The rate structure eliminates mileage bands and implements new weekday peak/off-peak and weekend rate periods. The new rates are shown in the old rate structure for the purposes of comparison.

** = AT&T initiated a new rate structure for business customers on November 5, 1997. The rate structure eliminates mileage bands and peak/off-peak rate periods. The new rates are shown in the old rate structure for the purposes of comparison.

Source: AT&T Tariffs and Industry Analysis Division, *Reference Book of Rates, Price Indices, and Household Expenditures for Telephone Service.* Cited in Federal Communications Commission, Common Carrier Bureau, Industry Analysis Division. *Trends in Telephone Service, July 1998.* Washington, DC: Federal Communications Commission, Common Carrier Bureau, Industry Analysis Division, 1998. Table 14.4. http://www.fcc.gov/Bureaus/Common_Carrier/Reports/FCC-State_Link/IAD/

4.17. Telephone Service Expenditures, United States, 1980–97.

	Annual Expenditures (Average for All Households)		Monthly Expenditures (Households with Telephone Service)		
Year	Telephone Expenditures	Percentage of Total Expenditures	Basic Local Service Charge*	Toll and Other Telephone Expenditures**	Total Telephone Expenditures
1980	$325	1.9%	$8.74	$21	$30
1981	$360	2.1%	$9.71	$23	$33
1982	$375	2.1%	$10.75	$23	$34
1983	$415	2.1%	$11.58	$26	$38
1984	$435	2.0%	$13.35	$26	$40
1985	$455	1.9%	$14.54	$27	$41
1986	$471	2.0%	$16.13	$26	$43
1987	$499	2.0%	$16.66	$28	$45
1988	$537	2.1%	$16.57	$32	$48
1989	$567	2.0%	$17.53	$33	$51
1990	$592	2.1%	$17.79	$35	$53
1991	$618	2.1%	$18.66	$36	$55
1992	$623	2.1%	$18.70	$37	$55
1993	$658	2.1%	$18.94	$39	$58
1994	$690	2.2%	$19.07	$42	$61
1995	$708	2.2%	$19.49	$42	$62
1996	$772	2.3%	$19.63	$44	$64
1997	$809	2.3%	$19.52	$47	$67

Notes:

* = Monthly service charges for unlimited local service, taxes, and subscriber line charges.

** = Calculated as total monthly bill minus the cost of basic local service. Figures may not add due to rounding. The "Toll and Other" category is primarily toll, but also includes charges for equipment, additional access lines, connection, touch-tone, call waiting, 900 service, directory listings, etc.

Source: Bureau of Labor Statistics. Cited in Federal Communications Commission, Common Carrier Bureau, Industry Analysis Division. *Trends in Telephone Service, March 1999*. Washington, DC: Federal Communications Commission, Common Carrier Bureau, Industry Analysis Division, 1998. Table 4.1 .http://www.fcc.gov/Bureaus/Common_Carrier/Reports/FCC-State_Link/IAD/trend199.pdf

4.18. Area Code Assignments, United States, 1984–2000.

Location	Date	Previous Code	Added Code
California	Jan-84	213	818
New York	Sep-84	212	718
Colorado	Mar-88	303	719
Florida	Apr-88	305	407
Massachusetts	Jul-88	617	508
Illinois	Nov-89	312	708
New Jersey	Nov-90	201	908
Texas	Nov-90	214	903
California	Sep-91	415	510
Maryland	Oct-91	301	410
California	Nov-91	213	310
New York	Jan-92	212	917
New York	Jan-92	718	917
Georgia	May-92	404	706
New York	Jul-92	212	718
Texas	Nov-92	512	210
California	Nov-92	714	909
North Carolina	Nov-93	919	910
Michigan	Dec-93	313	810
Pennsylvania	Jan-94	215	610
Alabama	Jan-95	205	334

4.18. Area Code Assignments, United States, 1984–2000 *(continued)*.

Location	Date	Previous Code	Added Code
Washington	Jan-95	206	360
Texas	Mar-95	713	281
Arizona	Mar-95	602	520
Colorado	Apr-95	303	970
Florida (Tampa)	May-95	813	941
Virginia	Jul-95	703	540
Georgia (Atlanta)	Aug-95	404	770
Connecticut	Aug-95	203	860
Florida (Miami)	Sep-95	305	954
Tennessee	Sep-95	615	423
Oregon	Nov-95	503	541
South Carolina	Dec-95	803	864
Florida (North)	Dec-95	904	352
Missouri	Jan-96	314	573
Illinois (Chicago)	Jan-96	708	847
Puerto Rico	Mar-96	809	787
Ohio	Mar-96	216	330
Minnesota	Mar-96	612	320
Florida (Southeast)	May-96	407	561
Virginia	Jul-96	804	757
Illinois (Chicago)	Aug-96	708	630
Texas (Dallas)	Sep-96	214	972
Ohio	Sep-96	513	937
Illinois	Oct-96	312	773
Texas (Houston)	Nov-96	713	281
California (Southern)	Jan-97	310	562
Indiana	Feb-97	317	765
California	Mar-97	619	760
Arkansas	Apr-97	501	870
Washington State	Apr-97	206	253
Washington State	Apr-97	206	425
Michigan	May-97	810	248
Texas	May-97	817	254
Texas	May-97	817	940
Maryland	Jun-97	809	868
Maryland	Jun-97	301	240
New Jersey	Jun-97	201	973
New Jersey	Jun-97	908	732
U.S. Virgin Islands	Jun-97	809	340
California	Jun-97	818	626
Guam	Jul-97	NA	671
Commonwealth of the Northern Mariana Islands	Jul-97	NA	670
Texas	Jul-97	210	830
Texas	Jul-97	210	956
Kansas	Jul-97	913	785
Wisconsin	Jul-97	414	920
California	Aug-97	415	650
Ohio	Sep-97	216	440
Massachusetts	Sep-97	617	781
Massachusetts	Sep-97	508	978
Tennessee	Sep-97	615	931
Mississippi	Sep-97	601	228
Utah	Sep-97	801	435
Missouri	Oct-97	816	660
California	Nov-97	916	530
Ohio	Dec-97	614	740

4.18. Area Code Assignments, United States, 1984–2000 *(continued)*.

Location	Date	Previous Code	Added Code
Michigan	Dec-97	313	734
North Carolina	Dec-97	910	336
Georgia (Atlanta)	Jan-98	770	678
Pennsylvania	Feb-98	412	724
California	Mar-98	510	925
South Carolina	Mar-98	803	843
Alabama	Mar-98	205	256
California	Apr-98	714	949
California (Los Angeles)	Jun-98	213	323
California	Jul-98	408	831
California	Nov-98	209	559
Nevada	Dec-98	702	775
California	Feb-99	805	661
California	Jun-99	619	858
California	Jun-00	619	935

Notes: Includes U.S. and its territories and possessions.

1999 and 2000 are only partially covered.

Source: Bell Communications Research. Cited in Federal Communications Commission, Common Carrier Bureau, Industry Analysis Division. *Trends in Telephone Service.* Washington, DC: Federal Communications Commission, Common Carrier Bureau, Industry Analysis Division, March 1999. Table 21.

4.19 New Area Code Assignments by Year, 1984–98.

1984	2
1985	0
1986	0
1987	0
1988	3
1989	1
1990	2
1991	3
1992	6
1993	2
1994	1
1995	14
1996	12
1997	34
1998	10

Note:
Calculated by the author based on the above table.

4.20. Telephone Numbers Assigned for 800 Service, United States, 1993–98.

Year	Month	Working 800 Numbers	Misc. 800 Numbers*	Total 800 Numbers Assigned	Spare 800 Numbers Still Available
1993	April	2,448,985	642,725	3,091,710	4,618,290
	May	2,511,933	708,192	3,220,125	4,489,875
	June	2,589,123	722,006	3,311,129	4,398,871
	July	2,675,483	705,416	3,380,899	4,329,101
	August	2,738,259	701,009	3,439,268	4,270,732
	September	2,818,262	639,547	3,457,809	4,252,191
	October	2,891,994	660,544	3,552,538	4,157,462
	November	3,083,250	728,514	3,811,764	3,898,236
	December	3,155,955	731,438	3,887,393	3,822,607
1994	January	3,257,540	580,216	3,837,756	3,872,244
	February	3,381,646	731,005	3,837,756	3,872,244
	March	3,516,620	743,813	4,260,433	3,449,567
	April	3,659,129	699,212	4,358,341	3,351,659
	May	3,793,865	738,767	4,532,632	3,177,368
	June	3,933,037	792,698	4,725,735	2,984,265
	July	4,099,174	699,803	4,798,977	2,911,023
	August	4,312,486	807,881	5,120,367	2,589,633
	September	4,506,014	841,381	5,347,395	2,362,605
	October	4,611,014	871,684	5,482,698	2,227,302
	November	4,817,854	875,416	5,693,270	2,016,730
	December	4,948,605	763,235	5,711,840	1,998,160
1995	January	5,096,646	807,294	5,903,940	1,806,060
	February	5,278,800	811,221	6,090,021	1,619,979
	March	5,528,723	793,771	6,322,494	1,387,506
	April	5,741,780	797,902	6,539,682	1,170,318
	May	5,980,848	843,093	6,823,941	886,059
	June	6,340,534	481,633	6,822,167	887,833
	July	6,402,785	443,717	6,846,502	863,498
	August	6,428,120	442,270	6,870,390	839,610
	September	6,503,018	437,215	6,940,233	769,767
	October	6,583,344	396,605	6,979,949	730,051
	November	6,647,880	310,043	6,957,923	752,077
	December	6,700,576	286,487	6,987,063	722,937
1996	January	6,766,607	297,001	7,063,608	646,392
	February	6,861,093	335,557	7,196,650	513,350
	March	6,907,098	293,244	7,200,342	509,658
	April	6,934,085	280,927	7,215,012	494,988
	May	6,943,620	333,140	7,276,760	433,240
	June	6,986,821	324,899	7,311,720	398,280
	July	7,022,309	339,900	7,362,209	347,781
	August	7,074,772	311,273	7,386,045	323,955
	September	7,119,167	310,562	7,429,729	280,271
	October	7,185,135	325,088	7,510,223	199,777
	November	7,242,377	337,502	7,579,879	130,121
	December	7,272,819	343,905	7,616,724	93,276
1997	January	7,333,632	323,804	7,657,436	52,564
	February	7,388,696	318,571	7,707,267	2,733
	March	7,402,769	305,362	7,708,131	1,869
	April	7,411,118	296,925	7,708,043	1,957
	May	7,411,291	294,320	7,705,611	4,389
	June	7,415,591	293,802	7,709,393	607
	July	7,421,288	283,794	7,705,082	4,918
	August	7,430,733	276,024	7,706,757	3,243
	September	7,427,717	280,668	7,708,385	1,615
	October	7,433,483	276,490	7,709,973	27
	November	7,423,662	276,490	7,700,238	9,762
	December	7,429,160	267,429	7,696,589	13,411

4.20. Telephone Numbers Assigned for 800 Service, United States, 1993–98 *(continued)*.

1998				
	January	7,431,789	264,143	7,695,932	14,068
	February	7,445,338	257,493	7,702,831	7,169
	March	7,455,240	249,964	7,705,204	4,796
	April	7,464,692	232,462	7,697,154	12,846
	May	7,476,270	228,409	7,704,679	5,321
	June	7,480,468	227,041	7,707,509	2,491
	July	7,485,866	221,078	7,706,944	3,056
	August	7,483,417	224,242	7,707,659	2,341
	September	7,489,271	219,080	7,708,351	1,649
	October	7,479,005	229,889	7,708,894	1,106
	November	7,478,913	228,892	7,707,805	2,195
	December	7,487,529	215,267	7,702,796	7,204

Notes:

* = Miscellaneous numbers include those in the 800 service management system maintained by Data Service Management, Inc., and categorized as reserved, assigned but not yet activated, recently disconnected, or suspended.

Source: Federal Communications Commission, Common Carrier Bureau, Industry Analysis Division. *Trends in Telephone Service, March 1999*. Washington, DC: Federal Communications Commission, Common Carrier Bureau, Industry Analysis Division, 1999. Table 21.2. http://www.fcc.gov/Bureaus/Common_Carrier/Reports/FCC-State_Link/IAD/trend199.pdf

4.21. Telephone Penetration in the United States, 1983–98.

		Households (millions)	Households with Telephones (millions)	Percentage with Telephones	Households without Telephones (millions)	Percentage without Telephones
1983	November	85.8	78.4	91.4%	7.4	8.6%
1984	March	86.0	78.9	91.8%	7.1	8.2%
	July	86.6	79.3	91.6%	7.3	8.4%
	November	87.4	79.9	91.4%	7.5	8.6%
1985	March	87.4	80.2	91.8%	7.2	8.2%
	July	88.2	81.0	91.8%	7.2	8.2%
	November	88.8	81.6	91.9%	7.2	8.1%
1986	March	89.0	82.1	92.2%	6.9	7.8%
	July	89.5	82.5	92.2%	7.0	7.8%
	November	89.9	83.1	92.4%	6.8	7.6%
1987	March	90.2	83.4	92.5%	6.8	7.5%
	July	90.7	83.7	92.3%	7.0	7.7%
	November	91.3	84.3	92.3%	7.0	7.7%
1988	March	91.8	85.3	92.9%	6.5	7.1%
	July	92.4	85.7	92.8%	6.7	7.2%
	November	92.6	85.7	92.5%	6.9	7.5%
1989	March	93.6	87.0	93.0%	6.6	7.0%
	July	93.8	87.5	93.3%	6.3	6.7%
	November	93.9	87.3	93.0%	6.6	7.0%
1990	March	94.2	87.9	93.3%	6.3	6.7%
	July	94.8	88.4	93.3%	6.4	6.7%
	November	94.7	88.4	93.3%	6.3	6.7%
1991	March	95.3	89.2	93.6%	6.1	6.4%
	July	95.5	89.1	93.3%	6.4	6.7%
	November	95.7	89.4	93.4%	6.3	6.6%
1992	March	96.6	90.7	93.9%	5.9	6.1%
	July	96.6	90.6	93.8%	6.0	6.2%
	November	97.0	91.0	93.8%	6.0	6.2%
1993	March	97.3	91.6	94.2%	5.7	5.8%
	July	97.9	92.2	94.2%	5.7	5.8%
	November	98.8	93.0	94.2%	5.8	5.8%
1994	March	98.1	92.1	93.9%	6.0	6.1%
	July	98.6	92.4	93.7%	6.2	6.3%
	November	99.8	93.7	93.8%	6.2	6.2%

4.21. Telephone Penetration in the United States, 1983–98 *(continued).*

		Households (millions)	Households with Telephones (millions)	Percentage with Telephones	Households without Telephones (millions)	Percentage without Telephones
1995	March	99.9	93.8	93.9%	6.1	6.1%
	July	100.0	94.0	94.0%	6.0	6.0%
	November	100.4	94.2	93.9%	6.2	6.1%
1996	March	100.6	94.4	93.8%	6.2	6.2%
	July	101.2	95.0	93.9%	6.1	6.1%
	November	101.3	95.1	93.9%	6.2	6.1%
1997	March	102.0	95.8	93.9%	6.2	6.1%
	July	102.3	96.1	93.9%	6.2	6.1%
	November	102.8	96.5	93.8%	6.3	6.2%
1998	March	103.4	97.4	94.1%	6.1	5.9%

Source: Federal Communications Commission, Common Carrier Bureau, Industry Analysis Division. *Trends in Telephone Service, July 1998.* Washington, DC: Federal Communications Commission, Common Carrier Bureau, Industry Analysis Division, 1998. http://www.fcc.gov/Bureaus/Common_Carrier/Reports/FCC-State_Link/IAD/

4.22. Historical Telephone Penetration Estimates.

Year	Percentage of Households with Telephones	Access Lines per 100 Population
1920	35.0%	9.6
1930	40.9%	12.5
1940	36.9%	12.7
1950	61.8%	21.7
1960	78.3%	27.6
1970	90.5%	35.0
1980	92.9%	46.2
1990	94.8%	54.8

Sources: FCC staff estimates based on data from the Bureau of the Census, *Historical Statistics of the United States, Colonial Times to 1970, Part 2,* page 783, for all percentage data except 1980 and 1990, which are from the decennial censuses. Access line data for 1920 through 1970 are estimated by multiplying the number of telephones by the proportion of main plus equivalent main stations to total telephones for the Bell System. Prior to 1950, the 1950 proportion is used. For 1980 and 1990, access lines reported by USTA are used. Cited in Federal Communications Commission, Common Carrier Bureau, Industry Analysis Division. *Trends in Telephone Service, July 1998.* Washington, DC: Federal Communications Commission, Common Carrier Bureau, Industry Analysis Division, 1998. Table 17.3. http://www.fcc.gov/Bureaus/Common_Carrier/Reports/FCC-State_Link/IAD/

4.23. Dial Equipment Minutes, United States, 1980–97 (billions).

Year	Local	Intrastate Toll	Interstate Toll	Total
1980	1,458	141	133	1,733
1981	1,492	151	144	1,787
1982	1,540	158	154	1,853
1983	1,587	166	169	1,923
1984	1,639	198	208	2,045
1985	1,673	222	250	2,145
1986	1,699	237	270	2,207
1987	1,713	253	295	2,261
1988	1,795	269	321	2,384
1989	1,829	286	344	2,459
1990	1,846	298	353	2,497
1991	1,859	302	366	2,527
1992	1,926	311	381	2,618
1993	2,027	316	396	2,739
1994	2,126	327	420	2,873
1995	2,227	343	451	3,021
1996	2,405	370	487	3,262
1997	2,683	404	525	3,612

Increase Over Prior Years

Year	Local	Intrastate Toll	Interstate Toll	Total
1981	2%	7%	8%	3%
1982	3%	5%	7%	4%
1983	3%	5%	10%	4%
1984	3%	19%	23%	6%
1985	2%	12%	20%	5%
1986	2%	7%	8%	3%
1987	1%	7%	9%	2%
1988	5%	6%	9%	5%
1989	2%	6%	7%	3%
1990	1%	4%	3%	2%
1991	1%	1%	4%	1%
1992	4%	3%	4%	4%
1993	5%	2%	4%	5%
1994	5%	3%	6%	5%
1995	5%	5%	7%	5%
1996	8%	8%	8%	8%
1997	12%	9%	8%	11%

Percent Distribution

Year	Local	Intrastate Toll	Interstate Toll	Total
1980	84%	8%	8%	100%
1981	83%	8%	8%	100%
1982	83%	9%	8%	100%
1983	83%	9%	9%	100%
1984	80%	10%	10%	100%
1985	78%	10%	12%	100%
1986	77%	11%	12%	100%
1987	76%	11%	13%	100%
1988	75%	11%	13%	100%
1989	74%	12%	14%	100%
1990	74%	12%	14%	100%
1991	74%	12%	14%	100%
1992	74%	12%	15%	100%
1993	74%	12%	14%	100%
1994	74%	11%	15%	100%
1995	74%	11%	15%	100%
1996	74%	11%	15%	100%
1997	74%	11%	15%	100%

Source: National Exchange Carrier Association. Cited in Federal Communications Commission, Common Carrier Bureau, Industry Analysis Division. *Trends in Telephone Service, March 1999.* Washington, DC: Federal Communications Commission, Common Carrier Bureau, Industry Analysis Division, 1999. Table 12.1. http://www.fcc.gov/Bureaus/Common_Carrier/Reports/FCC-State_Link/IAD/trend199.pdf

4.24. International Message Telephone Service for 1997 (figures rounded to the nearest million).

International Point	Traffic Billed in the United States				Traffic Billed in Foreign Countries		
	Number of Messages	Number of Minutes	U.S. Carrier Revenue	Owed to Foreign Carriers	Retained Revenue	Number of Messages	Number of Minutes
Africa	124	621	$610	$382	$227	29	100
Asia	865	4,653	$3,822	$2,591	$1,232	254	1,061
Caribbean	221	1,358	$1,015	$637	$378	87	343
Eastern Europe	96	592	$603	$280	$324	30	127
Middle East	111	655	$689	$480	$210	51	231
North and Central America	1,365	7,292	$3,660	$1,755	$1,905	985	4,182
Oceania	107	607	$319	$155	$164	37	209
South America	350	1,815	$1,478	$921	$558	101	457
Western Europe	1,005	5,078	$2,941	$798	$2,143	510	2,389
Other regions	3	11	$39	$28	$11	Less than 1	1
Total for foreign points	**4,233**	**22,611**	**$15,135**	**$8,016**	**$7,119**	**2,078**	**9,062**
Total for U.S. points	**14**	**70**	**$43**	**$10**	**$32**	**5**	**37**
Total for all international points	**4,247**	**22,682**	**$15,178**	**$8,026**	**$7,152**	**2,083**	**9,100**

Notes:
The region totals include all traffic reported by carriers serving Alaska, Hawaii, Puerto Rico, and the coterminous United States, and include traffic between these points and offshore U.S. points such as Guam and the U.S. Virgin Islands. This traffic is shown separately as the total for U.S. points, and also is included in the total for all international points.

Source: Federal Communications Commission, Industry Analysis Division, Section 43.61 International Telecommunications Data. Cited in Federal Communications Commission, Common Carrier Bureau, Industry Analysis Division. *Trends in Telephone Service, February 1999.* Washington, DC: Federal Communications Commission, Common Carrier Bureau, Industry Analysis Division, 1999. Table 7.3. http://www.fcc.gov/Bureaus/Common_Carrier/Reports/FCC-State_Link/IAD/trend199.pdf

4.25. International Service from United States to Foreign Points, 1980–97 (minute, message, and revenue amounts shown in millions).

	Minutes	Messages	Total	Per minute*	Per call	Telex	Telegraph	Private Line	Misc.
1980	1,569	199	$2,097	$1.34	$10.53	$325	$63	$115	
1981	1,857	233	$2,239	$1.21	$9.61	$350	$62	$126	
1982	2,187	274	$2,382	$1.09	$8.70	$363	$56	$138	
1983	2,650	322	$2,876	$1.09	$8.92	$379	$54	$154	
1984	3,037	367	$3,197	$1.05	$8.71	$394	$46	$158	
1985	3,350	411	$3,435	$1.03	$8.37	$415	$45	$172	
1986	3,917	482	$3,891	$0.99	$8.07	$390	$42	$175	
1987	4,480	570	$4,559	$1.02	$8.00	$360	$35	$191	
1988	5,190	687	$5,507	$1.06	$8.02	$310	$30	$194	
1989	6,109	835	$6,517	$1.07	$7.80	$243	$27	$208	
1990	7,215	984	$7,626	$1.06	$7.75	$196	$24	$201	
1991	8,986	1,371	$9,096	$1.01	$6.63	$200	$15	$303	$23
1992	10,156	1,643	$10,179	$1.00	$6.20	$155	$16	$313	$24
1993	11,393	1,926	$11,353	$1.00	$5.89	$135	$12	$365	$23
1994	13,393	2,313	$12,255	$0.92	$5.30	$123	$12	$432	$55
1995	15,837	2,821	$13,990	$0.88	$4.96	$119	$6	$432	$55
1996	19,119	3,485	$14,079	$0.74	$4.04	$119	$5	$649	$26
1997	22,611	4,233	$15,135	$0.67	$3.58	$110	$4	$840	$36

Notes:
* = Billed revenue per minute data is based on traffic for domestic U.S. points only. The domestic U.S. includes Puerto Rico but excludes American Samoa, Guam, the Northern Mariana Islands, and the U.S. Virgin Islands.

Sources: Federal Communications Commission, Industry Analysis Division. *Trends in the International Telecommunications Industry and Section 43.61 International Telecommunications Data.* Cited in Federal Communications Commission, Common Carrier Bureau, Industry Analysis Division. *Trends in Telephone Service, March 1999.* Washington, DC: Federal Communications Commission, Common Carrier Bureau, Industry Analysis Division, 1999. Table 7.1. http://www.fcc.gov/Bureaus/Common_Carrier/Reports/FCC-State_Link/IAD/trend199.pdf

4.26. Cellular Telephone Subscribers, United States, 1984–98.

Year		Number of Systems	Subscribers
1984	December	32	91,600
1985	June	65	203,600
	December	102	340,213
1986	June	129	500,000
	December	166	681,825
1987	June	206	883,778
	December	312	1,230,855
1988	June	420	1,608,697
	December	517	2,069,441
1989	June	559	2,691,793
	December	584	3,508,944
1990	June	592	4,368,686
	December	751	5,283,055
1991	June	1,029	6,390,053
	December	1,252	7,557,148
1992	June	1,483	8,892,535
	December	1,506	11,032,753
1993	June	1,523	13,067,318
	December	1,529	16,009,461
1994	June	1,550	19,283,506
	December	1,581	24,134,421
1995	June	1,581	28,154,415
	December	1,627	33,785,661
1996	June	1,629	38,195,466
	December	1,740	44,042,992
1997	June	2,005	48,705,553
	December	2,228	55,312,293
1998	June	2,300	60,831,431

Source: Cellular Telecommunications Industry Association. Cited in Federal Communications Commission, Common Carrier Bureau, Industry Analysis Division. *Trends in Telephone Service, January 1999.* Washington, DC: Federal Communications Commission, Common Carrier Bureau, Industry Analysis Division, 1999. Table 2.1. http://www.fcc.gov/Bureaus/Common_Carrier/ Reports/FCC-State_Link/IAD/trend199.pdf.

4.27. Cellular Telephone Service Average Monthly Bill, United States, 1987–98.

Year	Month	Monthly Bill
1987	December	$96.83
1988	June	$95.00
	December	$98.02
1989	June	$85.52
	December	$89.30
1990	June	$83.94
	December	$80.90
1991	June	$74.56
	December	$72.74
1992	June	$68.51
	December	$68.68
1993	June	$67.31
	December	$61.48
1994	June	$58.65
	December	$56.21
1995	June	$52.42
	December	$51.00
1996	June	$48.84
	December	$47.70
1997	June	$43.86
	December	$42.78
1998	June	$39.88

Source: Cellular Telcommunications Industry Association. Cited in Federal Communications Commission, Common Carrier Bureau, Industry Analysis Division. *Trends in Telephone Service, January 1999.* Washington, DC: Federal Communications Commission, Common Carrier Bureau, Industry Analysis Division, 1999. Table 2.2 http://www.fcc.gov/Bureaus/Common_Carrier/Reports/FCC-State_Link/IAD/trend199.pdf

4.28. Paging Industry Numbers, United States, 1993–97.

Year	Paging units	Revenues	Average Monthly Revenue per Unit	Number of Rental Paging Units	Number of Customer-owned and -maintained Paging Units
1993	19,300,000	$2,686,000,000	$13.76	11,190,000	8,110,000
1994	26,164,300	$3,233,000,000	$12.31	12,620,000	13,680,000
1995	34,426,868	$3,841,000,000	$10.51	15,870,000	18,630,000
1996	42,484,220	$4,404,000,000	$9.77	18,100,000	25,000,000
1997	49,800,000	$5,079,000,000	$9.11	18,960,000	30,940,000

Source: The Strategis Group. *The State of the U.S. Paging Industry: 1997.* Cited in Federal Communications Commission, Wireless Bureau. *Third Annual CMRS Competition Report.* Washington, DC, 1998. Page C-2.

4.29. Mobile Telephone Industry—U.S. National Penetration, 1993–97.

Year	Percentage
1993	6.2%
1994	9.3%
1995	12.8%
1996	16.6%
1997	20.7%

Notes:
Figures are for December of each year.

Source: Cellular Telecommunications Industry Association. Calculated using U.S. population estimates from Dennis Leibowitz et al. "The Wireless Communications Industry." Donaldson, Lufkin & Jenrette (DLJ Report). Fall 1997. Cited in Federal Communications Commission, Wireless Bureau. *Third Annual CMRS Competition Report.* Washington, DC: Federal Communications Commission, Wireless Bureau, 1998.Adapted from Figure 3.

BROADCASTING

4.30. Broadcast Stations Licensed in the United States as of May 31, 1998.

Station Type	Number
AM Radio	4,724
FM Radio	5,615
FM Educational	1,980
Total	12,322
UHF Commercial TV	652
VHF Commercial TV	559
UHF Educational TV	243
VHF Educational TV	125
Total	1,579
FM Translators & Boosters	3,011
UHF Translators	2,727
VHF Translators	2,197
Total	7,935
UHF Low Power TV	1,530
VHF Low Power TV	559
Total	2,089

Source: Federal Communications Commission, Mass Media Bureau, News Release at http://www.fcc.gov/Bureaus/Mass_Media/News_Releases/1998/nmm8019.txt

4.31. FCC Auctions Summary—Service Design, as of 1998.

Auction Number and Name	Number of Licenses[1]	Geographic License Scheme[2]	Spectrum per License	Total Spectrum (in megahertz)	Service Description
1　Nationwide Narrowband PCS	11[3]	Nationwide	11 blocks, 5=50/50KHz, 3=50/12.5KHz, 2=50KHz	0.7875 MHz	Advanced paging/data
2　Interactive Video and Data Service	594	MSA	2 blocks of 500KHz	1 MHz	Interactive data
3　Regional Narrowband PCS	30	Regional	6 blocks, 2=50/50KHz, 4=50/12.5 KHz	0.45 MHz	Advanced paging/data
4　A & B block Broadband PCS	102[4]	MTA	2 blocks of 30 MHz	60 MHz	Mobile voice and data
5　10 C block Broadband PCS[5]	493	BTA	1 block of 30 MHz	30 MHz	Mobile voice and data
6　Multichannel Distribution Service	493	BTA	Max of 13 channels of 6 MHz	78 MHz[6]	Wireless cable
7　900 MHz Specialized Mobile Radio	1020	MTA	20 blocks of 25 KHz	5 MHz	Mobile voice and data
8　Digital Broadcast Service[7]	1	Full U.S. coverage	500 MHz	437.5 MHz	Multichannel video
9　Digital Broadcast Service[7]	1	Partial U.S. coverage	Uses same spectrum as full coverage license	375 MHz	Multichannel video
11　D, E, & F block Broadband PCS	1479	BTA	3 blocks of 10 MHz	30 MHz	Mobile voice and data
12　Cellular Unserved	14	MSA/RSA	2 blocks of 25 MHz	50 MHz	Mobile voice and data
13　Interactive Video and Data Service	981	MSA/RSA	2 blocks of 500 KHz	1 MHz	Interactive data[8]
14　Wireless Communications Service	128	MEA/REAG	4 blocks, 2=10 MHz, 2=5 MHz	30 MHz	
15　Digital Audio Radio Service	2	Full U.S. coverage	2 blocks of 12.5 MHz	25 MHz	Multichannel audio
16　Upper 800 MHz Specialized Mobile Radio	525	EA	3 blocks, 1 MHz, 3 MHz, and 6 MHz	10 MHz	Mobile voice and data
17　Local Multipoint Distribution Service	986[9]	BTA	2 blocks, 1150 MHz and 150 MHz	1300 MHz	Fixed voice, data, and video

Notes:
[1] This is the total number of licenses initially granted in each service. It does not take into account any partitioning and disaggregation activity. Some of these licenses have not yet been granted.
[2] MTAs = Major Trading Areas; BTAs = Basic Trading Areas; MSAs = Metropolitan Statistical Areas; RSAs = Rural Service Areas; MEAs = Major Economic Areas; REAGs = Regional Economic Area Groups.
[3] Includes one pioneer preference license.
[4] Includes three pioneer preference licenses.
[5] To date, two auctions have been completed for C block PCS, the original and one reauction.
[6] To be precise, Multipoint Distribution Service (MDS) total spectrum should be 76 MHz because Channel 2 was originally 16 MHz only in the top 50 markets. In the rest of the markets, it was Channel 2A with 4 MHz. As noted in the MDS Auction Procedure Terms and Conditions: "In 1992, the 2160-2162 MHz frequency was reallocated to emerging technologies, and thus any subsequent MDS use of these 2 MHz will be secondary."
[7] There is a total of 500 MHz of DBS downlink spectrum available. The same spectrum can be reused at each of the eight DBS orbital slots. The figures in the table are (28/32) x 500 and (24/32) x 500 respectively, but they each refer to portions of the same 500 MHz spectrum.
[8] WCS is permitted to implement a wide range of services, subject to FCC engineering requirements, including fixed, mobile, radio location, and broadcasting-satellite (sound) service.
[9] Cellularvision, Inc., has been granted a pioneer preference for a portion of the 1150 MHz New York BTA.

Source: Federal Communications Commission, Wireless Bureau. *Third Annual CMRS Competition Report.* Washington, DC: Federal Communications Commission, Wireless Bureau. 1998. Page A-2. http://www.fcc.gov/Bureaus/Wireless/Reports/fcc98091.pdf

4.32. FCC Auctions Summary—Auction Results, July 1994—February 1998.

Auction Number and Name	Total Winning Bids[1]	Bid Price (dollars per person per MHz)	Auction Duration Began	Auction Duration Ended	No. Rounds	Number of Winning Bidders
1 Nationwide Narrowband PCS	$650,306,674	$3.10	7/25/94	7/29/94	47	6
2 Interactive Video and Data Service	$213,892,375	$0.85	7/28/94	7/29/94	Oral Outcry	178
3 Regional Narrowband PCS	$392,706,797	$3.46	10/26/94	11/8/94	105	9
4 A & B block Broadband PCS	$7,721,184,171	$0.52	12/5/94	3/13/95	112	18
5 10 C block Broadband PCS[2]	$10,102,121,394	$1.33	12/18/95	5/6/96	184	90
			7/3/96	7/16/96	25	
6 Multichannel Distribution Service	$216,239,603	$0.067[3]	11/13/95	3/28/96	181	67
7 900 MHz Specialized Mobile Radio	$204,267,144	$0.24[3]	12/5/95	4/15/96	168	80
8 Digital Broadcast Service	$682,500,000	$0.01	1/24/96	1/25/96	19	1
9 Digital Broadcast Service	$52,295,000	$0.00	1/25/96	1/26/96	25	1
11 D, E, & F block Broadband PCS	$2,517,439,565	$0.33	8/26/96	1/14/97	276	125
12 Cellular Unserved	$1,842,533	N/A	1/13/97	1/21/97	36	10
13 Interactive Video and Data Service	[4]	[4]	[4]	[4]	[4]	[4]
14 Wireless Communications Service	$13,638,940	$0.00	4/15/97	4/25/97	29	17
15 Digital Audio Radio Service	$173,234,888	$0.03	4/1/97	4/2/97	25	2
16 Upper 800 MHz Specialized Mobile Radio	$96,232,060	$0.04	10/28/97	12/8/98	235	14
17 Local Multipoint Distribution Service	$578,663,029	$0.00	2/18/98	3/25/98	127	104

Notes:
[1] Total Winning Bids includes high bids from the auction (net of any bidding credits) plus the price paid for any prior preference licenses.
[2] C block broadband PCS was auctioned in two auctions. The Total Winning Bids, Bid Price, and Number of Winning Bids have been combined to avoid duplication.
[3] Estimated to adjust for encumbered spectrum.
[4] The second IVDS auction was postponed on January 29, 1997.

Source: Federal Communications Commission, Wireless Bureau. *Third Annual CMRS Competition Report.* Washington, DC: Federal Communications Commission, Wireless Bureau, 1998. Page A-3. http://www.fcc.gov/Bureaus/Wireless/Reports/fcc98091.pdf

4.33. Value of Shipments of Radio, Television, and Studio Equipment, United States, 1996–97 (thousands of dollars).

Product Code		Product Description	1996		1997	
			Number of Companies	Value	Number of Companies	Value
		Radio station type communication systems and equipment except broadcast:[1]				
36631	15	Amateur	12	r/ $14,163	11	b 8,917
36631	17	Broadcast (sound and TV)[2]	8	D	5	D
36631	19	Earth exploration[2]	1	D	1	D
36631	21	Meteorological	4	b $12,113	5	b 11,981
36631	23	Fixed[2]	24	$2,037,990	29	$2,561,864
36631	25	Intersatellite	15	$2,330,099	15	D
36631	27	Telecommand	5	$53,494	5	a 19,626
36631	29	Telemetry	18	$111,652	15	b 72,123
36631	31	Radionavigational and locational	9	a $258,117	9	$536,535
36631	33	Aeronautical communications[2]	9	D	9	D
36631	46	Other[2]	D	D	14	D
36632	—	Broadcast, studio, and related electronic equipment	174	$3,000,164	NA	$3,350,487
		Audio equipment, excluding consumer and public address types:				
36632	12	Amplifiers and preamplifiers	29	b 305,617	26	a 53,435
36632	13	Control consoles and switchers	24	$92,306	21	a 96,578
36632	17	Other (power supplies, terminal equipment, broadcast recorders, etc.)	37	b $325,751	35	b 365,231
		Video equipment, excluding consumer and public address types:				
36632	22	Amplifiers[3]	5	D	7	D
36632	24	Television cameras[3]	4	D	4	D
36632	25	Videotape recorders[3]	3	r/ $71,547	3	$77,594
36632	28	Other (power supplies, synchronization equipment, terminal equipment, monitors, telecine chains, control consoles and switchers, film equipment, television outside vans)	23	r/ $ 477,966	23	$489,208
		Transmitters, translators, radio frequency power amplifiers, and related equipment:				
36632	31	AM and FM transmitters	13	$63,573	11	$67,792
36632	34	Television transmitters	8	$122,676	9	$126,093
36632	37	Other (broadcast transmission line equipment, phasing equipment, TV boosters and repeaters, translators, etc.)	21	r/ $347,951	23	$614,685

4.33. Value of Shipments of Radio, Television, and Studio Equipment, United States, 1996–97 (thousands of dollars) *(continued)*.

Product Code	Product Description	1996 Number of Companies	Value	1997 Number of Companies	Value
36632 39	Studio transmission links (STL) and remote pickup equipment	5	r/ $64,700	5	$78,628
	Cable TV (master antennae and CATV equipment):				
36632 41	Head end equipment (antennae baluns, carrier generators, head end control units, single and broadband preamplifiers and strip amplifiers, converters, modulators and demodulators; splitting filters and traps, power supplies, switches, etc.)	20	$304,715	22	a 265,455
36632 46	Subscriber converters[4]	0	0	0	0
36632 49	Subscriber equipment (decoders, switchers; wall outlet taps; distribution amplifiers; power suppliers; directional couplers, splitters, alternators, and equalizers)[4][5]	4	D	5	D
36632 43	Broadcasting transmitting antennae and community antennae systems	10	$29,925	8	$59,218
36632 44	Closed circuit television systems and equipment, excluding broadcast and consumer products (specially-designed cameras, monitors, video recorders, receivers, scan converters and control consoles [4][5]	24	$529,075	0	$612,619
36632 45	Other broadcast, studio, theater, and commercial sound quipment, sold separately, excluding studio lighting equipment, radiating and supporting towers	7	a$250,218	5	a$ 227,741

Notes:

[1] Of this total, space-based (satellite) accounted for $4.3 billion in 1996 and $3.3 billion in 1995; airborne and marine-based accounted for $548.5 million in 1997, $471.1 million in 1996, and $474.5 million in 1995; earth-based (fixed) accounted for $4.0 billion in 1997, $3.1 billion in 1996, and $2.4 billion in 1995; earth-based portable (mobile) accounted for $9.5 billion in 1996 and $9.8 billion in 1995. Space based (satellite) and earth (mobile) based systems amounts are withheld in 1997 to avoid disclosing data for individual companies.

[2] Product codes 36631 17 and 36631 19 are included with product code 36631 23 to avoid disclosing data for individual companies. Product codes 36631 17, 36631 19, 36631 25, 36631 33, and 36631 46 have a combined total of $5,111,768 for 1997 and r/ $4,733,920 for 1996.

[3] Product codes 36632 22 and 36632 24 are included with product code 36632 25 to avoid disclosing data for individual companies.

[4] For 1995, product codes 36632 46 and 36632 49 are included with product code 36632 44 to avoid disclosing data for individual companies.

[5] For 1996 and 1997, product code 36632 49 is included with product code 36632 44 to avoid disclosing data for individual companies.

D = Withheld to avoid disclosing data for individual companies.

r/ = Revised by 5 percent or more from previously published data.

a = 10 to 25 percent of this item has been estimated.

b = 25 to 50 percent of this item has been estimated.

c = Over 50 percent of this item has been estimated.

NA = Not available.

Source: Bureau of the Census. Current Industrial Reports, Series MA36P: *Communication Equipment, Including Telephone, Telegraph, and Other Electronic Systems and Equipment, 1997.* Washington, DC: U.S. Bureau of the Census, 1998. Table 2. http://www.census.gov/cir/www/ma36p.html

Chapter 5. Computers

Few technologies have been adopted more rapidly than personal computers. But can anyone think of a technology that inspires a more intense love/hate relationship? Is there a consumer or business technology that's harder to use or more vital?

Love them or hate them, everyone has to deal with computers constantly, even those of us who don't own one. So it is important to understand the lingo, the subfields, and the economics of computers. That's not easy because the computer world involves lots of jargon, difficult concepts, complexity, and rapid change. However, the following list divides the field into useful categories (see the glossary for definitions):

Software
  Operating systems
  Application software
  Utilities, device managers, database management systems, search engines
  Programming languages
  Security

Hardware
  Central processors/chips
  Storage devices and media (hard disks, zip drives, CD-ROM drives)
  Peripherals (printers, scanners, monitors, modems, graphics controllers)
  Systems (servers, workstations)

Networks and data communications
  Local area networks
  Wide area networks
  Internet/intranets/extranets
  Computer telephony
  Videoconferencing equipment

Databases, knowledge management, and content
  Data mining and online analytical processing (OLAP, tools for processing decision support and executive information systems by organizing data into multidimensional hierarchies)
  Digital text and numeric databases
  Images, sound, video, animation
  Online services and electronic commerce

Miscellaneous
  Voice and speech recognition
  Artificial intelligence
  Virtual reality

Obviously overlap among these divisions exists. Does "online service" belong under "networks," "content," or some other heading, for example? It matters little as long as you know what an online service is.

Computers have become so pervasive in our lives that it is almost unfair to segregate them into their own chapter. In fact, you will find computer-related information throughout this book, in the chapters dealing with agriculture, business and manufacturing, construction, consumer products and entertainment, education and libraries, law enforcement and military, medicine, and space. But here you will find a core of information about hardware, software, digital databases, and the use of computer networks.

Take hardware. Tables 5.1 through 5.3 provide a taste of microprocessor history, summing up a generation of computing power. Table 5.3 depicts something you don't often see: the relative sizes of Intel chips, those tiny workhorses that power the monster on your desk. Tables 5.4, 5.5, and 5.7 reinforce the information in Chapter 3's Table 3.1 showing how our use of personal and laptop computers has skyrocketed. Do not look for the numbers in all the tables to coincide, however. Table 3.1 covers the United States and Canada, while Tables 5.4 and 5.7 treat the United States alone. Also, the definitions and survey methodologies differ. Table 5.7 reports sales through consumer channels only. Printer figures differ similarly among Tables 3.1, 5.5, and 5.8. Tables 5.4 and 5.5 show the size of the computer hardware and peripherals industry—$50 billion per year—broken down by various types of equipment. Check out Table 5.11 for revenues and growth rates of data communications equipment, 1996 through 1998.

Our use of online services and networks has surged as well. Table 5.12 shows that we are enthusiastically embracing intranets as means of intra-organizational communication and information sharing. Tables 5.14 through 5.16 reinforce our suspicions that we are spending more and more time sending and reading e-mail.

Table 5.23 shows us that Java—the popular programming language invented by Sun Microsystems—is indeed hot. And Table 5.17 proves that despite fears of credit card hacking, we are purchasing ever more goods and services online.

COMPUTER EQUIPMENT

5.1. History of Microprocessors, 1971–98.

Selected Intel Processor Specifications	4004	8008	8080	8086	80286
Processor	4004	8008	8080	8086	80286
Introduction date	11/15/71	4/1/72	4/74	6/8/78	2/82
Clock speed	108 kilohertz	200 kilohertz	2 MHz	5 MHz, 8 MHz, 10 MHz	6 MHz, 10 MHz, 12 MHz
MIPS (million instructions per second)	0.06	0.06	0.64	0.33, 0.66, 0.75	0.9, 1.5, 2.66
Number of transistors	2,300 (10 microns)	3,500 (10 microns)	6,000 (6 microns)	29,000 (3 microns)	134,000 (1.5 microns)
Bus width	4 bits	8 bits	8 bits	16 bits	16 bits
Addressable memory	640 bytes	15 Kbytes	64 Kbytes	1 megabyte	16 megabytes
Virtual memory	—	—	—	Portable computing	1 gigabyte
Typical use	Busicom calculator	Dumb terminals, general calculators, bottling machines,	Traffic light controller, Altair computer (first PC)		Standard microprocessor for all PC clones at the time
Comments	First microcomputer chip, arithmetic manipulation	Data/character manipulation	Ten times the performance of the 8008 at 200 kilohertz	Ten times the performance of the 8080	Three to six times the performance of the 8086.

Processor	Intel386 (TM) DX CPU	Intel486 (TM) DX CPU	Pentium (R) Processor	Pentium (R) Processor
Introduction date	10/17/85	4/10/89	3/22/93	6/10/96
Clock speed	16, 20, 25, 33 MHz	25, 33, 50 MHz	60, 66 MHz	200 MHz
MIPS (million instructions per second)	5-6, 6-7, 8.5, 11.4	20, 27, 41	100, 112	
Number of transistors	275,000 (1.5 microns, now 1 micron)	1.2 million (1 micron, with 50 MHz at .8 micron)	3.1 million (.8 micron, BiCMOS)	3.3 million (.35 micron BiCMOS)
Bus width	32 bits	32 bits	64 bits (external data bus), 32 bits (address bus)	64 bits (external data bus), 32 bits (address bus)
Addressable memory	4 gigabytes	4 gigabytes	4 gigabytes	4 gigabytes
Virtual memory	64 terabytes	64 terabytes	64 terabytes	64 terabytes
Typical use	Desktop computing	Desktop computing and servers	Desktops	High-performance desktops and servers
Comments	Can address enough memory to manage an 8-page history of every person on earth	50 times the performance of the 8088.	This is a 32-bit microprocessor	

Processor	Pentium (R) Processor with MMX (TM) Technology	Pentium (R) Pro Processor	Pentium (R) II Processor	Pentium (R) II Processor
Introduction date	1/8/97	11/1/95	5/7/97	4/15/98
Clock speed	166 MHz	150 MHz	233 MHz	350, 400 MHz
MIPS (million instructions per second)				
Number of transistors	4.5 million (.35 micron CMOS)	5.5 million (.6 micron), 256K L2; 15.5 million (.6 micron)	7.5 million (.35 micron), 512K L2	7.5 million (.25 micron process), 512K L2 cache Bus speed is 100 MHz
Bus width	64 bits (external data bus), 32 bits (address bus)	64 bits front side; 64 bits to L2 cache	64 bit System Bus w/ECC; 512K L2	64 bit System Bus w/ECC; 64 bit Cache Bus w/opt.ECC
Addressable memory	4 gigabytes	4 gigabytes	4 gigabytes	4 gigabytes
Virtual memory	64 terabytes	64 terabytes	64 terabytes	64 terabytes
Typical use	High-performance desktops and servers	High-end desktops, workstations, and servers	High-end business desktops, workstations, and servers	Business and consumer PCs, one and two-way servers and workstations
Comments	This is a 32-bit microprocessor			

Source: Intel Corp. Santa Clara, CA. Available at Intel's online Processor Hall of Fame at http://www.intel.com/intel/museum/25anniv/Hof/hof_main.htm

5.2. Computing Milestones, 1971 to 1995.

1971, Intel 4004. 2,300 transistors.
Designed by Intel's Ted Hoff, Stan Mazor, and Federico Faggin, along with Busicom's Masatoshi Shima, the Intel 4004 was the world's first general-purpose microprocessor. It consisted of only 2,300 transistors in a 4-bit architecture, supported 45 instructions, and ran at under 1 MHz. Laughably underpowered by today's standards, this seminal chip literally changed the world.

1972, Intel 8008. 3,500 transistors.
The 8008, which had 3,500 transistors, was the first 8-bit microprocessor. Eight-bit data allowed the 8008 to manage alphanumeric data.

1974, Intel 8080. 6,000 transistors.
Intel introduced the 8080, a 2-MHz, 6,000-transistor microprocessor with 16-bit addressing that was eventually to become the heart of the MITS Altair, the first microcomputer. A small software startup, Microsoft, was launched by Bill Gates and Paul Allen, who wrote a BASIC interpreter for the Altair system as the company's first project.

1974, Motorola 6800. 4,000 transistors.
The Motorola 6800, with 4,000 transistors, was designed by Chuck Peddle and Charlie Melear. It was mainly used in automotive controls and small-business machines.

1975, Zilog Z80. 8,500 transistors.
The Z80, designed by Faggin and Shima, was seen as an improved version of the 8080. An 8-bit, 8,500-transistor processor with 16-bit addressing that ran at 2.5 MHz, it hosted CP/M, the first standard microprocessor operating system. The Z80 was the choice of many pioneer system vendors including Osborne and Kaypro, and, in many ways, brought PCs into business.

1976, MOS Technologies 6502. Approximately 9,000 transistors.
MOS Technologies introduces the 6502, an 8-bit processor with very few registers and a 16-bit address bus, developed by Peddle and colleagues. It sold for around $25 in the mid-seventies, a price that appealed to Steve Wozniak for his Apple II design. The 6502 launched the notion of personal computing in the public imagination and was used in several other popular PCs, including the Commodore PET, the Commodore 64, and the early Atari machines. Essentially an enhanced Motorola 6800, it made graphics effects easier to program and faster to execute, setting the computer gaming phenomenon in motion.

1978, Intel 8086. 29,000 transistors.
The 8086 was a 16-bit chip with 29,000 transistors. It introduced the *x*86 instruction set that's still present on *x*86-compatible chips today. Its segmented memory addressing was quite ingenious, but it ultimately proved to be a millstone that hung around the industry's neck for years.

1979, Intel 8088. 29,000 transistors.
Intel's 8088 was based on the earlier 8086. Like the earlier processor, the 8088 had a 16-bit internal architecture, but it communicated with other components through an 8-bit bus. IBM chose this cost-saving design for its first PC, the direct ancestor of today's dominant personal computing platform. This was the chip that would launch DOS, Lotus 1-2-3, and other groundbreaking software.

1979, Motorola 68000. 68,000 transistors.
Motorola's 68000, with a new 32-bit instruction set, was the platform for some of the early Unix systems. More important, Apple chose it to implement the Lisa and later the Macintosh, the system that featured the first commercially successful graphical user interface.

1982, Intel 286. 134,000 transistors.
Intel introduces its 286, the first *x*86 processor to support general protection and virtual memory. Used in the IBM PC AT (whose 16-bit AT extension-bus design is still in use for slower peripherals), the 286 ran at speeds of 8 to 12 MHz and delivered up to six times the power of the 8086. The 286 could support up to 16MB of physical memory.

1985, Intel386. 275,000 transistors.
The 386 was a pivotal chip that enabled the transition to the modern era of personal computing. This 32-bit design—with over a quarter of a million transistors and a 4GB address space—was the first mainstream Intel chip to support linear addressing. It was on this platform that graphical operating environments, such as MS Windows and OS/2, began to seem workable. And it was with this chip that we stopped thinking about IBM compatibility and started to focus on the processor and operating system as the true platform.

1986, MIPS R2000. 185,000 transistors.
MIPS ships its R2000, the first commercial RISC microprocessor.

1987, Sun SPARC. 50,000 transistors.
Sun introduces its first SPARC microprocessor. This chip and its offspring defined several generations of RISC-based workstations.

1989, Intel486. 1.2 million transistors.
Intel ships its 486 processor, an enhanced 386 design. Its more than 1 million transistors included a built-in floating point unit and 8K of internal RAM cache.

5.2. Computing Milestones, 1971 to 1995 *(continued)*.

1993, Intel Pentium. 3.1 million transistors.
Intel ships its Pentium, incorporating a superscalar architecture whose dual-pipeline design could execute two instructions at once. With dual-integer units and a single FPU, and running at relatively high speeds, this chip (with its 3.1 million transistors) forms the basis for today's mass-market computer industry. It became the platform of choice for running Windows 95 and a host of PC applications, while at the same time bringing *x86*-based servers into direct competition with non-Intel machines.

1993, IBM/Motorola PowerPC 601. 2.8 million transistors.
IBM and Motorola's PowerPC 601 brought RISC technology to mass-market computers. It was one of the first microprocessors to implement out-of-order execution of instructions. This processor and its successors have been adopted by Apple for its Power Macintosh line.

1995, Intel Pentium Pro. 5.5 million transistors.
The Pentium Pro is the most powerful Intel processor in production today. It uses an aggressively superscalar design that can execute up to three instructions simultaneously. The core CPU, with 5.5 million transistors, is paired with a second chip containing a Level 2 cache. Mounted in a single package, these dies are connected by an ultra-high-speed bus. The Pentium Pro debuted at the high end of the Intel-based server market but has since found its way into high-end and even mainstream workstations.

Source: Reprinted from *PC Magazine,* December 17, 1996. Copyright © 1996 Ziff Davis.

5.3. Intel Chip Sizes.

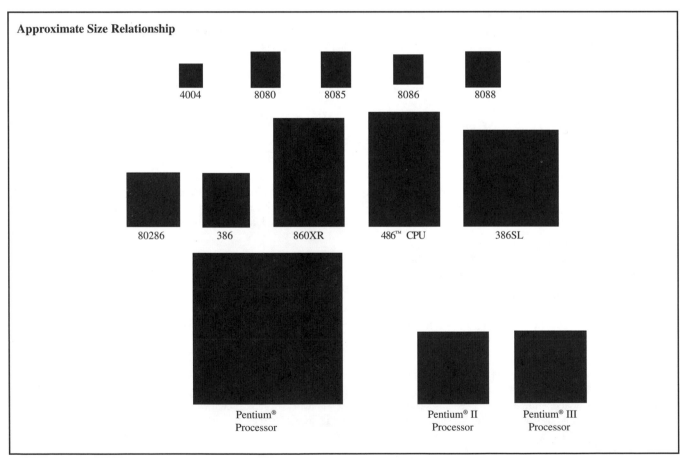

Source: © 1998 Intel Corporation.

5.4. Computers, Value of Shipments, United States, 1989–97.

Product Code	Product Description	1989	1990	1991	1992	1993	1994	1995	1996	1997*
3571	Electronic Computers [1]	$25,076.9	$25,630.1	$26,274.1	$28,571.2	$29,659.1	$38,260.7	$49,038.1	$50,681.5	$49,605.5
35713	Large-scale processing equipment [1]	N/A	N/A	N/A	$5,267.1	$4,104.4	$4,458.5	$5,288.8	$5,962.3	N/A
35714	Medium-scale and small-scale processing equipment [1]	N/A	N/A	N/A	$3,347.3	$2,442.3	$2,815.5	r/ $2,278.6	$2,401.6	N/A
35715	Personal computers and workstations [1]	N/A	N/A	N/A	$17,059.6	$18,888.8	$24,298.0	r/ $34,349.3	$35,767.9	N/A
35716	Portable computers [1]	N/A	N/A	N/A	$1,644.2	$2,575.6	$5,028.3	$5,774.3	$5,290.8	N/A
35717	Other general purpose digital processing units [1]	N/A	N/A	N/A	$493.5	$946.2	$958.8	r/ $457.9	$432.0	N/A
35718	Other computers, typically specialized for an application, including array, database, and image processors, computer chassis, and other analog, hybrid, or special purpose computers [1]	N/A	N/A	N/A	$759.4	$701.8	$701.6	r/ $889.3	$826.8	N/A
36798	Loaded computer processor boards and board subassemblies [2]	$2,854.3	$2,247.2	$12,590.4	$13,745.8	$15,087.1	$17,514.8	r/ $24,447.7	$24,133.8	$25,283.4
35721	Computer storage devices and equipment	$7,215.3	$7,488.2	$5,609.7	$6,282.4	$5,730.7	$5,555.6	$7,903.0	$8,160.7	$8,617.2
35722	Parts for computer storage devices and subassemblies	$686.5	$955.2	$780.4	$1,357.1	$1,496.4	$1,951.9	$2,235.6	$1,775.8	$2,142.5
35751	Computer terminals	$2,759.2	$2,066.7	$1,729.5	$1,707.9	$1,531.3	$1,243.5	$1,086.0	$1,062.3	$791.7
35752	Parts for computer terminals	$208.3	$362.9	$417.0	$192.1	$275.9	(D)	(D)	$165	(D)
35771	Computer peripheral equipment, n.e.c.	$8,271.1	$7,696.6	$7,763.6	$8,505.6	$9,810.2	$11,943.6	$12,331.0	$12,783.9	$13,686.2
35772	Parts for input/output equipment	$3,647.1	$3,705.5	$4,186.9	$3,053.2	$2,554.3	$2,498.7	$2,391.0	$5,441.6	$2,487.5
36950	Magnetic and optical recording media	$3,672.8	$3,695.4	$3,894.4	$4,336.8	$4,483.3	$4,777.1	$5,106.2	$5,155.8	$5,242.8

Notes:

D=Withheld to avoid disclosing data for individual companies.

NA=Not available.

n.e.c.=Not elsewhere classified.

r/=Revised by 5 percent or more from previously published data

[1] Prior to 1992, product class separation for computers is not available.

[2] These data are also collected for the Current Industrial Report Series MA36Q, Semiconductors, Printed Circuit Boards, and Other Electronic Components.

*=In 1997, the computer industry data were revised. Previous years are not comparable.

Source: U.S. Bureau of the Census. Current Industrial Reports, Series MA35R: *Computer and Office Accounting Machines.* Washington, DC: U.S. Bureau of the Census, 1997 and 1998. Table 1. http://www.census.gov/cir/www/ma35r.html

5.5. Computers, Quantity and Value of Shipments, United States, 1995–97.

Product Code	Product Description	1996 Number of Companies	1996 Quantity	1996 Value	1997 Number of Companies	1997 Quantity	1997 Value
3571—	Electronic computers (automatic data processors)	201	19,938,816	$50,681,536	172	20,124,027	$49,605,501
	Host computers (multi-users):						
35711 10	Large-scale systems, including mainframes and Supercomputers	NA	NA	NA	16	376,693	$6,065,320
35711 20	Medium-scale systems and Unix servers	NA	NA	NA	29	2,414,401	$8,327,930
35711 30	PC servers	NA	NA	NA	16	160,954	$644,394
35711 90	Other host computers	NA	NA	NA	6	33,038	$290,520
	Single-user computers:						
35712 10	Personal computers	NA	NA	NA	44	13,473,265	$21,931,109
35712 20	Workstations	NA	NA	NA	23	475,951	$5,137,371
35712 30	Laptops (AC/DC)	NA	NA	NA	10	502,939	$1,256,536
35712 40	Notebooks, subnotebooks (battery operated)	NA	NA	NA	12	1,899,263	$4,755,233
35712 50	Personal digital assistants [1]	NA	NA	NA	1	D	D
35712 60	Other portable computers [1]	NA	NA	NA	4	127,597	$122,756
35712 90	Other single user computers	NA	NA	NA	4	175,239	$104,858
35713—	Large-scale processing equipment (64 megabytes in MINIMUM main memory configuration)	29	101,083	$5,962,313	NA	NA	NA
35714—	Medium-scale and small-scale processing equipment (up to 64 megabytes in MINIMUM main memory configuration), excluding personal computers or workstations	39	267,502	$2,401,566	NA	NA	NA
35715—	Personal computers and workstations, excluding portables	86	14,422,799	$35,767,946	NA	NA	NA
35716—	Portable computers (typically with attached display, e.g., laptops, notebooks, palmtops)	39	4,373,641	$5,290,819	NA	NA	NA
35717—	Other general purpose digital processing units	21	546,837	$432,044	NA	NA	NA
35718—	Other computers, typically specialized for an application, including array, database, and image processors, computer chassis, and other analog, hybrid, or special purpose computers (excluding supercomputers, LAN servers, and engineering workstations)	66	a/ 226,954	a/ $826,848	NA	NA	NA
35719—	Other computers, including array, analog, hybrid, or specialty purpose computers	NA	NA	NA	45	484,687	$969,474
35721—	Computer storage devices and equipment	91	X	r/ $8,909,234	86	X	$8,617,175
	Rigid magnetic disk drives:						
35721 35	Less than 2 ½-inch	3	D	D	1	D	D
35721 37	2 ½-inch up to, but not including, 3 ½-inch	7	D	D	4	D	D
35721 39	3 ½-inch up to, but not including, 5 ¼-inch	26	2,564,485	$2,167,206	21	1,272,936	$1,530,804

5.5. Computers, Quantity and Value of Shipments, United States, 1995–97 (continued).

Product Code	Product Description	1996			1997		
		Number of Companies	Quantity	Value	Number of Companies	Quantity	Value
35721 41	5 ¼-inch and greater	17	207,027	$401,746	8	101,395	$362,803
35721 45	Disk subsystems and disk arrays for multiuser computer systems	28	119,924	$1,267,634	24	323,002	$1,175,704
35721 47	Flexible magnetic disk drives	11	D	D	8	D	D
	Optical disc drives, including CD-ROM and magneto-optical equipment:						
35721 51	CD-ROM juke-box capable of handling multiple discs	13	1,177,962	$144,580	8	D	D
35721 53	Single-disk CD-ROM equipment	7	164,663	$595,609	8	1,883,565	$810,812
35721 55	WORM (write once, read many times)	6	12,959	$48,237	4	D	D
35721 57	Rewritable CD-ROM	10	r/ 585,653	r/ $206,253	7	864,987	$331,919
35721 59	Optical subsystems for multiuser computer systems	7	b/ 24,341	$176,775	8	15,755	$220,431
	Auxiliary storage, not disc- or tape-based, for multiuser computer systems:						
35721 61	Encased or enclosed in a housing, enclosure, or cabinet	11	D	D	10	D	D
35721 63	Other	5	D	D	6	575,082	$347,576
35721 65	Other direct access storage equipment	14	288,784	$268,187	12	274,432	$177,973
	Serial access storage equipment (e.g., tape drives):						
35721 71	1/2-inch tape drives	13	316,391	$1,234,766	10	461,481	$1,515,458
35721 73	1/4-inch tape drives	13	2,378,892	$393,088	13	1,555,833	$269,198
35721 75	Helical scan tape drives	13	D	D	12	232,894	$396,441
35721 77	Cassette tape drives	7	D	D	7	26,689	$154,235
35721 79	Other serial access storage equipment	10	1,193,439	$459,710	7	D	D
35722 11	Computer storage parts and subassemblies	21	X	$1,719,880	16	X	$2,142,516
35751 —	Computer terminals	54	X	$1,103,897	51	X	$791,737
35751 80	Remote batch terminals	8	a/ 19,032	b/ $32,151	6	a/ 16,163	b/ $24,319
35751 61	Teleprinters	7	r/ 31,452	r/ $39,421	7	34,620	a/ $60,565
	Display terminals, including graphics type, whether or not incorporating a printing mechanism:						
35751 63	X-terminals	13	r/ 178,308	r/ $202,322	10	278,160	$262,409
	Other than X-terminals:						
35751 65	13=inch or less display	8	57,075	a/ $38,808	9	55,804	$79,896
35751 67	More than 13=inch but less than 19=inch display	22	1,761,025	$751,713	17	a/ 649,637	a/ $335,408
35751 69	19=inch or more display	9	b/ 22,941	b/ $39,482	6	b/ 8,460	b/ $29,140
35752 41	Computer terminal parts and subassemblies	13	X	r/ $199,169	9	X	D

5.5. Computers, Quantity and Value of Shipments, United States, 1995–97 (continued).

Product Code	Product Description	1996			1997		
		Number of Companies	Quantity	Value	Number of Companies	Quantity	Value
35771 —	Computer peripheral equipment, n.e.c.	260	X	r/ $11,462,756	253	X	$13,686,202
	Keying equipment:						
35771 33	Keyboards	26	20,810,401	$560,224	23	a/ 20,350,506	a/ $612,843
35771 35	Other keying equipment	5	D	D	4	D	D
35771 02	Mouse devices	12	D	D	11	D	D
35771 04	Digitizers and light pen tablets	9	r/ 176,279	r/ $51,534	10	551,647	$123,967
35771 06	Other manual input devices (joysticks, trackballs, touchscreens, etc.)	10	a/ 561,783	a/ $33,475	10	a/ 500,581	a/ $40,396
35771 08	Computer output to microfilm (COM) equipment	4	D	D	5	D	D
35771 12	Media copying and/or conversion equipment	7	b/ 17,391	b/ $60,621	9	a/ 14,522	$59,390
	Optical scanning devices:						
35771 14	Bar code devices	31	2,090,463	$850,428	29	2,510,684	a/ $909,933
35771 16	OCR equipment	15	b/ 635,428	a/ $237,433	11	215,135	a/ $259,624
	Other than bar code or OCR devices:						
35771 18	Flat bed scanners	10	D	D	6	b/ 43,563	a/ $55,777
35771 22	Hand-held scanners	3	D	D	4	46,476	$52,968
35771 24	Other	17	349,142	r/ $112,030	17	382,861	$169,326
35771 26	Voice recognition equipment	8	D	D	7	D	D
35771 28	Magnetic strip and ink recognition equipment	12	311,513	$205,810	12	D	D
35771 29	Other document entry equipment	3	D	D	3	D	D
	Computer printers:						
	Impact printers:						
35771 74	Line type (typically with a Centronics interface with output measured in lines per minute)	15	329,468	$386,361	12	255,630	$303,154
35771 72	Serial type (typically with a serial and/or parallel interface with output measured in characters per second)	27	569,705	$254,343	24	a/ 512,806	b/ $203,741
	Nonimpact printers:						
35771 40	Laser	31	r/ 1,941,839	r/ $1,946,660	30	2,427,245	$2,547,672
35771 42	Inkjet	10	15,523,120	r/ $1,320,686	13	7,486,107	$1,511,483
35771 45	Other (thermal, ion deposition, etc.)	25	596,686	r/ $638,451	23	1,278,689	$480,769
35771 47	Peripheral sharing devices	13	a/ 181,453	a/ $59,407	10	a/ 190,759	b/ $68,934
35771 49	Font cartridges	1	D	D	3	D	D
35771 50	Plotters, including electrostatic	14	47,443	a/ $235,734	13	37,740	b/ $217,440
	Monitors, excluding terminals:						
35771 55	Flat panel displays	6	D	D	5	D	D
	Other than flat panel displays (e.g., CRT):						

5.5. Computers, Quantity and Value of Shipments, United States, 1995–97 (continued).

Product Code	Product Description	1996			1997		
		Number of Companies	Quantity	Value	Number of Companies	Quantity	Value
35771 56	Less than 19=inch display	26	3,656,804	b/ $1,813,417	22	D	D
35771 58	19=inch or more display	13	228,693	$224,383	11	232,940	$181,241
35771 62	Monitor screen projection devices, e.g., LCD panels	7	D	D	7	D	D
35771 63	All other input/output devices	61	X	r/ $898,458	58	X	a/ $1,288,509
35771 68	Accessories for computer peripherals (e.g., device supports, ergonomic aids, etc.)	41	X	r/ $1,227,314	40	X	$2,378,630
35772 —	Parts and subassemblies for computer peripherals and input/output equipment	68	X	$5,504,761	66	X	$2,487,482
36950 —	Magnetic and optical recording media	57	X	r/ $5,739,079	57	X	$5,242,777
	Rigid disks:						
36950 12	Less than 2 1/2 inch	1	D	D	1	D	D
36950 14	2 1/2 inch up to but not including 3 1/2 inch	7	81,089	$236,727	6	D	D
36950 16	3 1/2 inch up to but not including 5 1/4 inch	14	18,005,756	$1,072,171	11	27,234,720	$1,151,126
36950 18	5 1/4 inch and greater	9	12,980,963	$226,714	5	3,225	$16,790
	Flexible discs:						
36950 22	3 1/2 inch up to but not including 5 1/4 inch	11	2,359,779	$564,942	13	782,694	$717,747
36950 24	5 1/4 inch and greater	10	1,536,169	$117,079	9	b/ 128,467	b/ $17,893
36950 09	Optical discs	11	198,689	r/ $648,527	12	229,754	$548,763
36950 26	Bulk magnetic tape	2	D	D	4	133,255	$24,424
	Packaged magnetic tape:						
36950 11	In reels for computer use	5	D	D	5	D	D
36950 13	In cassette and cartridge for computer use	15	862,420	$657,596	14	44,723	$742,739
36950 28	In forms suitable for audio use	9	431,461	r/ $310,039	10	b/ 433,078	b/ $315,805
36950 21	In video cassette, 8 mm and 1/2 inch	13	697,108	r/ $910,457	15	755,300	$868,228
36950 23	In video cassette, 3/4 (19mm) and 2 inch (51 mm)	4	38,046	r/ $99,089	4	9,909	$83,588
36950 34	Other packaged magnetic tape	3	181,958	r/ $346,519	4	199,452	$363,624
36950 39	Other magnetic recording media, including parts	15	X	r/ $309,414	16	X	b/ $183,634
	Nonmanufacturing revenue of manufacturing establishments:						
50650 00	Supplies and accessories for use with computers or business machines (resales)	55	X	$1,415,878	50	X	$1,442,467
99980 49	Research and development, testing and evaluation of systems and components	18	X	$47,949	18	X	$96,493
	Software sold with computer systems shipped by manufacturers:						
73721 00	Operating system software	21	X	$121,773	14	X	$94,358
73722 00	Utilities and device managers	16	X	$5,476	12	X	$2,806
73723 00	Application software	42	X	$437,019	36	X	$425,435
73724 00	Other software	16	X	$126,546	10	X	$66,031

5.5. Computers, Quantity and Value of Shipments, United States, 1995–97 (continued).

Product Code	Product Description	1996			1997		
		Number of Companies	Quantity	Value	Number of Companies	Quantity	Value
73730 00	Systems integration revenues, including configuration, installation and site preparation, and custom programming	31	X	$299,960	26	X	$408,653
73700 00	Other nonmanufacturing revenues associated with computer systems	80	X	$1,547,807	72	X	$1,517,293

Notes:
D=Withheld to avoid disclosing data for individual companies.
n.e.c.=Not elsewhere classified.
r/=Revised by 5 percent or more from previously published data.
/a=10 to 25 percent of this item has been estimated.
/b Over 25 percent of this item has been estimated.
X=Not applicable.
NA=Not available.
[1] Receipts for personal digital assistants (PDAs) are combined with other portable computers to avoid disclosing data for individual companies.

Source: U.S. Bureau of the Census. Current Industrial Reports, Series MA35R: *Computer and Office Accounting Machines.* Washington, DC: U.S. Bureau of the Census, 1997 and 1998. Table 2. http://www.census.gov/cir/www/ma35r.html

5.6. Modems/Fax Modems—U.S. Sales to Dealers, 1993–96, Estimated for 1997, and Projected for 1998.

Year	Unit Sales (thousands)	Dollar Sales (millions)	Household Penetration	Average Unit Price
1993	2,220	$460	NA	$207
1994	2,900	$550	NA	$190
1995	4,670	$770	16%	$165
1996	6,350	$960	18%	$151
1997 est.	7,800	$1,170	21%	$150
1998 proj.	9,100	$1,274	24%	$140

Notes:
NA=Not available.

Source: Consumer Electronics Manufacturers Association (CEMA). *CEMA Consumer Electronics U.S. Sales & Forecasts.* Arlington, VA: CEMA, 1998.

5.7. Personal Computers—U.S. Sales to Dealers, 1982–96, Estimated for 1997, and Projected for 1998.

Year	Unit Sales (thousands)	Dollar Sales (millions)	Household Penetration	Average Unit Price
1982	1,550	$1,375	NA	NA
1983	3,750	$2,070	7%	NA
1984	3,975	$2,385	13%	NA
1985	3,200	$2,175	15%	NA
1986	2,950	$3,060	16%	NA
1987	3,125	$3,100	18%	NA
1988	3,500	$3,340	20%	NA
1989	3,900	$3,711	21%	NA
1990	4,000	$4,187	22%	NA
1991	3,900	$4,287	25%	NA
1992	4,875	$6,825	27%	NA
1993	5,850	$8,190	30%	$1,400
1994	6,725	$10,088	33%	$1,500
1995	8,400	$12,600	36%	$1,500
1996	9,400	$15,040	40%	$1,600
1997 est.	11,000	$15,950	41%	$1,450
1998 proj.	12,600	$17,640	42%	$1,400

Notes:
Sales through consumer channels, includes notebooks. Does not include TV/PC.
NA=Not available.

Source: Consumer Electronics Manufacturers Association (CEMA). *CEMA Consumer Electronics U.S. Sales & Forecasts.* Arlington, VA: CEMA, 1998.

5.8. Computer Printers—U.S. Sales to Dealers, 1993–96, Estimated for 1997, and Projected for 1998.

Year	Unit Sales (thousands)	Dollar Sales (millions)	Household Penetration	Average Unit Price
1993	4,320	$1,793	NA	$415
1994	5,160	$2,018	NA	$391
1995	6,480	$2,430	33%	$375
1996	8,400	$2,999	38%	$357
1997 est.	10,400	$3,900	39%	$375
1998 proj.	11,600	$4,234	NA	$365

Notes:

Sales through consumer channels. Includes laser, inkjet, bubblejet, and dot matrix printers sold through retail channels including direct to consumer channels.

NA=Not available.

Source: Consumer Electronics Manufacturers Association (CEMA). *CEMA Consumer Electronics U.S. Sales & Forecasts.* Arlington, VA: CEMA, 1998.

5.9. Value of Shipments of Semiconductors, Printed Circuit Boards, and Other Electronic Components by Class of Product, 1991–97 (Quantity in thousands of units. Values in thousands of dollars.).

Product Code	Product Description	1991	1992	1993	1994	1995	1996	1997
36713	Transmittal, industrial, and special-purpose electron tubes (except x-ray)	$1,072.5	$868.4	$677.6	$854.9	r/ $854.9	$611.4	$699.1
36714	Receiving type electron tubes and cathode ray picture tubes	$1,925.1	$2,047.4	$2,258.6	$2,627.1	$2,907.0	$3,272.0	$3,433.6
36715	Electron tube parts	$120.3	$129.5	$160.1	$114.9	$120.2	$153.1	$161.0
36720	Printed circuit boards	$6,275.3	$5,721.7	$6,273.0	$6,812.3	$8,367.3	$8,216.8	$8,784.6
36741	Integrated microcircuits (semiconductor networks)	$19,150.9	$20,065.4	$23,636.1	$36,020.4	r/$48,437.9	$52,639.3	$56,866.1
36742	Transistors	$736.9	$674.8	$704.5	$834.6	$942.5	$945.9	$801.2
36743	Diodes and rectifiers	$653.5	$652.2	$640.7	$829.5	$1,066.7	$861.4	$786.5
36749	Other semiconductor devices	$5,760.8	$5,899.2	$6,908.2	$9,915.4	$12,639.4	$10,976.1	$10,271.5
36750	Capacitors for electronic applications	$1,224.0	$1,288.0	$1,294.3	$1,512.3	$1,785.3	$1,653.4	$2,006.6
36760	Resistors	$721.5	$708.3	$716.5	$869.5	r/ $953.2	$911.9	$992.8
36770	Coils, transformers, reactors, and chokes for electronic applications	$1,006.6	$1,095.3	$1,121.4	$1,250.7	$1,411.8	$1,435.6	$1,450.8
36781	Coaxial connectors	$384.4	$424.5	$419.0	$642.4	r/ $731.6	$656.9	$624.5
36782	Cylindrical connectors	$483.4	$503.9	$482.8	$511.3	$552.9	$642.5	$586.2
36783	Rack and panel connectors	$484.9	$461.9	$427.6	$545.6	$540.5	$530.7	$673.3
36784	Printed circuit connectors	$843.8	$855.2	$859.4	$923.4	$1,026.4	$1,095.3	$1,276.2
36785	Other connectors, including parts	$1,068.9	$1,123.3	$1,152.8	$1,377.2	$1,401.9	$1,617.1	$1,795.5
36791	Filters (except microwave) and piezoelectric devices	$451.5	$448.8	$534.3	$674.1	r/ $729.2	$719.3	$692.6
36793	Microwave components and devices	$1,178.5	$1,258.4	$1,136.3	$1,227.2	$1,233.4	$1,251.4	$1,238.5
36795	Transducers, electrical/electronic input or output	$799.5	$810.3	$832.5	$970.5	$1,111.3	$1,104.8	$1,209.2
36796	Switches, mechanical types for electronic circuitry	$545.1	$541.7	$572.6	$621.5	$666.3	$738.0	$828.3
36798	Printed circuit assemblies	$12,590.4	$13,745.8	$15,087.1	$17,514.8	r/ $24,447.7	$24,937.3	$25,283.4
36799	All other electronic components, n.e.c.	$5,997.3	$5,260.3	$5,257.6	$6,149.7	r/ $6,978.1	$7,199.1	$7,637.6

Notes:

n.e.c.=Not elsewhere classified.

r/=Revised by 5 percent or more than previously published data.

Source: Bureau of the Census. Current Industrial Reports, Series MA36Q: *Semiconductors, Printed Circuit Boards, and Other Electronic Components.* Washington, DC: U.S. Bureau of the Census, 1997 and 1998. Table 1. http://www.census.gov/cir/www/ma36q.html

5.10. Worldwide Market Size, Speech and Voice Recognition, 1993 and Forecast for 1998.

Total	1993 $255.5 million	1998 (Forecast) $810.7 million
Data entry	19.8%	14.4%
Computer control	10.2%	12.6%
Consumer	12.1%	8.1%
Speech-to-text	22.7%	27%
Voice verification	7.1%	5.8%
Telephone	28.1%	32.1%

Source: Voice Information Associates. Lexington, MA. Cited in "Speech Finally Recognized." *Electronic Engineering Times,* n. 833 (1995): 30.

5.11. The U.S. Data Communications Market, 1996–97 and Projected for 1998 (in millions of dollars).

Products	1996 Revenues	1997 Revenues	1997 Growth Rate	1998 Revenues (Projected)	1998 Growth Rate (Projected)
LAN switches	$2,461	$4,297	75%	$5,927	38%
Routers	$2,763	$3,095	12%	$3,436	11%
Hubs	$2,240*	$1,789	-20%	$1,573	-12%
NICs (network interface cards)	$2,643	$2,115	-20%	$1,820	-14%
Servers	$15,935*	$18,370	15%	$21,325	16%
Wiring	$2,194	$2,685	22%	$3,100	15%
Network operating systems	$1,700	$1,921	13%	$2,209	15%
Internet/intranet applications	$665	$965	45%	$1,447	50%
Web servers	$46	$62	35%	$82	32%
Groupware	$338	$422	25%	$549	30%
Network security	$408*	$685	68%	$1,164	70%
Mainframe peripherals	$252*	$262	4%	$271	3%
SNA gateways	$201	$240	19%	$285	19%
Remote access devices	$1,186	$1,604	35%	$2,044	27%
Modems	$3,302	$3,077	-7%	$3,489	13%
Frame relay switches	$628*	$851	36%	$1,033	21%
Frame relay access devices	$277*	$380	37%	$494	30%
ATM (asynchronous transfer mode) enterprise switches	$257	$415	61%	$618	49%
Packet-switching equipment	$592	$520	-12%	$439	-16%
Multiplexers	$612	$590	-4%	$570	-3%
CSU/DSUs'	$717	$803	12%	$867	8%
PBXs (private branch exchanges)	$4,954*	$4,805	-3%	$6,117	27%
Computer telephony integration	$272*	$729	168%	$1,260	73%
Videoconferencing equipment	$715*	$930	30%	$1,160	25%
VSAT (very small aperture terminals) equipment	$390	$460	18%	$529	15%
Network and systems management	$2,475*	$3,094	25%	$3,868	25%
Diagnostic and test equipment	$507	$558	10%	$607	9%
Products Total	$48,730	$55,726	14%	$66,283	19%
Network Support Services	$6,912	$7,880	14%	$9,220	17%
Data and Network Services					
Leased lines	$8,423	$9,750	16%	$10,700	10%
ISDN (integrated services digital network)	$770*	$1,036	35%	$1,658	60%
Frame relay	$1,130	$2,330	106%	$4,720	103%
X.25	$893*	$818	-8%	$753	-8%
ATM (asynchronous transfer mode)	$63	$160	154%	$320	100%
Commercial Internet services	$1,066*	$1,517	42%	$3,087	103%

5.11. The U.S. Data Communications Market, 1996–97 and Projected for 1998 (in millions of dollars) *(continued).*

	1996 Revenues	1997 Revenues	1997 Growth Rate	1998 Revenues (Projected)	1998 Growth Rate (Projected)
Content hosting	$161	$400	148%	$880	120%
Data and Network Services Total	$12,506	$16,011	28%	$22,119	38%
Products and Services Total	$68,148	$79,617	17%	$97,625	23%

Notes:
* Revised from figures provided in 1996 Data Communications Market Forecast.
[1] A device that incorporates the functions of a channel service unit (CSU) and a data service unit (DSU). It interfaces between a dedicated digital service line and a user's data equipment.

Source: Data Communications magazine, December 1997. Copyright 1998 The McGraw-Hill Companies, Inc. All rights reserved.

NETWORKS AND THE INTERNET

5.12. Worldwide Intranet Market for 1996–2000.

Year	Value
1996	$22,733,616,000
1997	$35,178,546,000
1998	$46,066,293,000
1999	$57,313,542,000
2000	$69,545,541,000

Note:
Intranet figures are calculated to be 65% of Adjusted Gross Internet Product (AGIP), which is the value of all goods and services sold by the Internet.

Source: Zona Research. *Internet and Intranet: 1997 Markets, Opportunities, and Trends* (3rd ed.). Redwood City, CA, 1997. Based on Figure 3.2.

5.13. Worldwide Intranet Market by Category, 1996–2000 (dollars in millions).

	1996	1997	1998	1999	2000
Creation Authoring Tools Application Development Tools Application Development Services Web Site Development Page Development	$565,766,500	$731,815,500	$927,361,500	$1,053,188,500	$1,198,359,500
Content Ad-supported Web Content Subscription-supported Content	$2,211,475,500	$4,198,525,500	$6,067,984,000	$7,721,233,000	$9,061,175,500
Containment Hardware - Servers Software - Servers Software - Database Management Systems Software - Retrieval Services - Maintenance Services - Site Hosting	$5,029,024,000	$7,473,310,000	$8,804,926,000	$10,779,561,000	$12,533,118,000

5.13. Worldwide Intranet Market by Category, 1996–2000 (dollars in millions) *(continued)*.

	1996	1997	1998	1999	2000
Communication	$13,234,474,000	$20,361,055,000	$27,113,658,000	$33,460,264,000	$40,874,483,000
Medium - Bandwidth					
Medium - Managed Services					
Mechanism - Services - Internet Service Providers					
Mechanism - Services - Consulting					
Mechanism - Hardware					
Mechanism - Software					
Consumption	$853,203,000	$1,369,699,500	$1,844,927,500	$2,643,231,500	$3,766,308,000
Internet Clients - Commercial Thin Clients and Network Computers					
Internet Clients - Consumer Internet Access Devices					
Applications - Vertical/ Horizontal					
Browsers					
Other software					
Control	$839,566,000	$1,044,127,500	$1,307,436,000	$1,656,063,500	$2,112,103,500
Firewall/virtual private network					
Encryption/authentication					
Recovery/backup					
Digital IDs/certificates					
Intrusion/Analysis					
Consulting					
Total	$22,733,509,000	$35,178,533,000	$46,066,293,000	$57,313,541,500	$69,545,547,500

Notes:
Intranet figures are calculated to be 65% of Adjusted Gross Internet Product (AGIP), which is the value of all goods and services sold by the Internet. An intranet is a private Internet-based network used for internal organizational communication.

Source: Zona Research. *Internet and Intranet: 1997 Markets, Opportunities, and Trends* (3rd ed.). Redwood City, CA, 1997. Based on Table 3.1.

5.14. The Total Electronic Mail Market, United States, 1994–2000.

	Number of Electronic Mail Users (millions)	Number of Transmissions (millions)	Number of Messages (millions)	Value ($millions)
1994	31.5	197,780	812,669	13,497
1995	39.2	259,899	1,109,056	14,839
1996	51.0	382,225	1,738,182	18,828
1997	66.9	574,140	2,718,389	24,702
1998	81.4	802,538	3,958,850	30,118
1999	95.6	1,047,499	5,371,817	36,025
2000	108.3	1,287,773	6,862,322	41,795

Source: Electronic Messaging Association. *EMA Market Research Survey* prepared by Wilkofsky Gruen Associates. Arlington, VA.

5.15. Electronic Mail at Work (Usage), United States, 1994–2000.

	Total Number of Electronic Mail Users (millions)	Transmissions per User per Week	Aggregate Number of Transmissions per Year (millions)	Average Number of Recipients per Transmission (millions)	Total Number of Electronic Mail Messages Sent (millions)
1994	23.0	160	184,000	4.2	776,480
1995	28.2	170	239,785	4.4	1,055,054
1996	37.1	190	352,450	4.7	1,656,515
1997	49.2	215	528,900	4.9	2,591,610
1998	60.6	245	742,718	5.1	3,787,859
1999	72.0	270	972,000	5.3	5,151,600
2000	82.8	290	1,200,600	5.5	6,603,300

Source: Electronic Messaging Association. *EMA Market Research Survey* prepared by Wilkofsky Gruen Associates. Arlington, VA.

5.16. Electronic Mail at Home (Usage), United States, 1994–2000.

	Total Number of Electronic Mail Users (millions)	Transmissions per User per Week	Aggregate Number of Transmissions per Year (millions)	Average Number of Recipients per Transmission	Total Number of Electronic Mail Messages Sent (millions)
1994	7.5	34	13,260	2.69	35,669
1995	9.8	38	19,365	2.75	53,253
1996	12.6	44	28,829	2.8	80,721
1997	16.3	52	44,075	2.85	125,614
1998	19.4	58	58,510	2.9	169,680
1999	22.3	64	74,214	2.95	218,932
2000	24.3	68	85,925	3.0	257,774

Source: Electronic Messaging Association. Arlington, VA. *EMA Market Research Survey* prepared by Wilkofsky Gruen Associates.

5.17. Online Retail Revenues, U.S. and Canada, Projected 1997–2001 (in millions of dollars).

Category	1997	1998	1999	2000	2001
PC hardware and software	$863	$1,616	$2,234	$2,901	$3,766
Travel	$654	$1,523	$2,810	$4,741	$7,443
Entertainment	$298	$591	$1,143	$1,921	$2,678
Books and music	$156	$288	$504	$761	$1,084
Gifts, flowers, and greetings	$149	$264	$413	$591	$802
Apparel and footwear	$92	$157	$245	$361	$514
Food and beverages	$90	$168	$250	$354	$463
Jewelry	$38	$56	$78	$107	$140
Consumer electronics	$19	$34	$60	$93	$143
Sporting goods	$20	$29	$43	$63	$84
Toys and hobbies	$13	$21	$32	$47	$71
Health, beauty, and drugs	$11	$16	$25	$36	$50
Tools and gardening	$10	$22	$31	$44	$59
Home furnishings	$9	$15	$21	$28	$38
Other (pets, photo, etc.)	$22	$28	$35	$42	$52
Total	$2,444	$4,828	$7,924	$12,090	$17,387

Source: Forrester Research, Inc. Cambridge, MA. Unpublished.

DATABASES, PROGRAMMING LANGUAGES, AND SECURITY

5.18. Number of Database Records by Year, 1975–97.

Year	Records (millions)
1975	52
1977	72
1979	148
1981	250
1983	310
1984	1,000
1985	1,680
1987	2,065
1988	2,255
1989	2,694
1990	3,569
1991	4,060
1992	4,527
1993	5,572
1994	6,319
1995	8,160
1996	10,757
1997	11,270

Notes:
The author of the data explains:
"The statistics presented in this article . . . cover the worldwide database industry and are independent of the media in which databases are distributed and accessed unless specifically noted." By "database," she means an electronic collection of commercially available structured data; the term does not refer to data internal to companies or other organizations. Databases may include words, numbers, images, sound, and other types of information.
"The entities counted as database records vary widely but generally range from 200 to 2,000 words (or, in the case of non-word-oriented records, they require a comparable number of bytes for storage). Records may be citations, abstracts, news stories, magazine articles, biographical records, unique names of chemicals, unique chemical structures, property data, recipes, time series, software programs, images, or descriptions or listings of virtually anything."

Source: Martha E. Williams. "The State of Databases Today: 1998." Foreword in Erin E. Braun and Lisa Krumar, eds. *Gale Directory of Databases, Volume I: Online Databases* and *Volume 2: CD-ROM, Diskette, Magnetic Tape, Handheld, and Batch Access Database Products.* Gale Research. Detroit, 1998. Page xvii.

5.19. Growth in Number of Database Vendors, Producers, Database Entries, and Databases, 1975–97.

Year	Vendors	Producers	Entries	Databases
1975	105	200	301	301
1979	263	316	528	528
1982	311	422	773	773
1985	614	1,210	2,700	3,010
1988	750	1,733	4,042	4,200
1989	770	1,950	4,786	5,578
1990	850	2,224	5,689	6,750
1991	933	1,372	6,261	7,637
1992	1,438	3,007	6,998	7,907
1993	1,629	2,744	7,538	8,261
1994	1,691	2,778	7,979	8,776
1995	1,810	2,860	8,525	9,207
1996	1,805	2,938	9,290	10,033
1997	2,115	3,216	9,662	10,338

Note:
Databases and entries (in *Gale Directory of Databases*) that are provided in more than one medium are counted only once.

Source: Martha E. Williams. "The State of Databases Today: 1998." Foreword in Erin E. Braun and Lisa Krumar, eds. *Gale Directory of Databases, Volume I: Online Databases* and *Volume 2: CD-ROM, Diskette, Magnetic Tape, Handheld, and Batch Access Database Products.* Gale Research. Detroit, 1998. Page xvii.

5.20. Database Entries Classed by Form of Representation of Data, 1988–97 (Normalized to One Class per Database Entry).

DB Classes (multiple per database)	1985 ** Number	Percent	1988 Number	Percent	1989 Number	Percent	1990 Number	Percent	1991 Number	Percent	1992 Number	Percent	1993 Number	Percent	1994 Number	Percent	1995 Number	Percent	1996 Number	Percent	1997 Number	Percent
Word-oriented	1,728	64%	2,797	69%	3,370	70%	4,080	72%	4,491	72%	4,925	70%	5,421	72%	5,729	72%	6,044	71%	6,467	70%	6,764	70%
Number-oriented	972	36%	1,136	28%	1,236	26%	1,298	23%	1,370	22%	1,533	22%	1,437	19%	1,428	18%	1,407	17%	1,410	15%	1,450	15%
Image/Video			14	<1%	34	<1%	113	2%	145	2%	272	4%	340	4%	431	5%	592	7%	846	9%	864	9%
Audio			1	<1%	2	<1%	16	<1%	28	<1%	83	1%	106	1%	152	2%	235	3%	304	3%	290	3%
Electronic Services			90	2%	134	<3%	170	<3%	172	3%	146	2%	203	3%	207	3%	204	2%	204	2%	193	2%
Software			4	<1%	10	<1%	12	<1%	55	1%	39	<1%	31	<1%	32	<1%	43	<1%	59	1%	96	1%
Total	2,700		4,042		4,786		5,689		6,261		6,998		7,538		7,979		8,525		9,290		9,662	

Notes:

*=Number is number of database entries in *Gale Directory of Databases*; an entry represents a database on one or more media. Estimates for individual databases in 1997 can be calculated by multiplying the percentages by 10,338.

**=In 1985 there were only two basic classes of databases.

Percentages are rounded.

Source: Martha E. Williams. "The State of Databases Today: 1998." Foreword in Erin E. Braun and Lisa Krumar, eds. *Gale Directory of Databases, Volume 1: Online Databases* and *Volume 2: CD-ROM, Diskette, Magnetic Tape, Handheld, and Batch Access Database Products.* Gale Research. Detroit, 1998. Page xvii.

5.21. Media for Database Distribution/Access, 1989–97 (Number* and Percent by Year).

Medium for Access and Distribution	1989 Number	Percent	1990 Number	Percent	1991 Number	Percent	1992 Number	Percent	1993 Number	Percent	1994 Number	Percent	1995 Number	Percent	1996 Number	Percent	1997 Number	Percent
Online	3,524	57%	4,018	53%	4,170	51%	5,486	65%	5,564	61%	5,646	57%	5,801	55%	5,950	52%	5,926	49%
CD-ROM	433	7%	715	10%	1,019	12%	1,321	15%	1,648	18%	2,016	20%	2,371	23%	2,953	26%	3,626	30%
Diskette	478	8%	626	8%	695	9%	676	8%	781	8%	956	10%	1,049	10%	1,110	10%	1,246	10%
Magnetic Tape	787	12%	906	12%	954	12%	584	7%	600	7%	686	7%	701	7%	719	6%	772	6%
Batch	999	16%	1,252	17%	1,321	16%	389	5%	481	5%	500	5%	489	5%	508	5%	466	4%
Handheld	0	0%	0	0%	0	0%	39	<1%		<1%	65	1%	73	1%	97	1%	105	1%
Total	6,221		7,517		8,159		8,495		9,131		9,869		10,484		11,337		12,141	

Notes:

*=Number is number of *Gale Directory of Databases* entries where a database can be on one or more media.

Percentages are rounded.

Source: Martha E. Williams. "The State of Databases Today: 1998." Foreword in Erin E. Braun and Lisa Krumar, eds. *Gale Directory of Databases, Volume I: Online Databases* and *Volume 2: CD-ROM, Diskette, Magnetic Tape, Handheld, and Batch Access Database Products.* Gale Research. Detroit, 1998. Page xvii.

5.22. Data Mining Market Forecast, 1996 and 2000.

	1996	Percent	2000	Percent
Micromining [1]	$33,000,000	1%	$84,000,000	1%
Data visualization [2]	99,000,000	3%	168,000,000	2%
Macromining [3]	165,000,000	5%	672,000,000	8%
Packages [4]	132,000,000	4%	1,092,000,000	13%
Systems integrators [5]	231,000,000	7%	1,680,000,000	20%
Data service providers [6]	2,640,000,000	80%	4,704,000,000	56%
Total	$3.3 billion	100%	$8.4 billon	100%

Notes:

Data mining is a process that involves using artificial intelligence techniques to look for hidden or unexpected patterns in a database. Data mining can yield vital strategic business information.

[1] Macromining is an expensive but flexible type of tool that operates on high-end servers and covers all steps of a data mining process. Macromining tools offer many types of algorithms, such as statistical, decision tree, and neural network.

[2] Data visualization allows users to see vast amounts of data in graphical form.

[3] Micromining is an inexpensive type of tool that features short learning curves and usually runs on PCs. The usefulness of micromining tools may be limited because these programs offer only a single algorithm that may not work equally well in different business applications.

[4] Compilations of products that treat horizontal and/or vertical markets.

[5] Vendors and consultants who offer unifying frameworks for delivering information.

[6] Furnishers of industry and other comparative data that can be used to enrich internal company data before applying data mining techniques.

Source: META Group. *Data Mining Market Trends: A Multiclient Study, 1997–1998.* Stamford, CT, 1997. Based on Figure 4, page 24.

5.23. U.S. Java Software Market Size Projections, 1997–2000 (sales in millions of dollars).

Year	
1997	$ 58.9
1998	$ 98.3
1999	$ 142.4
2000	$ 179.5

Source: Zona Research. Redwood City, CA. Cited in "Java Market Still in Infancy, Survey Shows." *Network World 14* (December 8, 1997): 39.

5.24. 1998 Security Revenue, United States, Projected.

	Percent	Revenues
Security firewalls	34%	$ 394,400,000
Authentication software and devices	24%	$ 278,400,000
Encryption software and hardware	17%	$ 197,200,000
Certificate servers and authorities	1%	$ 11,600,000
Other software and applications	24%	$ 278,400,000
Total		$ 1.16 billion

Notes:

Authentication. A way of insuring that digital information is delivered to the person for whom it is intended, often through the use of a user name and password. Authentication also provides a pedigree that assures the receiver of the message's integrity and source.

Certificate. A packet of encrypted information used on the Internet or on an intranet to establish a secure connection between parties. Certificates contain information about the owner, issuer, and valid dates as well as a unique identifier and an encrypted fingerprint that helps verify the contents. A certificate authority is a third party that issues security certificates; a certificate server is a software program that allows the user to create, distribute, and manage security certificates without going through a certificate authority. A certificate must be recognized by the user's browser software or the transaction will be rejected.

Encryption. The conversion of information to a code or cipher so that it is not readable without a special key.

Firewall. A software- or hardware-based barrier in a network that lets through only data intended and authorized to reach the other side.

Source: Data Communications magazine, December 1997. Copyright 1998 The McGraw-Hill Companies, Inc. All rights reserved.

Chapter 6. Construction and Infrastructure

Humans have built structures for millennia, but it wasn't until relatively recently that our buildings and public works provided real comfort and healthy living conditions. Romantic stories entice us with images of royal castles and palaces, but in reality, those homes were drafty, uncomfortable places, far less hospitable than today's average tract home. Elizabeth I couldn't even take a hot shower! However, as lucky as we are to know indoor plumbing, air conditioning, and fiberglass insulation, sophistication in structural engineering is anything but new. From the pyramids to intelligent homes, building has inspired people to accomplish near-impossibilities. Ancient, medieval, and Renaissance peoples built structures that seem way ahead of their times. Just look at a Gothic cathedral, Stonehenge, the Great Wall of China, the Egyptian or Mayan pyramids, Easter Island, or Roman aqueducts, and you know that our ancestors possessed sophisticated engineering know-how.

Civilization breeds variety in building form, function, and design. Not only do we build houses, office buildings, shopping malls, schools, hotels, theaters, and religious buildings, but we also construct huge airplane hangars, sports arenas, scientific laboratories, libraries, factories, and museums. Table 6.2 shows how many new square feet for each general type we are building year by year, while Table 6.3 shows how tall we are making our structures. While all buildings perform certain common functions, some structures require specialized engineering. The library floor must be able to support considerable weight; the laboratory must be capable of stringent environmental control; the hangar must be large enough to accommodate all kinds of aircraft, even those not yet designed; the theater must offer excellent acoustics and a sufficient number of women's restrooms. We also construct sewers, bridges, roads, pipelines, aqueducts, canals, freeways, power plants, and communications networks. These must be stable, safe, and capable of handling increasing loads. Table 6.7 and 6.10 dem-

onstrate that the quality of our roads and bridges is improving, while Table 6.9 lists roads and how they are distributed throughout the 50 states.

Buildings need to perform a number of functions besides the obvious one—shelter. They must provide services, such as power, plumbing, and telecommunications. They should be comfortable, with efficient heating, cooling, and sometimes air filtering systems. (See Table 6.6 for a glimpse at the heating/air conditioning/dehumidifying market.) They need to resist the loads caused by their own weight as well as those from rain, snow, and ice, and in some areas, they need to resist strong winds and/or seismic vibrations. Buildings also must protect inhabitants and their possessions from fire. At the least they should allow for quick emergency exit; at best they won't burn at all, or will provide early warning systems. Tables 6.4 and 6.5 illustrate how we are addressing the issue. Fire detection and prevention is an almost $800 million industry, with 16 million smoke detectors shipped in 1996 alone.

Hot topics in construction include

- Smart buildings, that is, buildings that provide for and integrate advanced technologies like computer-controlled environmental, security, and telecommunications systems. (See Table 6.1, which indicates that 71% of building owners and facilities managers surveyed intend to purchase intelligent controls or systems for their new construction projects.)
- Earthquake-resistant design
- Energy-efficient design and materials. (Table 6.1 also shows that many builders are thinking along these lines.)
- The use of recycled and special materials

The statistics that follow give you a taste of how we're dealing with these needs and issues.

BUILDINGS AND COMPONENTS

6.1. Building Technologies: Plans for Modernization and/or New Construction, United States, 1998.

	Modernization Projects	New Construction Projects
Types of buildings currently involved in construction		
Office	55%	42%
Education	15%	22%
Retail	14%	14%
Apartment	12%	13%
Government	11%	15%
Hospital/Healthcare	11%	16%
Hotel	4%	7%
Other	20%	20%
Use of CAD[1] and CAFM[2] systems		
CAD	81%	46%
CAFM	52%	17%
Use of electronic media in conjunction with projects		
Catalogs on CD-ROM	50%	56%
Catalogs on diskette	26%	33%
Online services	42%	46%
Web sites	52%	58%
Percent of projects including resolution of laws/regulations/issues		
Fire codes	79%	N/A
Americans with Disabilities Act	78%	N/A
Energy conservation	77%	N/A
Indoor air quality	43%	N/A
Hazardous waste	43%	N/A
Energy conservation technologies to be included		
Update HVAC systems[3]	71%	N/A
Relighting	65%	N/A
Reroofing	52%	N/A
Windows	52%	N/A
Energy management systems	50%	N/A
Building automation/integration	43%	N/A
Boilers/boiler controls	37%	N/A
Water heaters	31%	N/A
Exterior insulation	24%	N/A
Window film	17%	N/A
Solar control	9%	N/A
Products intended for purchase		
Paints/coating material	81%	N/A
Carpet	80%	N/A
HVAC[3]	75%	N/A
Ceilings	70%	N/A
Lighting fixtures, indoor	70%	N/A
Signage	68%	N/A
Doors/entrances	67%	N/A
Building hardware	65%	N/A
Restroom accessories	65%	N/A
Plumbing fixtures	64%	N/A
Roofing	64%	N/A
Fire detection/prevention systems	61%	N/A
Lighting controls	59%	N/A
Resilient/hard surface floors	59%	N/A
Security systems	59%	N/A
Sprinkler systems	59%	N/A
Wallcoverings	57%	N/A
Wiring/cabling/fiber optics	57%	N/A

6.1. Building Technologies: Plans for Modernization and/or New Construction, United States, 1998 *(continued)*.

	Modernization Projects	New Construction Projects
Concrete/masonry repair	55%	N/A
Energy management systems/ temperature control	54%	N/A
Furniture	54%	N/A
Lamps	54%	N/A
Windows	54%	N/A
Fixed interior walls	53%	N/A
Sealants	51%	N/A
Waterproofing	51%	N/A
Wire/cable distribution	51%	N/A
Telecommunications equipment	50%	N/A
Window treatments	45%	N/A
Water heaters	44%	N/A
Venetian/vertical blinds	43%	N/A
Workstations	43%	N/A
Building automation systems	42%	N/A
Insulation	41%	N/A
Electronic access control	39%	N/A
Seating	39%	N/A
Air filtration	38%	N/A
Elevators/escalators	38%	N/A
UPS/standby power[4]	38%	N/A
Boiler/boiler controls	36%	N/A
Moveable walls partitions	36%	N/A
Structural bracing	34%	N/A
Occupancy sensors	33%	N/A
Water coolers	32%	N/A
Pavers	31%	N/A
Filing systems	30%	N/A
Landscape management	30%	N/A
Access flooring	28%	N/A
Curtainwall (glass, marble, etc.)[5]	25%	N/A
Emergency communications equip.	25%	N/A
Exterior insulation	22%	N/A
Window film/solar control	18%	N/A
Inclusion of automated/intelligent/ integrated controls or systems in new construction	N/A	71%
Technologies to be included in new construction		
HVAC[3]	N/A	94%
Life and fire safety	N/A	83%
Energy management systems	N/A	70%
Lighting management/control	N/A	53%
Facility management systems	N/A	37%
Vertical transportation	N/A	33%

Notes:
[1] CAD = computer-aided design
[2] CAFM = computer-aided facilities management
[3] HVAC = heating, ventilating, and air conditioning
[4] UPS = uninterruptible power supply
[5] Curtainwall. An external skin on a building, usually made from light aluminum or steel and filled with glass or metal panels. A curtainwall bears no portion of the load but does protect the structure against weather. It also provides visual elements.

Methodology:
On February 2, 1998, a six-page questionnaire concerning modernization was mailed to 6,000 building owners and facilities management professionals chosen at random from the subscriber list of *Buildings* magazine. On the cut-off date of April 10, 1998, there were 757 usable replies, a return of 13%. On March 9, 1998, a four-page questionnaire concerning new construction was mailed to 3,000 building owners and facilities management professionals chosen at random from the subscriber list of *Buildings* magazine. On the cut-off date of May 15, 1998, there were 324 usable replies, a return of 11%.

Source: Market research report. Cedar Rapids, IA: *Buildings* magazine, April 1998. Report is distributed to the industry.

6.2. Building Types: Percent of Total Square Feet for Each, New Construction, United States, 1993–97.

	1993	1994	1995	1996	1997
Industrial	10.7	11.6	12.3	11.4	12.0
Commercial[1]	31.0	33.6	35.1	32.9	29.0
Office	9.0	9.0	8.8	10.4	12.4
Parking	7.0	7.0	6.7	7.0	8.0
Assembly[2]	24.7	21.7	20.7	20.0	20.6
Residential[3]	4.0	4.0	5.2	6.0	6.6
Medical	7.5	6.3	5.5	6.0	6.0
Public	3.0	4.0	3.1	3.3	3.0
Miscellaneous	2.6	2.6	2.6	2.5	2.3

Notes:
[1] Includes warehouses.
[2] Includes schools (but not dormitories), sports facilities, and churches.
[3] Includes dormitories.

Source: American Institute of Steel Construction. Chicago, IL. Unpublished.

6.3. Buildings, Percent of Square Feet by Number of Stories, All Types, New Construction, United States, 1993–97 (square feet in thousands).

	1993 Sq. feet	1993 Percent	1994 Sq. Feet	1994 Percent	1995 Sq. Feet	1995 Percent	1996 Sq. Feet	1996 Percent	1997 Sq. Ft.	1997 Percent
One story	438,445	62.0%	719,823	64.3%	807,536	63.9%	556,554	61.1%	859,612	60.2%
2–4 stories	201,609	28.5%	300,972	27.0%	341,895	27.0%	259,968	28.8%	399,784	28.0%
5–19 stories	61,516	9.0%	87,845	7.8%	96,509	7.6%	77,222	8.1%	144,259	10.0%
20 stories and over	3,173	0.5%	12,018	1.0%	16,424	1.3%	15,234	1.6%	22,538	1.8%
Total	704,743	100%	1,120,658	100%	1,262,364	100%	908,978	100%	1,426,193	100%

Source: American Institute of Steel Construction. Chicago, IL. Unpublished.

6.4. Value of Shipments of Alarm Systems, and Fire Detection and Prevention Equipment, United States, 1996–97 (thousands of dollars).

Product Code		Product Description	1996		1997	
			Number of Companies	Value	Number of Companies	Value
36691	—	Alarm Systems	104	$1,824,165	NA	$1,921,083
		Intrusion detection:				
36691	48	Local	29	$123,798	23	$146,023
36691	49	Central station	24	$389,442	16	[a]$ 446,425
36691	50	Direct connect	12	$12,108	9	[b]$ 11,975
36691	51	Hold-up systems (commercial and industrial)[1]	8	D	4	D
		Fire Detection and Prevention				
		Smoke and heat detection alarms:				
36691	53	Ionization chamber type	9	$116,411	9	$123,945
36691	54	Other, including photo cell type	20	$320,115	17	$355,701
36691	56	Central station	14	$121,059	11	$117,185
36691	57	Direct connect	17	[a]$186,438	14	$179,828
36691	59	Other intercommunication and alarm systems, including electric sirens and horns (marine, industrial, and air raid) security locking systems[1]	38	$324,940	32	$335,066

Notes:
[1]Product code 36691 51 is included with product code 36691 59 to avoid disclosing data for individual companies.
D Withheld to avoid disclosing data for individual companies.
NA Not available.
[r]Revised by 5 percent or more from previously published data.
[a] 10 to 25 percent of this item has been estimated.
[b] 25 to 50 percent of this item has been estimated.

Source: Bureau of the Census. *Current Industrial Reports, Series MA36P: Communication Equipment, Including Telephone, Telegraph, and Other Electronic Systems and Equipment, 1997.* Table 1. Washington, DC: U.S. Bureau of the Census, 1998. http://www.census.gov/cir/www/ma36p.html

6.5. Carbon Monoxide and Smoke Detectors, Unit Shipments, United States, 1996.

Carbon Monoxide Detectors	6,300,000
Smoke Detectors	16,000,000

Source: "The Share-of-Market Picture for 1996: Security Appliances." *Appliance* 54 (September, 1997): 84. Reprinted by permission of *Appliance Magazine.* Copyright Dana Chase Publications, Inc., September 1997.

6.6. U.S. Market Size for Air Conditioners, Heat Pumps, Dehumidifiers, and Gas Furnaces in 1996 (unit shipments).

Comfort Conditioning	Units
Air conditioners (room)	4,825,000
ACs/Heat Pumps (unitary)	5,670,665
Dehumidifiers	976,900
Furnaces (gas)	2,871,256

Source: "The Share-of-Market Picture for 1996: Comfort Conditioning." *Appliance* 54 (September, 1997): 82. Reprinted by permission of *Appliance Magazine.* Copyright Dana Chase Publications, Inc., September 1997.

INFRASTRUCTURE

6.7. Highway Pavement Conditions, United States, 1994 and 1996 (percent).

Type of Road	Year	Poor	Mediocre	Fair	Good	Very Good	Total Miles Reported
Urban							
Interstates*	1994	13.0%	29.9%	24.2%	26.7%	6.2%	12,338
	1996	8.8%	28.2%	24.8%	30.6%	7.6%	12,419
Other freeways and	1994	5.3%	12.7%	58.1%	20.9%	2.9%	7,618
expressways	1996	3.4%	8.8%	55.1%	26.0%	6.7%	8,403
Other principal arterials	1994	12.5%	16.3%	50.8%	16.6%	3.8%	38,598
	1996	11.8%	14.2%	49.1%	17.4%	7.5%	44,469
Rural							
Interstates*	1994	6.5%	26.5%	23.9%	33.2%	9.9%	31,502
	1996	3.9%	19.4%	21.8%	38.4%	16.4%	31,298
Other principal arterials	1994	2.4%	8.2%	57.4%	26.6%	5.4%	89,506
	1996	1.5%	5.9%	49.3%	34.0%	9.0%	91,998
Minor arterials	1994	3.5%	10.5%	57.9%	23.6%	4.5%	124,877
	1996	2.3%	8%	50.7%	30.9%	7.8%	126,158

Notes:
Poor = Needs immediate improvement.
Mediocre = Needs improvement in the near future to preserve usability.
Fair = Will be likely to need improvement in the near future, but depends on traffic use.
Good = In decent condition; will not require improvement in the near future.
Very good = New or almost new pavement; will not require improvement for some time.
* = Interstates are held to a higher standard than other roads because of higher volume and speed.

Source: U.S. Department of Transportation, Federal Highway Administration. *Highway Statistics 1995 and 1996.* Washington, DC. Cited in U.S. Department of Transportation, Bureau of Transportation Statistics. *Transportation Statistics Annual Report 1998.* Washington, DC: Bureau of Transportation Statistics, 1998. Table 1-5, page 33. http://www.bts.gov/programs/transtu/tsar/prod.html

6.8. Funding for Highways, United States, 1980–97 (in millions of dollars).

Type	1980	1985	1990	1991	1992	1993	1994	1995 (preliminary)	1996 (estimated)	1997 (forecast)
Total receipts	$39,715	$61,506	$75,294	$82,379	$86,703	$88,380	$91,312	$95,312	$99,997	$101,850
Current income	$37,604	$54,957	$69,730	$75,452	$77,404	$80,610	$84,017	$87,693	$92,153	$93,750
Imposts on highway users[1]	$22,559	$35,599	$44,264	$50,349	$50,889	$54,401	$55,387	$59,495	$64,242	$65,173
Other taxes and fees	$11,808	$15,127	$19,827	$19,077	$19,933	$19,375	$21,598	$21,160	$20,599	$21,123
Investment income, other receipts	$3,237	$4,231	$5,639	$6,026	$6,582	$6,834	$7,032	$7,038	$7,312	$7,454
Bond issue proceeds[2]	$2,111	$6,549	$5,564	$6,927	$9,299	$7,770	$7,295	$7,619	$7,844	$8,100
Intergovernmental payments[3]	NA	NA	NA	NA	NA	NA	NA	NA	NA	NA
Funds from (+) or to (−) reserves	$2,080	$(4,058)	$114	$(3,768)	$(3,155)	$(1,955)	$(1,120)	$(2,808)	$(3,341)	$(1,591)
Total funds available	$41,795	$57,448	$75,408	$78,611	$83,548	$86,425	$90,192	$92,504	$96,656	$100,259
Total disbursements	$41,795	$57,448	$75,408	$78,611	$83,548	$86,425	$90,192	$92,504	$96,656	$100,259
Current disbursements	$40,084	$54,725	$72,457	$74,895	$78,959	$81,233	$85,645	$87,843	$91,780	$95,145
Capital outlay	$20,337	$27,138	$35,151	$36,638	$38,309	$39,528	$42,379	$43,097	$45,000	$46,597
Maintenance and traffic services	$11,445	$16,032	$20,365	$21,222	$22,223	$22,894	$23,553	$24,455	$25,662	$26,720
Administration and research	$3,022	$4,033	$6,501	$6,856	$7,718	$7,921	$8,376	$8,332	$8,624	$8,790
Law enforcement and safety	$3,824	$5,334	$7,235	$7,040	$7,088	$7,157	$7,673	$7,977	$8,270	$8,604
Interest on debt	$1,456	$2,188	$3,205	$3,139	$3,621	$3,733	$3,664	$3,982	$4,224	$4,434
Debt retirement[2]	$1,711	$2,723	$2,951	$3,716	$4,589	$5,192	$4,547	$4,661	$4,876	$5,114

Notes:

Data compiled from reports of state and local authorities.

NA = Not applicable.

[1] Excludes amounts later allocated for nonhighway purposes.

[2] Excludes issue and redemption of short-term notes or refunding bonds.

[3] Positive figures indicate net receipt of funds from other levels of government; negative figures, in parentheses, indicate net disbursement of funds to other levels.

Source: U.S. Bureau of the Census. *Statistical Abstract of the U.S., 1997.* Washington, DC.: U.S. Government Printing Office, 1997. Table 998.

6.9. Highway Mileage—Functional Systems and Urban/Rural, United States, 1995.

State	Functional Systems					Urban	Rural
	Total	Interstate[1]	Other arterial	Collector	Local		
U.S	3,912,226	54,714	376,405	793,124	2,687,983	819,706	3,092,520
AL	93,313	925	8,673	20,325	63,390	20,086	73,227
AK	13,486	1,086	1,511	2,444	8,445	1,781	11,705
AZ	54,561	1,250	4,701	8,721	39,889	16,322	38,239
AR	77,222	646	6,739	20,120	49,717	7,657	69,565
CA	170,389	3,750	26,754	32,206	107,679	82,520	87,869
CO	84,499	1,170	8,134	16,566	58,629	13,485	71,014
CT	20,500	542	2,851	2,945	14,162	11,627	8,873
DE	5,631	51	615	943	4,022	1,920	3,722
DC	1,421	31	248	152	990	1,421	0
FL	113,778	1,861	11,933	15,344	84,640	48,338	65,440
GA	111,273	1,413	12,998	23,093	73,769	26,402	84,871
HI	4,133	77	754	792	2,510	1,840	2,293
ID	59,733	613	3,580	9,647	45,893	3,546	56,187
IL	137,413	2,245	13,872	21,429	99,867	35,567	101,846
IN	92,780	1,303	7,912	22,598	60,967	19,467	73,313
IA	112,702	781	9,412	31,488	71,021	9,337	103,365
KS	133,323	1,008	9,167	33,201	89,947	9,691	123,632
KY	72,998	855	5,418	17,610	49,115	10,298	62,700
LA	60,119	929	5,289	12,545	41,356	13,919	46,200
ME	22,577	383	2,300	5,917	13,977	2,620	19,957
MD	29,680	711	3,551	4,987	20,431	13,899	15,781
MA	30,751	762	5,628	5,454	18,907	19,847	10,904
MI	117,611	1,458	12,150	25,847	78,156	28,083	89,528
MN	130,391	1,042	12,260	29,345	87,744	15,166	115,225
MS	73,102	726	7,018	15,554	49,804	7,919	65,183
MO	122,616	1,460	9,276	25,072	86,808	16,310	106,306
MT	69,537	1,190	6,003	16,428	45,916	2,399	67,138
NE	92,755	497	7,877	20,747	63,634	5,105	87,650
NV	44,936	586	2,729	4,895	36,726	4,453	40,483
NH	15,086	266	1,545	2,713	10,562	2,913	12,173
NJ	35,646	728	5,154	4,544	25,220	24,405	11,241
NM	61,289	1,003	4,509	6,811	48,966	6,102	55,187
NY	112,193	2,328	13,690	20,496	75,679	40,320	71,873
NC	96,809	1,237	8,850	17,788	68,934	22,149	74,660
ND	86,830	570	5,870	11,372	69,018	1,820	85,010
OH	114,563	1,937	10,616	22,097	79,913	33,139	81,424
OK	112,517	1,064	7,832	25,349	78,272	12,956	99,561
OR	83,944	780	6,510	18,429	58,225	10,155	73,789
PA	118,648	2,087	13,284	19,747	83,530	33,272	85,376
RI	5,893	137	751	818	4,187	4,572	1,321
SC	64,293	894	6,802	13,390	43,207	10,543	53,750
SD	83,360	681	6,284	19,272	57,123	1,938	81,422
TN	85,599	1,176	8,580	18,078	57,765	17,196	68,403
TX	296,186	4,474	28,121	63,174	200,417	82,201	213,985
UT	41,044	948	3,333	7,714	29,049	6,227	34,817
VT	14,184	339	1,300	3,110	9,435	1,334	12,850
VA	69,142	1,329	8,073	14,226	45,514	18,375	50,767
WA	79,710	1,079	7,282	16,791	54,558	17,612	62,098
WV	35,110	560	3,236	8,802	22,512	3,158	31,952
WI	111,489	830	11,767	21,421	77,471	15,982	95,507
WY	35,461	916	3,663	10,567	20,315	2,312	33,149

Notes:
[1] Also includes freeways and expressways.

Source: U.S. Department of Transportation, Federal Highway Administration. Highway Statistics, annual. Cited in U.S. Bureau of the Census. *Statistical Abstract of the U.S., 1997.* Washington, DC.: U.S. Government Printing Office, 1997. Table 995.

6.10. Condition of U.S. Bridges, 1990–96.

	1990	1991	1992	1993	1994	1995	1996
All Bridges, total	342,098	346,069	347,668	349,765	346,505	344,647	346,364
Urban Bridges, total	87,495	90,127	93,249	94,787	97,575	98,855	100,508
Rural Bridges, total	254,603	255,942	254,419	254,978	248,930	245,792	245,856
Deficient Bridges, total	111,321	117,151	102,207	99,956	94,791	93,094	93,197
Urban Deficient Bridges, total	34,726	38,578	34,394	34,361	34,358	34,489	34,867
Structurally	12,239	12,649	12,351	12,152	11,803	11,423	11,309
Functionally	22,487	25,929	22,043	22,209	22,555	23,066	23,558
Rural Deficient Bridges, total	76,595	78,573	67,813	65,595	60,433	58,605	58,330
Structurally	40,654	41,406	37,228	35,568	31,721	29,510	29,246
Functionally	35,941	37,167	30,585	30,027	28,712	29,095	29,084

Notes:
Structurally deficient bridges = those designated as needing significant maintenance attention, rehabilitation, or sometimes replacement.
Functionally deficient bridges = those that do not have the lane widths, shoulder widths, or vertical clearances adequate to serve the traffic demand; or the waterway of the bridge may be inadequate and may allow occasional flooding of the roadway.

Source: U.S. Department of Transportation, Federal Highway Administration. National Bridge Inventory Database. Cited in U.S. Department of Transportation, Bureau of Transportation Statistics. *National Transportation Statistics 1998.* Washington, DC: Bureau of Transportation Statistics, 1998. Table 1-39. http://www.bts.gov/programs/btsprod/nts/index.html.

6.11. Value of Some Public Utility and Infrastructure Construction*, United States, 1990–95 (in millions of dollars).

Type of Construction	Current Dollars						Constant Dollars					
	1990	1993	1994	1995	1996	1997 (prelim.)	1990	1993	1994	1995	1996	1997
Railroads	$2,600	$3,108	$3,340	$3,509	$4,398	$5,059	$2,633	$3,056	$3,186	$3,201	$3,894	$4,321
Electric light and power	$11,299	$15,567	$14,918	$14,049	$11,211	$12,144	$11,572	$15,096	$13,877	$12,656	$9,914	$10,545
Gas	$4,820	$5,645	$4,694	$6,279	$4,865	$4,390	$5,013	$5,536	$4,308	$5,637	$4,330	$3,820
Petroleum pipelines	$411	$986	$998	$929	$1,015	$969	$428	$965	$918	$834	$6903	$843
Highways and streets	$32,105	$34,299	$37,419	$38,498	$41,243	$45,197	$31,777	$34,164	$36,219	$35,303	$36,483	$38,605
Sewer systems	$10,276	$8,865	$8,700	$9,435	$10,433	$10,463	$10,670	$8,622	$8,199	$8,557	$9,260	$8,951
Water supply facilities	$4,909	$5,085	$4,647	$5,283	$5,964	$6,339	$4,987	$4,868	$4,237	$4,695	$5,187	$5,393

Notes:
NA = Not available.
* = Represents value of new construction put in place during year. Differs from building permit and construction contract data in timing and coverage. Includes installed cost of normal building service equipment and selected types of industrial production equipment (largely site fabricated). Excludes cost of shipbuilding, land, and most types of machinery and equipment.

Source: U.S. Bureau of the Census. Current Construction Reports, series C30, *Value of New Construction*, monthly. Washington, DC. Cited in U.S. Bureau of the Census, *Statistical Abstract of the United States, 1997 and 1998.* Washington, DC: U.S. Government Printing Office, 1997 and 1998. Tables 1180 and 1194 respectively.

Chapter 7. Consumer Products and Entertainment

Entertainment, fueled by technology, has become a huge business, but it wasn't always this way. Centuries ago, people either had to make their own music or be wealthy enough to commission musicians to perform. The same was true for visual entertainment. Even a couple of decades ago, if people missed their favorite TV show when it was on the air, they were out of luck. Just a few decades before that, TV didn't exist at all. No *Seinfeld*, no *60 Minutes*, no *Monday Night Football*. Now we can carry our own personal music, drama, comedy, dance, and sports with us wherever we go—even out into space. Have these developments changed lives in meaningful ways? What if we had to give up our CD players and VCRs right now? Worse yet, our televisions. What if we could never see another movie that used special effects?

Even though entertainment technologies may not affect us as critically as agriculture, energy, and transportation technologies, they still play an important part in our lives. And what about all the time-saving technologies we use in our day-to-day lives. Without appliances like refrigerators, gas and electric ranges, and washing machines, we couldn't maintain the frantic work and play schedules we do—we'd be too busy doing chores. And who wants to think about life without hot water heaters (a $1.3 billion per year industry and well worth every penny—see Table 7.4)? On the other hand, trash compactors, with a 4.4% saturation, and central cleaning systems, with a 4.9% saturation (see Table 7.5) fail to captivate us. (They obviously don't save us any work, or we'd be buying a lot more of them.)

Of course, some consumer and entertainment technologies are nothing but fluff. Who needs a yogurt maker or beard trimmer, for example? Video games are fun, but are they transforming? Or maybe we should ask, Who cares? It's fun and interesting to track these implements of pop culture and daily life. Besides, fortunes are made and lost over many of these gadgets.

Hot topics in consumer technologies include

- High-definition television
- Digital television

Other trends that you will find in this chapter include

- Refrigerators represent a $5 billion per year industry (Table 7.2)
- More than three times as many electric dryers as gas dryers are shipped per year (Table 7.3)
- Almost a third of U.S. homes are equipped with a camcorder (Table 7.11)
- Only about half of U.S. homes have CD players of any type, which makes one wonder what the rest are doing for music, if anything (Table 7.12)
- Direct to home satellite systems are still relatively rare, with an estimated household penetration of about 11% (Table 7.14). Projection TV is in about the same situation, with a 12% penetration (Table 7.16).
- Most Americans didn't have color TV in the sixties (Table 7.15)
- Computer animation for games isn't nearly as large a market as it is for advertising, film and video, and corporate communications (Table 7.21)

HOUSEHOLD APPLIANCES AND OTHER CONSUMER PRODUCTS

7.1. Quantity and Value of Shipments of Major Household Appliances, United States: Ranges, Ovens, Surface Cooking Units, Outdoor Cooking Equipment, 1991–97.

Product Code	Product Description	1991 Quantity (thousands of units)	1991 Value ($millions)	1992 Quantity (thousands of units)	1992 Value ($millions)	1993 Quantity (thousands of units)	1993 Value ($millions)	1994 Quantity (thousands of units)	1994 Value ($millions)	1995 Quantity (thousands of units)	1995 Value ($millions)	1996 Quantity (thousands of units)	1996 Value ($millions)	1997 Quantity (thousands of units)	1997 Value ($millions)
36311 —	Electric ranges, ovens, and surface cooking units	7400.1	$1,614.9	8267.6	$1,730.0	8057.9	$1,731.3	7883.2	$1,796.0	6439	$1,791.8	6,524.10	$1,865.8	7,286.2	$1,839.0
	Free-standing ranges:														
36311 11	Under 23" wide	94.5	$17.0	94.9	$16.8	109.6	$19.2	118.3	$20.7	113.9	$20.3	107.1	$19.3	(D)	(D)
36311 17*	23" to 32" wide	NA	NA	NA	NA	2547.7	$756.6	2782.1	$824.8	2727.1	$850	2,731.70	$906.6	2,734.5	$862.4
36311 35	33" wide and over	(D)	(D)	(D)	(D)	4.1	$2.1	4.2	$2	(D)	(D)	(D)	(D)	(D)	(D)
	Built-in ovens:														
36311 53	Single oven	r/256.0	$105.8	332.9	$128.8	309.0	$117.2	272.9	$106.7	303.0	$153.0	302.4	$157.5	(D)	(D)
36311 55	Two ovens	r/79.1	r/$47.3	80.6	$51.8	78.2	$50.3	97.9	$68.9	101.9	$83.6	102.8	$99.1	105.5	$99.1
36311 61	Surface cooking tops	366.4	$77.2	411.9	$87.2	395.1	$75.6	414.2	$82.2	365.1	$98.4	383.8	$109.0	331.7	$91.8
36311 65	Drop-in ranges	132.4	$51.4	129.7	$57.3	121.1	$51.4	124.7	$51.2	117.4	$64.4	(D)	(D)	(D)	(D)
36311 73	Microwave ranges and ovens, including combination microwave/electric ranges and built-in ovens that utilize microwave and electricthermal energy in one or more of the oven cavities	r/67.0	r/$47.8	77.6	$55.0	74.4	$50.1	80.9	$55.5	58.4	$57.4	53.1	$54.9	55.2	$64.0
36311 74	Microwave ovens, including countertop built-in and wall- or cabinet-hung ovens	3391.5	$438.5	3544.9	$455.6	3781.0	$459.3	3350.5	$422.8	2652.2	$301.3	2927.1	$342.6	(D)	(D)
36313 —	Gas household ranges, ovens, and surface cooking units	1954.3	$609.5	1990.6	$616.2	2077.3	$621.5	2158.4	$649.6	2033.3	$654.1	1,705.5	$544.6	1,107.3	$366.1
	Standard types, free-standing:														
36313 11	Over 32" cooking top and oven	r/71.3	r/$29.1	57.1	$24.9	58.6	$31.3	57.2	$34.9	61.7	$41.2	53.5	r/$33.7	29.1	$33.4
36313 13	24" to 32" cooking top and oven	1163.2	$350.8	1148.1	$344.6	1221.4	$349.4	1301.2	$364.7	1179.6	$359.8	993.6	$299.9	596.9	$188.6
36313 15	24" and under cooking top	179.6	$35.2	216.1	$40.9	173.1	$32.3	178.3	$32.2	212.6	$40.9	197.2	$38.2	(D)	(D)
36313 17	Standard type, for built-in installation	(D)	(D)	(D)	(D)	(D)	(D)	(D)	(D)	(D)	(D)	(D)	(D)	(D)	(D)
36313 19	Surface cooking tops	181.0	$38.3	225.3	$49.0	242.1	$51.1	211.9	$56.1	r/212.6	r/$69.4	205.9	$68.2	141.9	$47.7
36313 22	Nonstandard types, including wall-hung, slide-in, or drop-in	NA	NA	NA	NA	r/315.3	r/$82.7	342.4	$88.5	300.9	r/$70.6	220.8	$51.7	(D)	(D)
36314 —	Other household ranges and cooking equipment and outdoor cooking equipment, parts, and accessories	10404.8	$657.4	10611.9	$707.8	11804.2	$786.1	12300.1	$866.4	11662.7	$911.6	r/10,538.3	$1,004.0	9,846.7	$1,039.3
	Barbecuers, grills, stoves, braziers, etc. Portable units:														
35314 22	Gas fuel or gas/other hybrid	r/2649.8	r/$267.1	a/2837.2	$293.9	r/5487.5	r/$475.5	5674.4	$524.9	5245.6	r/$533.4	r/5,429.7	r/$575.1	5,509.3	$602.0
35314 24	Solid fuel (charcoal, wood, etc.)	5111.6	r/$93.9	5143.9	$98.0	5185.5	$113.8	5599.0	$129.1	4967.2	$126.7	r/3,694.6	$ a/141.3	3,416.8	$133.7
35314 26	Liquid fuel (oil, kerosene, etc.)	2494.7	$233.6	2489.2	$241.4	r/733.7	r/$76.3	801.1	$87.8	910.2	$105.5	877.6	$139.1	781.8	$164.8
35314 28	All other outdoor and other cooking equipment	NA	NA	NA	NA	b/397.5	a/$29.7	225.5	$22.2	r/539.2	r/$29.1	b/536.4	$ b/23.7	138.9	$13.9

Notes:
For 1995, product code 36311 35 is combined with 36311 17 to avoid disclosing data for individual companies.
D = Withheld to avoid disclosing data for individual companies.
r/ = Revised by 5% or more from previously published data.
a/ = 10 to 25% of this item is estimated.
b/ = Over 25% of this item is estimated.

Source: Current Industrial Reports, Series MA36F: Major Household Appliances. Washington, DC: U.S. Bureau of the Census. Table 2. http://www.census.gov/cir/www/ma36f.html

7.2. Quantity and Value of Shipments of Major Household Appliances, United States: Refrigerators and Freezers, 1993–97.

Product Code	Product Description	1993 Quantity (thousands)	1993 Value ($millions of units)	1994 Quantity (thousands)	1994 Value ($millions of units)	1995 Quantity (thousands)	1995 Value ($millions of units)	1996 Quantity (thousands)	1996 Value ($millions of units)	1997 Quantity (thousands)	1997 Value ($millions of units)
36321	—[1] Household refrigerators, including combustion refrigerator-freezers	10313.5	$4,310.3	11275.6	$4,969.0	11004.9	$4,739.4	11,118.8	$5,084.4	11,092.4	4,858.2
	Compression type, fitted with separate external doors:										
36321 02	9.4 cubic feet and under	[2]245.6	[2][5]64.8	[2]270.9	[2]$72.0	(D)	(D)	(D)	(D)	85.3	/a $19.1
36321 09	9.5 to 13.4 cubic feet					235.9	$60.7	203.7	$55.6	135.7	$39.2
36321 12	13.5 to 15.4 cubic feet	1419.3	$433.5	1520.7	$482.3	1506.3	$465.5	1,318.8	$436.0	1,275.1	$399.1
36321 14	15.5 to 17.4 cubic feet	691.3	$225.0	693.9	$229.9	679.3	$214.1	792.1	$268.7	745.4	$254.5
36321 19	17.5 to 19.4 cubic feet	2031.9	$806.8	2289.8	$932.4	2327.4	$927.8	2,332.0	$979.8	2,407.8	$936.1
36321 24	19.5 to 21.4 cubic feet	1408.8	$683.7	1557.4	$813.5	1572.3	$829.9	1,543.5	$852.0	1,432.8	$757.8
36321 25[4]	21.5 to 24.4 cubic feet			1864.5	$1,524.2	1705.1	$1,405.9	1,674.7	$1,422.0	1,331.8	$931.0
36321 28	Over 24.4 cubic feet	[3]1932.9	[3]$1481.4	230.2	$237.8	229.3	$269.2	556.2	$497.2	1,099.0	$1,008.2
	Compression type, fitted with only one external door:										
36321 63	6.4 cubic feet and under	790.8	$102.8	853.5	$115.2	881.6	$124.3	839.8	$121.1	734.9	$118.6
36321 64	6.5 cubic feet and over	(D)	(D)	(D)	(D)	(D)	(D)	(D)	(D)	(D)	(D)
36322	—[1] Food freezers, complete units, for freezing and/or storing frozen food (household type)										
36322 05	Upright type freezers	(D)	(D)	(D)	(D)	(D)	(D)	(D)	(D)	(D)	(D)
	Chest type freezers	(D)	(D)	(D)	(D)	(D)	(D)	(D)	(D)	(D)	(D)

Notes:

[1] Product classes 36321 00 are combined to avoid disclosing data for individual companies.
[2] Product code 36321 02 is included with product code 36321 09 to avoid disclosing data for individual companies.
[3] For 1993, product code 36321 26 is included with product code 36321 28 to avoid disclosing data for individual companies.
[4] Before 1995, product code 36321 25 was product code 36321 26.
D = Withheld to avoid disclosing data for individual companies.

Source: Current Industrial Reprots, Series MA36F: *Major Household Appliances.* Washington, DC: U.S. Bureau of the Census, Various years. Table 2. http://www.census.gov/cir/www/ma36f.html

7.3. Quantity and Value of Shipments of Major Household Appliances, United States: Washers and Dryers, 1991–97.

Product Code	Product Description	1991 Quantity (thousands of units)	1991 Value ($millions)	1992 Quantity (thousands of units)	1992 Value ($millions)	1993 Quantity (thousands of units)	1993 Value ($millions)	1994 Quantity (thousands of units)	1994 Value ($millions)	1995 Quantity (thousands of units)	1995 Value ($millions)	1996 Quantity (thousands of units)	1996 Value ($millions)	1997 Quantity (thousands of units)	1997 Value ($millions)
36330 —	Household laundry machines	NA	$2,919.0	NA	$3,070.6	11520.4	$3,389.4	12,112.7	$3,479.0	11,728.9	$3,095.4	12,082.9	$3,217.9	12,237.1	$3,216.0
	Washing machines, mechanical (electric):														
36330 11	Coin-operated	¹294.8	¹$154.3	¹462.7	¹$236.8	217.2	$92.6	219.7	$90.2	227.9	$95.4	227.4	$98.2	205.8	$92.9
36330 31	Noncoin operated	6108.9	$1,685.2	6102.7	$1,694.3	6282.5	$1,782.8	6,599.6	$1,841.1	6,378.0	$1,742.3	6,646.7	$1,828.3	6,617.3	$1,814.9
36330 37	Other washing machines, including those with a built-in centrifugal drier and gasoline-driven	¹	¹	¹	¹	239.2	r/ $158.4	281.2	$156.8	270.1	$155.2	276.1	$163.8	269.4	$151.8
	Dryers, mechanical (with heat):														
36330 51	Gas	988.3	$236.4	1105.8	$270.1	1207.7	$297.9	1,219.0	$310.7	1,187.6	$298.8	1,076.0	$281.0	1,098.6	$283.7
36330 55	Electric	3067.0	$630.6	3258.2	$672.8	3446.0	$741.5	3,679.2	$792.2	3,545.2	$758.6	3,735.3	$800.5	3,926.1	$826.8
36330 15	Coin operated	109.6	$44.3	132.4	$44.0	127.9	$52.1	114.0	$41.8	120.0	$45.0	121.5	$46.0	119.9	$45.8

Notes:

r/ = Data are revised 5 percent or more from previously published figures.

¹Data for product code 36330 37, "Other washing machines, n.e.c.," are included in product code 36330 11 to avoid disclosures of individual companies for 1991–92.

Source: Current Industrial Report, Series MA36F: *Major House Appliances.* Washington, DC: U.S. Bureau of the Census. Table 2. Various years. http://www.census.gov/cir/www/ma36f.html

7.4. Quantity and Value of Shipments of Major Household Appliances, United States: Water Heaters and Dishwashers, 1991–97.

Product Code	Product Description	1991 Quantity (thousands of units)	1991 Value ($millions)	1992 Quantity (thousands of units)	1992 Value ($millions)	1993 Quantity (thousands of units)	1993 Value ($millions)	1994 Quantity (thousands of units)	1994 Value ($millions)	1995 Quantity (thousands of units)	1995 Value ($millions)	1996 Quantity (thousands of units)	1996 Value ($millions)	1997 Quantity (thousands of units)	1997 Value ($millions)
36391 —	Water heaters, electric (permanent installation)	3,689.2	$446.7	3,211.6	$471.8	3,747.1	$473.4	4,003.9	$513.8	4,068.2	$513.0	4,384.9	$558.8	4,300.8	$550.1
36391 11	34 gallons and under	1,002.7	$98.3	721.6	$89.4	860.6	$91.8	721.6	$79.3	1,012.5	$116.0	/b 1,049.4	/b $122.4	/b 926.4	/b $ 110.7
36391 12	35 to 44 gallons	1,024.4	$122.0	882.8	$121.7	1,090.2	$128.3	1,325.1	$162.7	1,421.7	$170.1	/b 1,556.2	/b$187.1	/b 1,561.4	/b $ 188.4
36391 14	45 to 54 gallons	1,108.3	$142.4	1,026.9	$161.9	1,207.4	$153.7	1,330.1	$173.9	1,096.4	$146.0	/b 1,191.0	/b$160.0	/b 1,188.4	/b $ 158.3
36391 15	55 gallons and over	[1]	[1]	258.0	$58.3	350.8	$60.8	379.0	$65.8	291.7	$57.7	/a 312.7	/a$61.8	/a 332.5	/a $63.9
36391 98	Other types, including circulating and portable storage	[2]/r 553.8	[2]/r $ 84.0	322.3	$40.5	238.1	/r$ 38.7	248.1	/b $ 32.1	245.8	$23.3	275.5	$27.6	292.1	$28.7
36392 —	Water heaters, except electric Direct fired:	4,077.7	$616.1	3,716.5	$593.8	4,635.8	$700.0	4,841.8	$748.8	4366.0	/r$ 681.8	4,840.9	$764.6	4,798.7	$753.0
36392 12	Gas	3,941.6	$587.3	3,562.2	$566.4	4,493.9	$671.6	4,682.9	$715.4	4235.7	/r$653.0	b/ 4,732.7	/b $745.3	/b 4,691.4	/b $ 732.2
36392 24	Oil	[2]	[2]	[1]	[1]	/r 27.3	/r$9.9	28.1	$9.9	/r 22.1	/r$7.8	21.2	$6.8	(D)	(D)
36392 52	Indirect, including storage cast[1] or coil type (less tank); tanks (with or without generator coils); generator coils with tank collar (sold without tank); and instantaneous	136.1	[1]$28.8	[1]154.3	[1] $27.4	114.5	$18.6	130.7	$23.5	108.1	$21.0	r/ 108.2	r/ $ 19.2	107.3	$20.8
	Dishwashers:														
36395 12	Portable type, including convertible type	224.6	$61.0	186.0	$51.8	162.5	$46.4	206.8	$59.1	(D)	(D)	(D)	(D)	(D)	(D)
36395 14	Built-in type	3,399.7	$764.0	3,617.7	$810.6	/r 3,861.2	/r $882.3	4,448.1	$1,076.2	3,981.3	$1,020.2	4,369.0	$1,011.5	4,209.2	$974.3

Notes:
/r = Data are revised 5 percent or more from previously published figures.
/a = The percent of estimation is 10 to 25 percent.
/b = The percent of estimation is over 25 percent.
[1]For 1991, product code 36391 15 is included with 36391 98 to avoid disclosures of individual companies.
[2]Data for product code 36392 24, "oil direct fired water heaters," are included in product code 36392 52, "indirect water heaters," to avoid disclosures of individual companies.

Source: Current Industrial Reports, Series MA36F: *Major Household Appliances.* Washington, DC: U.S. Bureau of the Census. Table 2. Various years. http://www.census.gov/cir/rwww/ma36f.html

7.5. U.S. Households Owning Particular Appliances or Consumer Electronics Products, 1994–96 (percentages).

	Percent Saturation		
	1994	**1995**	**1996**
Major Appliances			
Compactors	3.9%	3.9%	4.4%
Standard Refrigerators	99.5%	99.7%	99.8%
Electric Housewares			
Clocks (Including Battery-operated)	99.0%	99.0%	98.0%
Pasta Makers	5.3%	5.2%	4.5%
Personal Care			
Hair Dryers, Handheld	90.0%	87.7%	88.0%
Mustache/Beard Trimmers	8.4%	8.2%	9.0%
Home Care			
Vacuum Cleaners	97.3%	98.1%	98.5%
Central Cleaning Systems	2.7%	4.7%	4.9%
Comfort Conditioning			
Fans (Except Ceiling)	67.5%	69.3%	69.7%
Electric Furnaces	11.0%	11.0%	11.8%
Consumer Electronics			
Calculators	97.0%	98.0%	98.0%
Radios	98.0%	98.3%	98.0%
Color Televisions	97.0%	98.0%	98.0%
Video Disc Players	1.0%	1.6%	2.0%
Outdoor Appliances			
Electric Grills	6.8%	8.5%	5.0%
Gasoline Power Motors	80.0%	81.9%	81.5%

Notes:
Figures represent the percentage of U.S. households that own at least one of the particular appliance.

Source: "The Saturation Picture." *Appliance* 54 (September 1997): 86+. Reprinted by permission of *Appliance Magazine.* Copyright Dana Chase Publications, Inc., September 1997.

7.6. Telephone Answering Devices—U.S. Sales to Dealers, 1982–96, Estimated for 1997, and Projected for 1998.

Year	Units (thousands)	Dollars (millions)	Household Penetration	Average Unit Price
1982	850	$110	NA	NA
1983	2,200	$190	3%	NA
1984	3,000	$230	4%	NA
1985	4,220	$325	7%	NA
1986	6,450	$464	12%	NA
1987	8,800	$607	17%	NA
1988	11,100	$755	20%	NA
1989	12,500	$838	28%	NA
1990	13,560	$827	35%	NA
1991	15,380	$1,000	43%	NA
1992	14,590	$934	50%	NA
1993	16,279	$1,026	54%	$63
1994	17,613	$1,153	58%	$65
1995	17,498	$1,077	57%	$62
1996	17,570	$1,004	63%	$57
1997 est.	19,000	$1,026	68%	$54
1998 proj.	19,200	$1,037	69%	$54

Source: Consumer Electronics Manufacturers Association (CEMA). *CEMA Consumer Electronics U.S. Sales & Forecasts.* Arlington, VA: CEMA, 1998.

7.7. Aftermarket Vehicle Security—Factory Sales, 1993–95, and Estimated for 1990–92 and for 1996–97.

Year	Dollars (millions)	Household Penetration
1990 est.	$190	NA
1991 est.	$210	NA
1992 est.	$240	NA
1993	$294	NA
1994	$401	24%
1995	$464	25%
1996 est.	$478	27%
1997 est.	$550	29%

Notes:
Aftermarket vehicle security includes car alarms and satellite trackers but not mechanical devices like "The Club."
NA = Not available.

Source: Consumer Electronics Manufacturers Association. *EIA-CEMA Consumer Electronics U.S. Sales.* Arlington, VA: CEMA, 1998.

ENTERTAINMENT PRODUCTS

7.8. Portable CD Equipment—U.S. Factory Sales, 1990–96, Estimated for 1997, and Projected for 1998.

Year	Units (thousands)	Dollars (millions)	Household Penetration	Average Unit Price
1990	3,186	$453	NA	NA
1991	4,681	$639	NA	NA
1992	8,341	$1,057	NA	NA
1993	11,276	$1,289	NA	$114
1994	15,262	$1,648	NA	$108
1995	17,849	$1,702	NA	$95
1996	16,970	$1,460	18%	$86
1997 est.	18,600	$1,414	20%	$76
1998 proj.	18,900	$1,418	22%	$75

Notes:
NA = Not available.

Source: Consumer Electronics Manufacturers Association (CEMA). *CEMA Consumer Electronics U.S. Sales & Forecasts.* Arlington, VA: CEMA, 1998.

7.9. Compact Audio Systems—U.S. Factory Sales, 1980–96, Estimated for 1997, and Projected for 1998.

Year	Units (thousands)	Dollars (millions)	Household Penetration*	Average Unit Price
1980	3,567	$729	NA	NA
1981	2,720	$631	NA	NA
1982	2,321	$428	NA	NA
1983	2,651	$452	NA	NA
1984	1,659	$246	NA	NA
1985	2,531	$325	NA	NA
1986	2,690	$320	NA	NA
1987	2,315	$326	NA	NA
1988	3,049	$336	NA	NA
1989	2,878	$470	NA	NA
1990	2,447	$466	NA	NA
1991	3,139	$597	NA	NA
1992	3,877	$756	NA	NA
1993	4,100	$919	NA	$224
1994	5,139	$1,108	NA	$216
1995	5,677	$1,162	31%	$205
1996	6,174	$1,157	35%	$187
1997 est.	7,200	$1,397	37%	$194
1998 proj.	7,400	$1,399	38%	$189

Notes:
*Household penetration includes rack or compact audio system.
Compact audio systems are all-in-one shelf units. Definition does not include portable or boombox-type systems.

Source: Consumer Electronics Manufacturers Association (CEMA). *CEMA Consumer Electronics U.S. Sales & Forecasts.* Arlington, VA: CEMA, 1998.

7.10. Rack Audio Systems—U.S. Factory Sales, 1988–96, Estimated for 1997, and Projected for 1998.

Year	Units (thousands)	Dollars (millions)	Household Penetration*	Average Unit Price
1988	1,666	$889	NA	NA
1989	1,438	$747	NA	NA
1990	1,557	$804	NA	NA
1991	1,415	$667	NA	NA
1992	1,341	$614	NA	NA
1993	1,116	$545	NA	$488
1994	1,143	$595	NA	$521
1995	944	$515	31%	$546
1996	695	$380	35%	$547
1997 est.	500	$268	37%	$536
1998 proj.	425	$221	38%	$519

Notes:
* = Household penetration includes rack or compact audio system.
Rack audio systems are defined as a single manufacturer's complete sound reproduction system, including rack or cabinet, packaged to sell as one unit. May be a compact or component system.

Source: Consumer Electronics Manufacturers Association (CEMA). *CEMA Consumer Electronics U.S. Sales & Forecasts.* Arlington, VA: CEMA, 1998.

7.11. Separate Audio Component Systems—U.S. Factory Sales, 1979–96, Estimated for 1997, and Projected for 1998.

Year	Dollar Sales (millions)	Household Penetration
1979	$1,178	NA
1980	$1,424	NA
1981	$1,363	NA
1982	$1,181	NA
1983	$1,283	NA
1984	$913	NA
1985	$1,132	NA
1986	$1,358	NA
1987	$1,854	NA
1989	$1,871	NA
1990	$1,935	NA
1991	$1,805	NA
1992	$1,586	NA
1993	$1,635	NA
1994	$1,686	NA
1995	$1,911	53%
1996	$1,808	54%
1997 est.	$1,625	55%
1998 proj.	$1,515	55%

Notes:
Separate audio component systems are those purchased individually and combined into a system by the consumer. Components can include radios, CD players, speakers. May be shelf or large stand-alone systems. For home use only.

Source: Consumer Electronics Manufacturers Association (CEMA). *CEMA Consumer Electronics U.S. Sales & Forecasts.* Arlington, VA: CEMA, 1998.

7.12. Camcorders—U.S. Sales to Dealers, 1985–96, Estimated for 1997, and Projected for 1998.

Year	Units (thousands)	Dollars (millions)	Household Penetration	Average Unit Price
1985	517	$793	NA	NA
1986	1,169	$1,280	2%	NA
1987	1,604	$1,651	4%	NA
1988	2,044	$1,972	5%	NA
1989	2,286	$2,007	8%	NA
1990	2,962	$2,260	11%	NA
1991	2,864	$2,013	15%	NA
1992	2,815	$1,841	18%	NA
1993	3,088	$1,958	19%	$634
1994	3,209	$1,985	21%	$625
1995	3,560	$2,130	22%	$598
1996	3,634	$2,084	25%	$598
1997 est.	3,675	$1,970	28%	$536
1998 proj.	3,800	$1,919	30%	$505

Note:
Includes digital camcorders.

Source: Consumer Electronics Manufacturers Association (CEMA). *CEMA Consumer Electronics U.S. Sales & Forecasts.* Arlington, VA: CEMA, 1998.

7.13. Total CD Players—U.S. Factory Sales, 1987–96, Estimated for 1983–86 and 1997, and Projected for 1998.

Year	Units (thousands)	Dollars (millions)	Household Penetration	Average Unit Price
1983 est.	35	NA	0%	NA
1984 est.	208	NA	0%	NA
1985 est.	1,000	NA	1%	NA
1986 est.	2,600	NA	4%	NA
1987	4,067	NA	8%	NA
1988	5,255	NA	12%	NA
1989	6,914	$1,696	18%	NA
1990	9,155	$2,016	24%	NA
1991	11,595	$2,391	30%	NA
1992	16,134	$3,005	39%	NA
1993	20,425	$3,552	42%	$174
1994	26,544	$4,368	43%	$165
1995	30,374	$4,532	47%	$149
1996	29,813	$4,118	49%	$138
1997 est.	33,000	$4,290	50%	$130
1998 proj.	33,900	$4,170	52%	$123

Notes:
Includes home CD players, audio systems sold with CD players, portable CD players including combinations, and autosound CD players. Excludes laserdisc combination players.
NA = Not available.

Source: Consumer Electronics Manufacturers Association (CEMA). *CEMA Consumer Electronics U.S. Sales & Forecasts.* Arlington, VA: CEMA, 1998.

7.14. Direct to Home (DTH) Satellite Systems, U.S. Sales to Dealers, 1986–89, Estimated for 1990–97, and Projected for 1998.

Year	Installations (thousands)	Dollar Sales (millions)	Household Penetration	Average Unit Price
1986	235	$256	NA	NA
1987	250	$304	NA	NA
1988	275	$331	NA	NA
1989	300	$365	NA	NA
1990 est.	330	$421	NA	NA
1991 est.	281	$370	NA	NA
1992 est.	303	$379	NA	NA
1993 est.	349	$408	NA	$1,169
1994 est.	1,320	$900	NA	$682
1995 est.	2,235	$1,265	4%	$566
1996 est.	3,480	$1,493	8%	$429
1997 est.	3,000	$990	11%	$330
1998 proj.	3,625	$1,088	11%	$300

Notes:
Includes DSS and DBS for 1994–98 only; all years include C-Band.
NA = Not available.

Source: Consumer Electronics Manufacturers Association (CEMA). *CEMA Consumer Electronics U.S. Sales & Forecasts.* Arlington, VA: CEMA, 1998.

7.15. Color TV Receivers—U.S. Sales to Dealers, 1954–96, Estimated for 1997, and Projected for 1998.

Year	Units (thousands)	Dollars (thousands)	Household Penetration	Average Unit Price
1954	5	$2,000	NA	NA
1955	20	$10,000	0.0%	NA
1956	100	$46,000	0.1%	NA
1957	85	$37,000	0.2%	NA
1958	80	$34,000	0.4%	NA
1959	90	$37,000	0.6%	NA
1960	120	$47,000	0.7%	NA
1961	147	$56,000	0.9%	NA
1962	438	$154,000	1.2%	NA
1963	747	$258,000	1.9%	NA
1964	1,404	$488,000	3.1%	NA
1965	2,694	$959,000	4.9%	NA
1966	5,012	$1,861,000	9.7%	NA
1967	5,563	$2,015,000	16.3%	NA
1968	6,215	$2,086,000	24.2%	NA
1969	6,191	$2,031,000	32.0%	NA
1970	5,320	$1,684,000	35.7%	NA
1971	7,274	$2,355,000	41.0%	NA
1972	8,845	$2,825,000	48.7%	NA
1973	10,071	$3,097,000	55.4%	NA
1974	8,411	$2,658,000	62.3%	NA
1975	6,485	$2,211,547	68.4%	NA
1976	7,700	$2,687,740	73.6%	NA
1977	9,107	$3,187,398	77.1%	NA
1978	10,236	$3,582,814	78.0%	NA
1979	9,846	$3,544,717	80.9%	NA
1980	10,897	$4,003,548	83.0%	NA
1981	11,157	$4,123,312	85.2%	NA
1982	11,366	$4,141,098	87.6%	NA
1983	13,986	$4,885,930	88.7%	NA
1984	16,083	$5,358,768	90.5%	NA
1985	16,995	$5,521,880	91.0%	NA
1986	18,204	$5,835,924	92.0%	NA
1987	19,330	$6,147,705	93.0%	NA
1988	20,216	$5,907,408	94.0%	NA
1989	21,706	$6,490,000	96.0%	NA
1990	20,384	$6,197,000	96.0%	NA
1991	19,474	$5,979,000	96.0%	NA
1992	21,056	$6,591,000	97.0%	NA
1993	23,005	$7,315,000	98.0%	$318
1994	24,715	$7,225,000	98.0%	$292
1995	23,231	$6,798,000	97.0%	$293
1996	22,384	$6,492,000	98.0%	$290
1997 est.	21,900	$6,198,000	98.0%	$283
1998 proj.	21,700	$6,098,000	98.0%	$281

Notes:
Includes portable, table, and console color televisions. Before 1990 includes TV/VCR combinations.
Excludes LCD, projection TV, and TV/VCR combinations. Sales Sources: *EIA Electronic Market Data Book* and *Consumer Electronics U.S. Sales.* Before 1975, statistics reported as factory sales (includes factory sales to distributors and dealers, and factory export sales.)
NA = Not available.

Penetration Sources: NBC, Nielsen, 1955–84, and *EIA Consumer Electronics U.S. Sales.*
Source: Consumer Electronics Manufacturers Association. Arlington, VA. Unpublished. January 1998.

7.16. Projection Television—U.S. Sales to Dealers, 1984–96, Estimated for 1997, and Projected for 1998.

Year	Units (thousands)	Dollars (millions)	Household Penetration	Average Unit Price
1984	195	$385	1%	NA
1985	266	$488	2%	NA
1986	304	$529	3%	NA
1987	293	$527	4%	NA
1988	302	$529	3%	NA
1989	265	$478	5%	NA
1990	351	$626	6%	NA
1991	380	$683	7%	NA
1992	404	$714	8%	NA
1993	465	$841	9%	$1,808
1994	636	$1,117	9%	$1,755
1995	820	$1,417	10%	$1,727
1996	887	$1,426	11%	$1,607
1997 est.	925	$1,378	12%	$1,490
1998 proj.	950	$1,368	12%	$1,440

Notes:
NA = Not available.

Source: Consumer Electronics Manufacturers Association (CEMA). *CEMA Consumer Electronics U.S. Sales & Forecasts.* Arlington, VA: CEMA, 1998.

7.17. VCR Decks—U.S. Sales to Dealers, 1979–96, Estimated for 1997, and Projected for 1998.

Year	Units (thousands)	Dollars (millions)	Household Penetration	Average Unit Price
1979	475	NA	0.7%	NA
1980	805	$621	1.3%	$771
1981	1,361	$1,127	2.2%	$828
1982	2,035	$1,303	3.5%	$640
1983	4,091	$2,162	5.7%	$528
1984	7,616	$3,585	9.9%	$471
1985	11,853	$4,738	18.0%	$400
1986	13,174	$5,258	30.0%	$331
1987	11,702	$3,442	40.0%	$294
1988	10,748	$2,848	52.0%	$265
1989	9,760	$2,625	64.0%	$269
1990	10,119	$2,439	69.0%	$241
1991	10,718	$2,454	74.0%	$229
1992	12,329	$2,947	78.0%	$239
1993	12,448	$2,851	80.0%	$229
1994	13,087	$2,869	81.0%	$218
1995	13,562	$2,767	87.0%	$204
1996	15,641	$2,815	88.0%	$180
1997 est.	16,800	$2,705	90.0%	$161
1998 proj.	16,600	$2,556	90.0%	$154

Notes:
NA = Not available.

Penetration Sources: RCA Corp., 1979–85; Electronic Industries Association, 1986+.
Sales Source: Consumer Electronics Manufacturers Association (CEMA). *CEMA Consumer Electronics U.S. Sales & Forecasts.* Arlington, VA: CEMA, 1998.

7.18. Laserdisc Players—U.S. Sales to Dealers, 1990–95, Estimated for 1985–89 and for 1996–97, and Projected for 1998.

Year	Units (thousands)	Dollars (millions)	Household Penetration	Average Unit Price
1985 est.	75	$23	NA	$300
1986 est.	85	$26	NA	$300
1987 est.	75	$26	NA	$340
1988 est.	90	$34	NA	$380
1989 est.	120	$50	NA	$420
1990	168	$72	NA	$429
1991	206	$81	NA	$394
1992	224	$93	1%	$417
1993	287	$123	1%	$429
1994	272	$122	1%	$447
1995	257	$108	2%	$419
1996 est.	155	$66	2%	$419
1997 est.	50	$25	2%	$500
1998 proj.	27	$14	2%	$500

Notes:
Includes Combination and Karaoke Players.

Source: Consumer Electronics Manufacturers Association (CEMA). *CEMA Consumer Electronics U.S. Sales & Forecasts.* Arlington, VA: CEMA, 1998.

7.19. Home Radios—U.S. Factory Sales, 1950–96, Estimated for 1997, and Projected for 1998.

Year	Units (thousands)	Dollars (millions)	Household Penetration	Average Unit Price
1950	9,218	NA	NA	NA
1951	6,445	NA	NA	NA
1952	7,232	NA	NA	NA
1953	7,283	NA	NA	NA
1954	6,119	NA	NA	NA
1955	7,327	NA	NA	NA
1956	8,951	NA	NA	NA
1957	9,952	NA	NA	NA
1958	10,797	NA	NA	NA
1959	15,772	NA	NA	NA
1960	18,031	NA	NA	NA
1961	23,654	NA	NA	NA
1962	24,781	NA	NA	NA
1963	23,602	NA	NA	NA
1964	23,558	NA	NA	NA
1965	31,689	NA	NA	NA
1966	34,779	NA	NA	NA
1967	31,684	NA	NA	NA
1968	34,332	NA	NA	NA
1969	39,414	NA	NA	NA
1970	34,049	NA	NA	NA
1972	42,149	NA	NA	NA
1973	36,968	NA	NA	NA
1974	33,076	NA	NA	NA
1975	25,434	NA	NA	NA
1976	28,198	NA	NA	NA
1977	41,430	NA	NA	NA
1978	31,760	NA	NA	NA
1979	27,684	NA	NA	NA
1980	28,062	NA	NA	NA
1981	29,415	NA	NA	NA

7.19. Home Radios—U.S. Factory Sales, 1950–96, Estimated for 1997, and Projected for 1998 *(continued).*

Year	Units (thousands)	Dollars (millions)	Household Penetration	Average Unit Price
1982	32,663	NA	NA	NA
1983	39,496	NA	98%	NA
1984	26,456	NA	98%	NA
1985	21,575	NA	98%	NA
1986	25,364	NA	98%	NA
1988	23,623	NA	98%	NA
1989	25,254	NA	98%	NA
1990	21,585	NA	98%	NA
1991	18,530	NA	98%	NA
1992	21,553	NA	98%	NA
1993	19,697	$307	98%	$16
1994	18,325	$306	98%	$17
1995	17,051	$284	98%	$17
1996	17,581	$291	98%	$17
1997 est.	17,800	$285	98%	$16
1998 proj.	17,600	$282	98%	$16

Notes:
NA = Not available.

Sources: Electronic Industries Association. *EIA Electronic Market Data Book.* Arlington, VA, 1998. Consumer Electronics Manufacturers Association (CEMA). *CEMA Consumer Electronics U.S. Sales & Forecasts.* Arlington, VA: CEMA, 1998.

7.20. The Computer Animation Industry, Market Size, Worldwide, 1991–96.

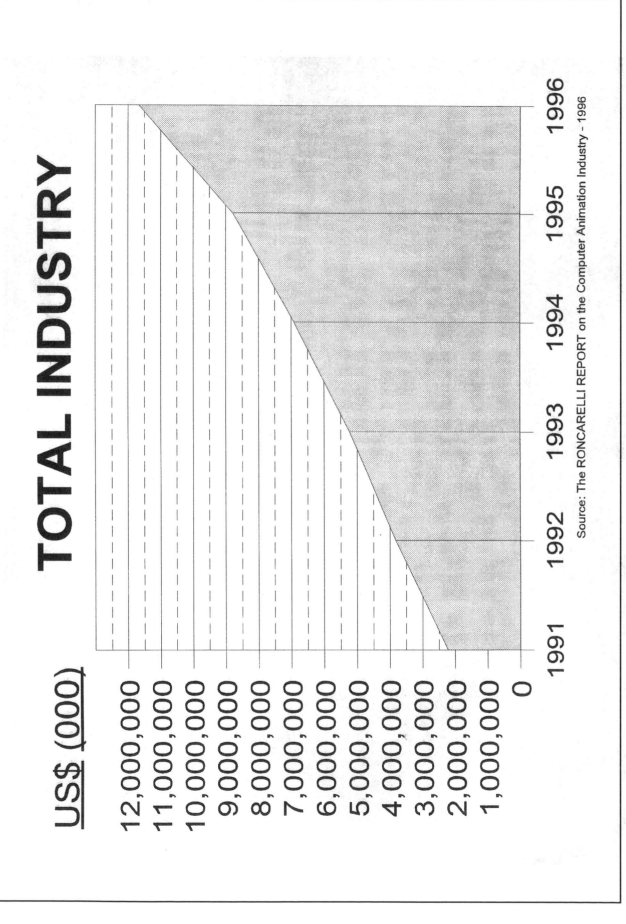

Source: The RONCARELLI REPORT on the Computer Animation Industry - 1996

7.21. The Computer Animation Industry, Total Use Analysis, Worldwide, 1996.

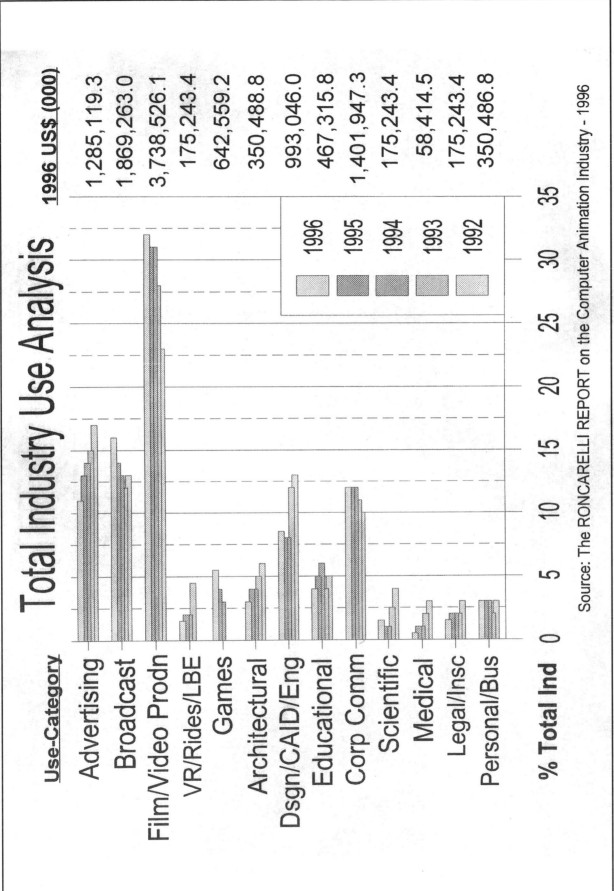

Total Industry Use Analysis

Use-Category	1996 US$ (000)
Advertising	1,285,119.3
Broadcast	1,869,263.0
Film/Video Prodn	3,738,526.1
VR/Rides/LBE	175,243.4
Games	642,559.2
Architectural	350,488.8
Dsgn/CAID/Eng	993,046.0
Educational	467,315.8
Corp Comm	1,401,947.3
Scientific	175,243.4
Medical	58,414.5
Legal/Insc	175,243.4
Personal/Bus	350,486.8

% Total Ind   0   5   10   15   20   25   30   35

Source: The RONCARELLI REPORT on the Computer Animation Industry - 1996

Source: The RONCARELLI REPORT on the Computer Animation Industry - 1996

Chapter 8. Education and Libraries

Education can be divided into four basic categories: learning by doing, learning by being told or shown, learning by reading, and learning by observing. Some of our education is formalized and some is ad hoc, but from clay tablets to books to computers to closed-circuit television, we have developed tools to help people learn in all four ways. At the same time, to facilitate learning and to preserve our cultures, we have constructed libraries, centralized repositories of knowledge, experience, and art, holding our collective successes and failures for all to see. Until recently, the primary technology used to store our culture and to impart knowledge was the low-tech book or magazine. Now we enjoy many more choices, and we face new issues like equality of access to and efficacy of technology, ease of use, and availability of funds for our gadgets and distribution networks. The statistics in this chapter help quantify our progress and deficiencies in these areas.

Tables 8.1 and 8.3 show how well our libraries are providing Internet access to patrons, while Table 8.12 reports progress in providing the same access through our public schools. According to the data, our schools are doing better than our libraries. Ironically, more libraries are hosting their own Web sites (Table 8.4) than are providing Web access for their patrons. Distance education is catching on, as Table 8.10 illustrates—at least with public higher education institutions; private institutions are holding back, with 61% of four-year colleges and 84% of two-year ones declining to offer distance education in the near future. Table 8.5 shows that we have made great progress in school library media center saturation since 1958, while Tables 8.6 and 8.7 tell us that public schools have again trumped private schools with respect to media center services and technologies like computers with modems, database searching with CD-ROMs, and televisions. On the other hand, Table 8.8 demonstrates that private schools hold more books, video materials, and microcomputer software per 100 students than public schools. Tables 8.13 and 8.14 show that most of our kids of all ages are using computers—mostly at school, but also at the library and at home.

Table 8.9 illustrates a steep rise in college and university library spending for machine readable materials and microforms between the early and mid-nineties, while spending for other categories of materials like serials, print materials, and audiovisual materials increased slowly. Tables 8.15 and 8.16 explore how our kids are using computers.

Educational institutions and libraries come in a variety of specialized types, both private and public. Most technologies are suitable for all institutions, but some are more widely used by certain types or age groups. You will find the statistics broken down into those categories:

> Primary schools
> Secondary schools
> Community colleges, colleges, and universities
> Vocational schools
> Adult education and extension schools
> Public libraries
> Academic libraries
> Special libraries (corporate, museum, nonprofit, etc.)

Hot topics relating to the educational and library use of technology include

- Possible loss of information due to changing technologies. As we migrate from one storage medium to another, can we afford to convert all our content and purchase the new equipment it requires? Will we have to make hard choices about what to keep and what to discard?
- Backup and disaster planning. Are we sufficiently backed up so that increasing centralization—especially of digital resources—doesn't leave us vulnerable when disasters strike?
- Efficacy of computers in the classroom. We have them, but are they helping kids learn effectively?
- The Internet in classrooms and libraries. With so much reference and educational material on the Net, the new medium is becoming indispensible. But

there are drawbacks for classrooms, like pornography. (See Tables 8.1, 8.2, and 8.3.)

- Equality of access to technology. If all these gadgets are helping kids and the public, then everyone should be able to use them. (See Tables 8.3, 8.5, 8.6, 8.7, 8.8, 8.10, 8.12, 8.13, 8.14, and 8.15.)
- Distance learning. Remote access to classrooms and teachers means that more people can learn more

conveniently. (See Tables 8.10 and 8.11.)

- Virtual libraries. When information is digital, many people can use the same item at the same time, and there is no waiting for interlibrary loan. But building virtual libraries takes time and money and requires that librarians and users master new skills (Tables 8.1 through 8.4).

LIBRARIES AND TECHNOLOGY

8.1. Percentage of U.S. Public Libraries Offering Access to the Internet, 1996–97.

	1996	1997
Public Access	27.8%	60.4%
No Public Access	72.2%	39.6%

Notes:
Based on a sample of 2,000 of the 8,921 U.S. public library systems. The sample was weighted to represent public libraries across various population service areas and central city, suburban, and rural locations. The survey achieved a response rate of 70.1%, for a total of 1,402 responding library systems. The responses were reweighted to compensate for nonrespondents. Thus, these data are national estimates.

Source: American Library Association, Office for Information Technology Policy. *1997 National Survey of U.S. Public Libraries and the Internet: Summary Results.* Washington, DC: American Library Association, 1998. http://www.ala.org/oitp/research/

8.2. Percentage of U.S. Public Libraries Connected to the Internet, 1996–97.

	1996	1997
Connected	44.4%	72.3%
Not Connected	55.6%	27.7%

Notes:
Based on a sample of 2,000 of the 8,921 U.S. public library systems. The sample was weighted to represent public libraries across various population service areas and central city, suburban, and rural locations. The survey achieved a response rate of 70.1%, for a total of 1,402 responding library systems. The responses were reweighted to compensate for nonrespondents. Thus, these data are national estimates.

Source: American Library Association, Office for Information Technology Policy. *1997 National Survey of U.S. Public Libraries and the Internet: Summary Results.* Washington, DC: American Library Association, 1998. http://www.ala.org/oitp/research/

8.3. Internet Access in U.S. Public Libraries Regionally, 1997.

Central City Access

No Web	36%
Web at main library	31%
Web at all outlets	23%
Web at some outlets	11%

Suburban Access

No Web	43%
Web at main library	39%
Web at all outlets	14%
Web at some outlets	4%

Rural Access

No Web	51%
Web at main library	40%
Web at all outlets	6%
Web at some outlets	3%

Notes:
Based on a sample of 2,000 of the 8,921 U.S. public library systems. The sample was weighted to represent public libraries across various population service areas and central city, suburban, and rural locations. The survey achieved a response rate of 70.1%, for a total of 1,402 responding library systems. The responses were reweighted to compensate for nonrespondents. Thus, these data are national estimates.

Source: American Library Association, Office for Information Technology Policy. *1997 National Survey of U.S. Public Libraries and the Internet: Summary Results.* Washington, DC: American Library Association, 1998. http://www.ala.org/oitp/research/

8.4. U.S. Public Libraries Hosting Their Own Web Sites, 1997 (Percentage Hosting a Web Site by Population of Legal Service Area).

1,000,000+	63.5%
500,000–999,999	66.4%
100,000–499,999	41.5%
25,000–99,999	22.1%
5,000–24,999	8.3%
Less than 5,000	3.6%

Notes:
Based on a sample of 2,000 of the 8,921 U.S. public library systems. The sample was weighted to represent public libraries across various population service areas and central city, suburban, and rural locations. The survey achieved a response rate of 70.1%, for a total of 1,402 responding library systems. The responses were reweighted to compensate for nonrespondents. Thus, these data are national estimates.

Source: American Library Association, Office for Information Technology Policy. *1997 National Survey of U.S. Public Libraries and the Internet: Summary Results.* Washington, DC: American Library Association, 1998. http://www.ala.org/oitp/research/

8.5. Percent of Schools with Library Media Centers and Percent of Pupils in Schools with Library Media Centers: Historical Summary, 1958–94.

Year	Percent of Schools with Library Media Centers		Percent of Pupils in Schools with Library Media Centers	
	Public	Private	Public	Private
1958[1]	50%	NA	68%	NA
1962[2]	59%	44%	74%	NA
1974[3]	85%	NA	NA	NA
1978/79[4]	85%	83%	93%	86%
1985[5]	93%	75%	98%	88%
1990–91	96%	87%	98%	95%
1993–94	96%	80%	98%	92%

Notes:

NA = Not available.

[1] The public school sample excluded schools with fewer than 150 pupils and is estimated to represent 97% of public school pupils.

[2] The public school sample excluded schools in districts with fewer than 150 pupils and is estimated to represent 98% of public school pupils. An extensive library study was also conducted in 1960–61 that represented all districts. This study found a pupil membership of 35,952,711, of which 25,300,243 pupils, or 70%, were in schools with library media centers. The survey also found that 46% of all schools overall had library media centers.

[3] The public school sample represented all districts.

[4] The public school sample represented all districts and included technical centers and special schools.

[5] The public school sample represented all districts and included regular public schools only. The private school sample in 1985 included a number of small and special/alternative schools that may not have been included in the universe listing used to draw the sample in 1978. This may account, in part, for the lower percentage of private schools reporting having a library media center in 1985 compared to 1979.

Source: U.S. Department of Education, National Center for Education Statistics, *Schools and Staffing Survey: 1990–91, School Questionnaire; Statistics of Public and Private School Library Media Centers:1985-86 (with historical comparisons from 1958 to 1985)*; and *Schools and Staffing Survey: 1993–94, School Questionnaire.* Cited in Bradford Chaney. *School Library Media Centers: 1993–94.* National Center for Education Statistics, U.S. Department of Education, August 1998.

8.6. School Library Media Centers Offering Selected Services and Equipment, Public and Private, Elementary and Secondary, United States, 1993–94.

Selected services and equipment	Public			Private		
	Total	Elementary	Secondary	Total	Elementary	Secondary
Percentage of school library media centers offering the following equipment:						
Telephone	61.2%	55.8%	74.7%	40.8%	30.7%	52.8%
Fax machine	7.8%	4.7%	15.1%	5.1%	3.1%	7.9%
Computer with modem	34.3%	28.3%	48.2%	19.5%	12.2%	30.2%
Automated catalog	24.0%	20.3%	32.8%	9.7%	5.5%	15.8%
Automated circulation system	37.9%	34.1%	47.7%	9.5%	5.4%	16.0%
Database searching with CD-ROM	31.2%	23.9%	48.1%	13.9%	6.3%	25.7%
Online database searching	9.4%	5.4%	18.9%	5.5%	0.7%	12.2%
Compact disc for periodical indices, etc.	46.7%	39.6%	63.5%	19.6%	12.3%	33.5%
Video laser disc	31.9%	30.3%	36.1%	6.3%	4.8%	9.9%
Connection to Internet	12.0%	9.5%	17.5%	5.3%	2.8%	9.2%
Cable television	76.2%	75.0%	80.6%	39.9%	42.6%	43.1%
Broadcast television	48.6%	48.0%	49.9%	39.9%	42.9%	39.7%
Closed circuit television	25.5%	22.0%	34.0%	8.8%	5.7%	19.2%
Satellite dish	22.9%	14.4%	41.1%	8.7%	5.6%	15.6%
Percentage of schools offering the following services:						
Microcomputers	90.1%	88.5%	94.0%	75.8%	72.8%	80.6%
Long-distance learning	19.0%	15.9%	24.8%	8.8%	8.3%	8.2%

Notes:

Percentages are based on schools that have library media centers. In school year 1990-91, 96% of public and 87% of private schools had library media centers.

Source: U.S. Department of Education, National Center for Education Statistics. *Schools and Staffing Survey, (Library Media Center Questionnaire), 1993–94.* Cited in U.S. Department of Education, National Center for Education Statistics. *The Condition of Education 1997.* Washington, DC: National Center for Education Statistics, 1997. Table 6.1. http://www.nces.ed.gov/pubs/ce/index.htm.

8.7. School Library Media Center Expenditures for Selected Services and Equipment, Public and Private, Elementary and Secondary, United States, 1993–1994.

Selected services and equipment	Public			Private		
	Total	Elementary	Secondary	Total	Elementary	Secondary
Expenditures per student for the 1992–93 school year on the following:						
Books	$ 8.52	$ 7.80	$ 10.06	$ 7.86	$ 6.94	$ 12.86
Current serial subscriptions (print and microfilm)	$ 2.18	$ 1.49	$ 4.15	$ 1.57	$ 0.95	$ 4.83
Video materials (tape and disc)	$ 1.24	$ 1.04	$ 1.77	$ 1.07	$ 0.87	$ 1.49
Other audiovisual materials	$ 0.82	$ 0.77	$ 0.87	$ 0.55	$ 0.55	$ 0.72
Microcomputer software	$ 1.09	$ 0.89	$ 1.48	$ 1.41	$ 1.25	$ 1.86
CD-ROM titles	$ 0.63	$ 0.38	$ 1.24	$ 0.69	$ 0.59	$ 1.69

Notes:

Locally budgeted expenditures exclude federal gifts and grants. Percentages are based on schools that have library media centers. In school year 1990–91, 96% of public and 87% of private schools had library media centers.

Source: U.S. Department of Education, National Center for Education Statistics. *Schools and Staffing Survey, (Library Media Center Questionnaire), 1993–94.* Cited in U.S. Department of Education, National Center for Education Statistics. *The Condition of Education 1997.* Washington, DC: National Center for Education Statistics, 1997. Table 6.1. http://www.nces.ed.gov/pubs/ce/index.htm.

8.8. School Library Media Centers, United States: Number of Selected Items Held per 100 Students at End of 1992–93 School Year.

Selected services and equipment	Public			Private		
	Total	Elementary	Secondary	Total	Elementary	Secondary
Books (number of volumes)	2,585	2,467	2,891	3,716	3,455	5,383
Current serial subscriptions (print and microfilm)	9	7	14	8	7	16
Video materials (tape and disc)	38	35	46	47	44	50
Other audiovisual materials	116	114	117	76	74	120
Microcomputer software	18	20	12	26	28	11
CD-ROM titles	1	1	2	1	0*	2

Notes:

* = Data less than 0.5 are rounded to 0.

Percentages are based on schools that have library media centers. In school year 1990–91, 96% of public and 87% of private schools had library media centers.

Source: U.S. Department of Education, National Center for Education Statistics. *Schools and Staffing Survey, (Library Media Center Questionnaire), 1993–94.* Cited in U.S. Department of Education, National Center for Education Statistics. *The Condition of Education 1997.* Washington, DC: National Center for Education Statistics, 1997. Table 6.1. http://www.nces.ed.gov/pubs/ce/index.htm.

8.9. College and University Library Hardware, Software, Audiovisual Materials, Microforms, and Machine-Readable Materials, 1991–92 and 1994–95.

Item	1991–1992	1994–1995
Number of libraries	3,274	3,639
Total enrollment, in thousands [1]	14,359	14,279
Full-time-equivalent enrollment, in thousands [1]	10,361	10,348
Microform titles at end of year (thousands of units)	—	160,188
Computer files at end of year (thousands of units)	—	480
Selected expenditures, total, in thousands:		
Furniture/equipment	—	$ 56,432
Computer hardware/software	—	$ 126,936
Utilities/networks/consortia	—	$ 81,686
Collections:	$ 1,197,293	$ 1,374,407
Print materials	$ 420,930	$ 451,988
Serial subscriptions	$ 639,128	$ 703,463
Microforms	$ 43,666	$ 61,702
Audiovisual materials	$ 23,879	$ 29,375
Machine readable materials	$ 29,093	$ 72,735
Collections, loans	—	$ 12,440
Other collection expenditures	$ 40,596	$ 42,704

Notes:
[1]Fall enrollment for the academic year specified.

Source: U.S. Department of Education, National Center for Education Statistics. *Library Statistics of Colleges and Universities*, various years; and Integrated Postsecondary Education Data System. "Academic Library Survey." July 1997. Cited in U.S. Department of Education, National Center for Education Statistics. *Digest of Education Statistics 1997*. Jessup, MD: Education Publications Center, 1998. Table 416.

TECHNOLOGY IN SCHOOLS

8.10. Distance Education in Higher Education Institutions, United States: Percent Distribution, by Current and Planned Use, 1995.

Institutional characteristics	Currently offering distance education courses	Planning to offer distance education courses in next 3 years	No plans to offer distance education in next 3 years
All institutions	33%	25%	42%
Institution type:			
Public 2-year	58%	28%	14%
Private 2-year	2%	14%	84%
Public 4-year	62%	23%	14%
Private 4-year	12%	27%	61%
Geographic region:			
Northeast	20%	27%	53%
Southeast	31%	28%	41%
Central	39%	24%	37%
West	40%	23%	37%
Enrollment:			
Less than 3,000	16%	27%	56%
3,000 to 9,999	61%	24%	15%
10,000 or more	76%	14%	10%

Notes:
Data are for higher education institutions in the 50 states, the District of Columbia, and Puerto Rico. Percents may not total 100 because of rounding.

Source: U.S. Department of Education, National Center for Education Statistics. Postsecondary Education Quick Information System. *Survey on Distance Education Courses Offered by Higher Education Institutions, 1995*. Cited in U.S. Department of Education, National Center for Education Statistics. *Issue Brief: Distance Education in Higher Education Institutions: Incidence, Audiences, and Plans to Expand*. Washington, DC: U.S. Department of Education, Office of Educational Research and Improvement, 1998. Table 1. http://www.nces.ed.gov/pubs98/98132.pdf.

8.11. Percent of U.S. Higher Education Institutions Using Various Technologies to Deliver Distance Education Courses, and Plans for Use by Level of Use, 1995.

Type of technology	Currently use — Percent of institutions [2]	Plan to use in the next 3 years [1] — Reduce or keep same number of courses	Start or increase number of courses	No plans
Two-way interactive video	57%	2%	79%	19%
Two-way audio, one-way video	24%	7%	35%	58%
One-way live video	9%	6%	28%	66%
One-way pre-recorded video	52%	8%	49%	43%
Audiographics	3%	3%	8%	89%
Two-way audio	11%	6%	20%	75%
One-way audio	10%	6%	11%	83%
Two-way online interactions	14%	(+)	71%	29%
Other computer-based technology	22%	1%	79%	20%

Notes:

(+) = Less than 0.5 percent.

[1] For plans, percentages are computed across each row, but may not total 100 because of rounding. Percentages are based on all institutions currently offering or planning to offer distance education courses in the next 3 years.

[2] Percentages are based on the number of all institutions that offered distance education courses in fall 1995. They add to more than 100 because an institution can use more than one type of technology. Data are for higher education institutions in the 50 states, the District of Columbia, and Puerto Rico.

Source: U.S. Department of Education, National Center for Education Statistics. Postsecondary Education Quick Information System. *Survey on Distance Education Courses Offered by Higher Education Institutions, 1995.* Cited in U.S. Department of Education, National Center for Education Statistics. *Issue Brief: Distance Education in Higher Education Institutions: Incidence, Audiences, and Plans to Expand.* Washington, DC: U.S. Department of Education, Office of Educational Research and Improvement, 1998. Table 2. http://www.nces.ed.gov/pubs98/98132.pdf.

8.12. Percentage of Public Schools Having Access to the Internet in Fall 1994, 1995, 1996, 1997 and 1998, by School Characteristics, United States.

School characteristics	Public schools having access to the Internet				
	1994	1995	1996	1997	1998
All public schools	35%	50%	65%	78%	89%
Instructional level*					
Elementary	30%	46%	61%	75%	88%
Secondary	49%	65%	77%	89%	94%
Size of enrollment					
Less than 300	30%	39%	57%	75%	87%
300 to 999	35%	52%	66%	78%	89%
1,000 or more	58%	69%	80%	89%	95%
Metropolitan status					
City	40%	47%	64%	74%	92%
Urban fringe	38%	59%	75%	78%	85%
Town	29%	47%	61%	84%	90%
Rural	35%	48%	60%	79%	92%
Geographic region					
Northeast	34%	59%	70%	78%	90%
Southeast	29%	44%	62%	84%	92%
Central	34%	52%	66%	79%	90%
West	42%	48%	62%	73%	86%
Minority enrollment					
Less than 6 percent	NA	52%	65%	84%	91%
6 to 20 percent	NA	58%	72%	87%	93%
21 to 49 percent	NA	54%	65%	73%	91%
50 percent or more	NA	40%	56%	63%	82%
Students eligible for free or reduced-price lunch					
Less than 11 percent	NA	62%	78%	88%	87%
11 to 30 percent	NA	59%	72%	83%	94%
31 to 70 percent	NA	47%	58%	78%	91%
71 percent or more	NA	31%	53%	63%	80%

Notes:
NA = Data not available.
* = Data for combined schools (those that span elementary and secondary grades) are included in the totals and in analyses by other school characteristics but are not shown separately.

Source: U.S. Department of Education, National Center for Education Statistics. "Advanced Telecommunications in U.S. Public Schools, K-12," NCES 95–731; "Advanced Telecommunications in U.S. Public Elementary and Secondary Schools, 1995," NCES 96–854; "Advanced Telecommunications in U.S. Public Elementary and Secondary Schools, Fall 1996," NCES 97–944; "Internet Access in Public Schools," NCES 98–031; and data from the "Survey on Internet Access in U.S. Public Schools, Fall 1998," FRSS 69, 1998. Cited in U.S. Department of Education, National Center for Education Statistics. *Issue Brief: Internet Access in Public Schools.* Washington, DC: National Center for Education Statistics, 1998. And U.S. Department of Education National Center for Education Statistics. *Issue Brief: Internet Access in Public Schools and Classrooms: 1994–98.* Washington, DC: U.S. Department of Education, Office of Research and Improvement, 1999. Table 1.

8.13. Students Who Reported Using a Computer, by Location of Use and Grade, United States, 1984–96.

Year	At home			At school [1]			At the library [2]		
	Grade 4	Grade 8	Grade 11	Grade 4	Grade 8	Grade 11	Grade 4	Grade 8	Grade 11
1984	44.8%	36.6%	30.3%	38.8%	33.3%	45.0%	25.2%	20.5%	22.2%
1988	45.5%	46.2%	39.7%	70.2%	58.2%	55.3%	27.6%	39.8%	37.4%
1990	43.3%	41.2%	42.8%	81.1%	59.5%	55.1%	34.5%	35.1%	46.2%
1992	43.0%	44.0%	50.7%	83.5%	62.4%	72.8%	45.6%	47.3%	62.1%
1994	50.0%	50.1%	51.0%	86.0%	72.3%	73.9%	48.1%	56.8%	61.3%
1996	62.6%	61.2%	63.3%	88.6%	76.7%	84.0%	61.4%	66.7%	72.9%

Notes:
[1] Based on the percentage of students who reported ever using a computer at school .
[2] School or public library.

Source: U.S. Department of Education, National Center for Education Statistics. *National Assessment of Educational Progress, Almanac: Writing, 1984 to 1996, 1998.* Cited in U.S. Department of Education, National Center for Education Statistics. *Condition of Education 1998.* Washington, DC: National Center for Education Statistics, 1998. Table 3.2. http://www.nces.ed.gov/pubs98/condition98.

8.14. Percentage of Students Who Used a Computer at School and/or Home, by Current Grade Level, Race-ethnicity, and Family Income, United States: 1984, 1989, 1993, and 1997.

Current grade level, race-ethnicity, and family income[1]	1984 Used a computer at			1989 Used a computer at			1993 Used a computer at			1997 Used a computer at		
	School	Home	Home or school	School	Home	Home or school	School	Home	Home or school	School	Home	Home or school
Total (Grades 1-12)	**29.7%**	**12.6%**	**36.2%**	**48.0%**	**18.4%**	**54.6%**	**62.0%**	**25.2%**	**68.3%**	**76.4%**	**45.2%**	**84.0%**
					Grades 1-6							
Total	**30.5%**	**11.8%**	**36.2%**	**52.4%**	**16.1%**	**56.9%**	**66.6%**	**23.0%**	**70.7%**	**79.1%**	**41.3%**	**83.8%**
Race-ethnicity												
White	35.5%	14.4%	42.3%	58.9%	20.4%	64.5%	71.6%	29.2%	76.6%	84.4%	52.3%	89.9%
Black	15.1%	5.1%	18.3%	34.3%	6.0%	36.2%	54.1%	8.3%	56.5%	70.1%	19.3%	72.9%
Hispanic	16.4%	3.5%	18.5%	41.1%	5.1%	42.3%	55.1%	6.8%	56.8%	67.7%	17.9%	70.5%
Family income												
Low income	18.5%	2.5%	20.0%	39.4%	5.7%	39.0%	57.4%	3.9%	58.1%	70.9%	12.4%	71.9%
Middle income	29.5%	9.7%	34.5%	52.3%	17.0%	49.9%	66.2%	18.0%	69.5%	78.6%	36.4%	82.8%
High Income	42.4%	24.4%	53.0%	62.5%	38.3%	63.9%	74.0%	48.5%	82.4%	86.5%	74.6%	95.0%
					Grades 7-12							
Total	**28.9%**	**13.4%**	**36.2%**	**43.0%**	**21.1%**	**52.1%**	**57.0%**	**27.7%**	**65.6%**	**75.5%**	**60.9%**	**89.0%**
Race-ethnicity												
White	31.9%	16.2%	40.8%	45.5%	25.6%	56.7%	59.6%	34.7%	70.2%	75.5%	60.9%	89.0%
Black	18.4%	4.9%	20.8%	36.5%	8.5%	39.7%	50.5%	10.2%	53.5%	74.2%	22.3%	77.9%
Hispanic	21.2%	3.6%	23.2%	34.4%	9.0%	38.3%	52.6%	9.5%	56.1%	65.4%	21.5%	69.4%
Family income												
Low income	20.0%	3.3%	22.3%	36.7%	5.7%	39.0%	49.0%	5.6%	50.4%	67.6%	14.9%	70.7%
Middle income	28.4%	10.1%	33.6%	42.6%	17.0%	49.9%	57.3%	22.2%	64.1%	74.1%	44.2%	83.5%
High Income	34.1%	24.8%	48.1%	47.2%	38.3%	63.9%	60.7%	51.2%	77.0%	75.4%	78.6%	93.3%

Notes:
[1] Low income is the bottom 20% of all family incomes; high income is the top 20% of all family incomes; and middle income is the 60% in between.

Source: U.S. Department of Commerce, Bureau of the Census, October Current Population Surveys. Cited in U.S. Department of Education, National Center for Education Statistics. *The Condition of Education 1999.* Washington, DC: National Center for Education Statistics, 1999. Page 38.

8.15. Percentage of Students Who Used a Computer at Home, by Purpose, Current Grade Level, Race-ethnicity, and Family Income, United States, 1997.

Current grade level, race-ethnicity, and family income[1]	Word processing	E-mail	Internet	School assignments	Databases	Graphics/ design
Total (Grades 1-12)	**33.9%**	**13.0%**	**17.5%**	**49.1%**	**1.6%**	**14.7%**
			Grades 1-6			
Total	**19.8%**	**6.8%**	**10.2%**	**34.0%**	**0.0%**	**12.0%**
Race-ethnicity						
White	21.7%	8.0%	11.5%	35.6%	0.0%	13.5%
Black	11.2%	2.5%	4.2%	27.4%	0.0%	6.1%
Hispanic	15.2%	2.2%	7.3%	28.4%	0.0%	7.6%
Family income						
Low income	12.5%	4.4%	4.7%	21.7%	0.0%	7.3%
Middle income	15.5%	4.8%	7.6%	29.7%	0.0%	10.3%
High income	27.5%	10.1%	15.1%	42.6%	0.0%	15.3%
			Grades 7-12			
Total	**47.5%**	**19.0%**	**24.6%**	**63.9%**	**3.1%**	**17.4%**
Race-ethnicity						
White	50.1%	20.9%	26.4%	65.6%	3.3%	18.8%
Black	31.7%	7.1%	12.8%	50.7%	1.1%	9.2%
Hispanic	37.6%	9.0%	16.6%	53.0%	2.1%	11.6%
Family income						
Low income	26.9%	8.0%	10.2%	44.6%	0.8%	9.7%
Middle income	41.4%	15.1%	19.2%	60.7%	3.1%	16.4%
High income	58.6%	25.9%	33.8%	70.8%	3.5%	19.7%

Notes:
[1]Low income is the bottom 20% of all family incomes; high income is the top 20% of all family incomes; and middle income is the 60% in between.

Source: U.S. Department of Commerce, Bureau of the Census, October Current Population Surveys. Cited in U.S. Department of Education, National Center for Education Statistics. *The Condition of Education 1999.* Washington, DC: National Center for Education Statistics, 1999. Table 18-1, Page 173.

8.16. Percent of Student Home Computer Users Using Specific Applications, by Selected Characteristics, United States, October 1997.

Selected Characteristics	Number of Students Using Computers at Home (thousands)	Percent of Computer Users Working with Specific Applications [1]										
		Home Bookkeeping	School Assignments	Games	Job-related	Home Connection to School or Work	Word Processing	Databases	Graphics	Desktop Publishing	Spreadsheets	Internet
Total, All Students	**32,459**	—	**65.9%**	**78.0%**	—	**7.4%**	**52.5%**	—	**21.4%**	—	—	**36.3%**
Preprimary	2,493	—	5.2%	89.5%	—	0.4%	4.6%	0.0%	10.8%	—	—	4.4%
1st to 8th Grade	14,056	—	58.3%	93.0%	—	1.3%	36.1%	0.0%	19.2%	—	—	25.1%
White, non-Hispanic	11,340	—	55.6%	89.1%	—	1.2%	35.7%	0.0%	19.5%	—	—	26.4%
Black, non-Hispanic	1,078	—	49.8%	84.1%	—	1.0%	23.6%	0.0%	9.5%	—	—	16.0%
Hispanic	863	—	51.0%	77.5%	—	2.0%	28.2%	0.0%	14.3%	—	—	19.4%
9th to 12th Grade	7,699	2.9%	84.8%	76.0%	5.2%	3.1%	65.9%	6.1%	22.3%	7.0%	6.5%	42.2%
White, non-Hispanic	6,265	2.8%	80.4%	72.5%	4.2%	3.1%	64.3%	6.0%	22.2%	7.1%	6.3%	43.6%
Black, non-Hispanic	534	1.7%	74.3%	69.7%	9.9%	1.4%	47.2%	2.4%	15.3%	4.0%	2.2%	27.6%
Hispanic	458	3.1%	73.0%	64.2%	4.4%	2.2%	55.2%	4.1%	13.4%	3.9%	4.5%	37.3%
Undergraduate	6,179	22.5%	82.0%	54.3%	17.8%	20.3%	79.7%	0.0%	25.4%	14.3%	24.6%	57.9%
Graduate	2,032	38.1%	73.4%	44.3%	43.4%	32.0%	88.0%	0.0%	32.9%	25.5%	36.6%	65.9%
Males	16,213	7.2%	62.5%	79.9%	7.2%	7.7%	47.0%	0.0%	20.3%	5.4%	8.9%	37.3%
Preprimary	1,308	—	5.5%	90.3%	—	0.8%	3.8%	0.0%	11.2%	—	—	5.9%
1st to 8th Grade	7,257	—	54.8%	89.8%	—	1.4%	31.3%	0.0%	16.3%	—	—	25.0%
9th to 12th Grade	3,971	3.0%	78.6%	78.8%	5.5%	3.4%	58.9%	0.0%	21.9%	6.6%	6.7%	44.5%
Undergraduate	2,841	23.5%	82.7%	59.8%	18.8%	23.7%	78.8%	0.0%	28.6%	13.8%	28.4%	63.6%
Graduate	834	44.2%	73.7%	50.4%	48.1%	39.0%	85.9%	0.0%	34.5%	26.0%	43.0%	68.8%
Females	16,246	7.5%	64.3%	70.2%	7.4%	6.6%	54.1%	0.0%	20.8%	6.4%	8.0%	35.4%
Preprimary	1,185	—	4.9%	88.6%	—	0.0%	5.3%	0.0%	10.4%	—	—	2.7%
1st to 8th Grade	6,799	—	55.4%	85.9%	—	1.1%	37.0%	0.0%	20.0%	—	—	25.1%
9th to 12th Grade	3,728	2.5%	81.3%	64.0%	4.1%	2.5%	65.5%	0.0%	20.0%	6.5%	5.5%	39.7%
Undergraduate	3,337	21.6%	81.5%	49.5%	17.0%	17.4%	80.6%	0.0%	22.7%	14.7%	21.3%	53.1%
Graduate	1,197	33.9%	73.2%	40.0%	40.2%	27.0%	89.4%	0.0%	31.8%	25.2%	32.2%	63.8%

Notes:

[1]Individuals may be counted in more than one computer activity.

— = Data not available or not applicable.

Data are based on a sample survey of households and are subject to sampling and nonsampling error.

Source: U.S. Department of Commerce, Bureau of the Census. *Current Population Survey.* October 1997. Unpublished data. This table was prepared October 1998. Cited in U.S. Department of Education, National Center for Education Statistics. *Digest of Education Statistics 1998.* Washington, DC: National Center for Education Statistics, 1999. Table 427.

Chapter 9. Energy and Environment

ENERGY

Energy makes technology work. Without power to make heat and light, run motors, and produce electricity, we would be lighting our caves or huts with candles, erecting our buildings using levers and pulleys, and cooking over open fires.

Our society is now so dependent on energy that we go to great lengths to obtain and preserve our access to it. So threatened do we feel by its possible loss that we fight wars, dam free-flowing rivers, construct pipelines across fragile ecosystems, and leak radioactive material into human settlements in its pursuit.

Since energy is so vital, we should know what kinds of energy we are using, what these energy options are costing us financially and in terms of pollution (more on the latter in Table 13.24, Chapter 13), how much work we can perform with the energy available to us, and what alternatives are available to us now and in the future. We also care about how we are spending our energy, and whether we can use it more wisely or frugally.

The statistics that follow illuminate these issues. Table 9.1 provides a broad overview of the types of fuels we have been consuming since 1949. Overall energy consumption has more than tripled in that time, with natural gas use increasing the most. Although the figures for renewable energy consumption seem to have jumped substantially from 1990 on, be careful: The way the Department of Energy measured renewable energy use changed that year, skewing the figures upward. Also, in this scheme, renewable energy is composed almost entirely of hydroelectric power, as you can see if you compare figures for renewable and hydroelectric. The types of energy we usually think of as renewable—solar, wind, and geothermal—barely register on the radar screen. Table 9.2 illustrates the renewable big picture even more clearly. Together, solar, wind, and geothermal comprise less than 1% of U.S. energy consumption. Table 9.5 shows that the residential commercial sector consumes most of the solar energy generated, while industry uses the largest share of geothermal energy. As

of 1996, the electric utility sector was barely using solar and wind energy, but with deregulation of the electricity industry and concerns about greenhouse gas emissions, that may change. Table 9.7 shows that electric generating capacity from wind is about six times as large as from solar energy. Tables 9.13 through 9.15 give us a greater in-depth look at the solar collector (for heating) and solar photovoltaic (for making electricity) markets. Table 9.9 illustrates the growth in nuclear power in the United States, though future tables should show decreases that reflect the industry's projected decline.

We are all acutely aware of gasoline prices and vehicle fuel efficiencies. Table 9.10. reminds us that gas prices peaked in 1981 at $2.05 per gallon, then declined steadily until 1990, when they surged back to $1.30 from an interim low of $1.12 in 1988. Between 1991 and 1997, prices fell again, with another low in 1994 and 1995 of $1.12 per gallon. Table 9.11 illustrates the ups and downs of fuel economy, which reached its highest levels in the 1990s.

Table 9.12 is one of the most interesting in the chapter. It shows prices for various types of fuel between 1970 and 1994. All figures are dollars per million Btu, so you cannot compare vehicle fuels directly with those used to generate heat and electricity, but you *can* compare within those categories, and you can see how much prices have risen over the period. Nuclear fuel prices have risen the least (38 cents), while natural gas rose by the largest factor—more than seven times—though the largest dollar increase was for electricity purchased by end users: $15.35. Even though prices for nuclear fuel are the lowest of all, the industry is unprofitable because of the high cost of operating the power plants. Obviously fuel costs represent only one factor in overall energy economics.

ENVIRONMENT

"The environment" did not spring into being with the publication of Rachel Carson's *Silent Spring* in 1962, but popular awareness of it did. By the first Earth Day in 1970, environmental degradation had become a white-hot issue, with people slashing their gasoline credit cards to protest oil pollution and worrying about the effects of overpopulation. In the years since that raucous day, we have developed devices and techniques for cleaning up pollution and, better yet, preventing it in the first place. Not that things are perfect. Our air, water, and land are still far from pristine, and we face a new possible danger—the ozone hole. We still use hazardous materials that can make people sick, cause birth defects, and kill all kinds of life, and we still don't dispose of those materials in ways that negate all their effects.

Sub-fields related to the environment include

- Air quality
- Water quality
- Soil quality
- Use and effects of hazardous materials, like chemicals and radioactive substances
- Biodiversity and endangered species
- Waste disposal and recycling
- Overpopulation and its effects, such as resource shortages, crowding, and noise
- Climate and global warming
- Ecology, that is, the balance of natural environments

This chapter contains tables that show some of the methods and techniques we use for cleaning up the environment or for stopping pollution before it starts. It also details quantities of hazardous waste by state and lists the top 100 most hazardous substances in use.

Tables 9.17 and 9.18 list the number and types of National Priorities List cleanup sites in the construction cleanup stage and the technologies used in the process. Most of the sites were polluted by industrial or landfill waste. Note that the total number of sites is smaller than the sum of the individual types and methods; multiple types of pollution and techniques apply to each site. Table 9.23 demonstrates how much wood and wood-fiber material we recycled in the United States in the latest year available, 1994. Table 9.19 lists the largest hazardous waste-generating states as California, Illinois, Louisiana, Michigan, New Jersey, Tennessee, Texas, and Washington, with Texas way out in front. South Dakota generates the least hazardous waste of the 50 states. Arsenic and lead are the highest priority hazardous substances (Table 9.20).

Tables 9.21 and 9.22 show the value of shipments of various air pollution control equipment from 1992 through 1997.

ENERGY

9.1. U.S. Energy Consumption, by Type, 1949–97.

Year	Biofuel Total Consumption Quadrillion Btu	Coal Consumption Quadrillion Btu	Fossil Fuels Consumption Quadrillion Btu	Geothermal Energy Consumption Quadrillion Btu	Conventional Hydroelectric Power Consumption Quadrillion Btu	Natural Gas Consumption Quadrillion Btu	Petroleum Consumption Quadrillion Btu	Renewable Energy Consumption Quadrillion Btu	Solar Energy Consumption Quadrillion Btu	Wind Energy Consumption Quadrillion Btu	Total Energy Consumption Quadrillion Btu
1949	0.0058	11.9812	29.0023	0	1.4486	5.1451	11.8827	1.4544	0	0	30.4568
1950	0.0055	12.3470	31.6318	0	1.4405	5.9684	13.3155	1.4459	0	0	33.0778
1951	0.0053	12.5528	34.0079	0	1.4536	7.0485	14.4280	1.4590	0	0	35.4668
1952	0.0064	11.3065	33.7999	0	1.4961	7.5496	14.9557	1.5026	0	0	35.3025
1953	0.0050	11.3727	34.8262	0	1.4387	7.9066	15.5558	1.4438	0	0	36.2699
1954	0.0032	9.7145	33.8771	0	1.3883	8.3302	15.8392	1.3915	0	0	35.2686
1955	0.0032	11.1673	37.4101	0	1.4074	8.9979	17.2550	1.4107	0	0	38.8208
1956	0.0017	11.3498	38.8882	0	1.4868	9.6140	17.9375	1.4886	0	0	40.3768
1957	0.0020	10.8207	38.9257	0	1.5565	10.1908	17.9317	1.5586	0	0	40.4843
1958	0.0019	9.5334	38.7168	0	1.6287	10.6632	18.5269	1.6307	0	0	40.3494
1959	0.0017	9.5184	40.5501	0	1.5875	11.7174	19.3226	1.5891	0	0	42.1414
1960	0.0015	9.8378	42.1367	0.0008	1.6568	12.3854	19.9192	1.6591	0	0	43.8018
1961	0.0013	9.6234	42.7582	0.0022	1.6805	12.9264	20.2164	1.6840	0	0	44.4619
1962	0.0013	9.9064	44.6807	0.0023	1.8218	13.7308	21.0490	1.8255	0	0	46.5326
1963	0.0013	10.4124	46.5092	0.0037	1.7724	14.4033	21.7008	1.7774	0	0	48.3247
1964	0.0015	10.9646	48.5433	0.0045	1.9068	15.2878	22.3013	1.9128	0	0	50.4959
1965	0.0028	11.5805	50.5764	0.0042	2.0576	15.7687	23.2457	2.0646	0	0	52.6841
1966	0.0035	12.1433	53.5142	0.0042	2.0729	16.9953	24.4005	2.0805	0	0	55.6589
1967	0.0033	11.9140	55.1271	0.0069	2.3435	17.9448	25.2837	2.3537	0	0	57.5693
1968	0.0039	12.3306	58.5023	0.0094	2.3421	19.2097	26.9794	2.3554	0	0	60.9993
1969	0.0033	12.3817	61.3619	0.0133	2.6592	20.6780	28.3383	2.6758	0	0	64.1915
1970	0.0037	12.2643	63.5221	0.0113	2.6541	21.7947	29.5207	2.6692	0	0	66.4306
1971	0.0033	11.5986	64.5959	0.0119	2.8611	22.4691	30.5613	2.8763	0	0	67.8851
1972	0.0034	12.0768	67.6958	0.0315	2.9436	22.6982	32.9467	2.9786	0	0	71.2581
1973	0.0034	12.9715	70.3164	0.0426	3.0098	22.5124	34.8399	3.0558	0	0	74.2823
1974	0.0026	12.6628	67.9060	0.0532	3.3091	21.7325	33.4546	3.3649	0	0	72.5430
1975	0.0020	12.6629	65.3549	0.0702	3.2190	19.9479	32.7306	3.2911	0	0	70.5458
1976	0.0028	13.5843	69.1043	0.0782	3.0656	20.3454	35.1747	3.1465	0	0	74.3619
1977	0.0050	13.9219	70.9892	0.0774	2.5150	19.9305	37.1222	2.5974	0	0	76.2884
1978	0.0035	13.7654	71.8558	0.0643	3.1414	20.0004	37.9653	3.2092	0	0	78.0892
1979	0.0052	15.0395	72.8916	0.0838	3.1412	20.6658	37.1234	3.2301	0	0	78.8975
1980	0.0045	15.4231	69.9846	0.1098	3.1175	20.3941	34.2024	3.2318	0	0	75.9555
1981	0.0038	15.9072	67.7500	0.1230	3.1054	19.9278	31.9310	3.2323	0	0	73.9899
1982	0.0034	15.3215	64.0365	0.1047	3.5720	18.5051	30.2316	3.6801	0	0	70.8478
1983	0.0040	15.8948	63.2899	0.1293	3.8989	17.3568	30.0539	4.0323	0	0	70.5247
1984	0.0092	17.0705	66.6173	0.1649	3.9741	18.5070	31.0513	3.9741	0	0.0001	74.1440
1985	0.0144	17.4781	66.2207	0.1983	3.3978	17.8339	30.9221	3.6107	0	0.0002	73.9803
1986	0.0123	17.2604	66.1476	0.2192	3.4462	16.7079	32.1961	3.6778	0	0.0002	74.2967
1987	0.0154	18.0081	68.6262	0.2291	3.1173	17.7443	32.8651	3.3619	0	0.0001	76.8941

9.1. U.S. Energy Consumption, by Type, 1949–97 *(continued)*.

Year	Biofuel Total Consumption Quadrillion Btu	Coal Consumption Quadrillion Btu	Fossil Fuels Consumption Quadrillion Btu	Geothermal Energy Consumption Quadrillion Btu	Conventional Hydroelectric Power Consumption Quadrillion Btu	Natural Gas Consumption Quadrillion Btu	Petroleum Consumption Quadrillion Btu	Renewable Energy Consumption Quadrillion Btu	Solar Energy Consumption Quadrillion Btu	Wind Energy Consumption Quadrillion Btu	Total Energy Consumption Quadrillion Btu
1988	0.0173	18.8455	71.6595	0.2173	2.6623	18.5524	34.2220	2.8969	0	0.0001	80.2177
1989	0.0203	18.9208	72.5459	0.1971	2.8805	19.3837	34.2111	3.0979	0	0	81.3206
1990	2.6320*	19.1009	71.9546	0.3450*	3.1041	19.2965	33.5525	6.1712*	0.0669	0.0233	84.0929
1991	2.6420	18.7699	71.2303	0.3535	3.1824	19.6062	32.8454	6.2730	0.0681	0.0270	83.9947
1992	2.7880	19.2085	72.8930	0.3675	2.8524	20.1307	33.5266	6.1056	0.0677	0.0300	85.5227
1993	2.7837	19.8286	74.5140	0.3806	3.1377	20.8266	33.8415	6.4027	0.0693	0.0314	87.3368
1994	2.8380	20.0178	76.0639	0.3814	2.9580	21.2875	34.7349	6.2816	0.0685	0.0358	89.2131
1995	2.9457	20.0846	76.9378	0.3250	3.4706	22.1633	34.6634	6.8466	0.0724	0.0329	90.9422
1996	3.0170	20.9885	79.2922	0.3542	3.9110	22.5870	35.7170	7.3933	0.0749	0.0362	93.8129
1997	2.7227	21.4400	80.3600	0.3700	3.9400	22.5900	36.3100	7.1400	0.0800	0.0400	94.2100

Notes:

Btu = British Thermal Unit. The amount of heat required to raise the temperature of a pound of water one degree Fahrenheit.

* = Beginning in 1990, coverage of nonutility use of removeable energy was expanded, resulting in a large increase in the figures for that year and later.

Source: Energy Information Administration, U.S. Department of Energy. Washington, DC. From the Department of Energy's query system at http://tonto.eia.doe.gov/er/, April 1998.

9.2. U.S. Energy Consumption by Energy Source, 1992–97 (quadrillion Btu).

Energy Source	1992	1993	1994	1995	1996	1997
			Fossil Fuels			
Coal	R 19.209	R 19.837	R 20.090	R 21.011	R 21.011	21.439
Coking Coal (Net Imports)	0.027	0.017	0.024	0.026	S	0.018
Natural Gas a/	20.131	20.827	R 21.288	R 22.163	22.560	22.588
Petroleum b/	33.527	33.841	34.735	R 34.663	35.864	36.314
Total Fossil Fuels	R 72.893	R 74.522	R 76.073	R 76.943	79.434	80.360
Nuclear Electric Power	6.607	6.519	6.837	R 7.177	7.168	6.686
Hydroelectric Pumped Storage c/	-0.043	-0.042	-0.035	R -0.028	-0.032	-0.042
			Renewable Energy			
Conventional Hydroelectric Power d/	2.852	R 3.147	R 2.969	R 3.472	R 3.914	3.932
Geothermal Energy	0.367	R 0.393	R 0.395	R 0.339	R 0.352	0.322
Biomass e/	2.788	2.784	R 2.838	R 2.846	R 2.938	2.723
Solar Energy f/, g/	0.068	R 0.071	R 0.072	0.073	0.075	0.074
Wind Energy	0.030	0.031	0.036	0.033	R 0.035	0.035
Total Renewable Energy	6.106	R 6.426	R 6.309	R 6.763	R 7.315	7.086
Total Energy Consumption g/	R 85.523	R 87.368	R 89.250	R 90.864	R 93.871	94.151

Notes:
a = Includes supplemental gaseous fuels.
b = Petroleum products supplied, including natural gas plant liquids and crude oil burned as fuel.
c = Represents total pumped-storage facility production minus energy used for pumping.
d = Hydroelectricity generated by pumped storage is not included in renewable energy.
e = Includes wood, wood waste, peat, wood sludge, municipal solid waste, agricultural waste, straw, tires, landfill gases, fish oils, and/or other waste.
f = Includes solar thermal and photovoltaic.
g = Includes off-grid photovoltaic electricity not included in Table 1.3 of the *Annual Energy Review 1996.*
R = Revised data.
S = Value less than 0.005 quadrillion BTU.
Totals may not equal sum of components due to independent rounding.

Sources: Cited in U.S. Department of Energy, Energy Information Administration. *Renewable Energy Annual 1997, Volume 1,* and *Renewable Energy Annual 1998 with data for 1997.* Washington, DC: Energy Information Administration, 1997 and 1998. Table 1.

9.3. U.S. and World Petroleum Consumption, 1949–96.

Year	U.S. Petroleum Consumption Million barrels per day	Percent Consumed by U.S.	World Petroleum Consumption Million barrels per day
1949	5.68	NA	NA
1950	6.36	NA	NA
1951	6.91	NA	NA
1952	7.18	NA	NA
1953	7.49	NA	NA
1954	7.64	NA	NA
1955	8.33	NA	NA
1956	8.67	NA	NA
1957	8.68	NA	NA
1958	8.99	NA	NA
1959	9.39	NA	NA
1960	9.68	45.37%	21.34
1961	9.83	42.75%	23.00
1962	10.25	41.48%	24.89
1963	10.59	39.33%	26.92
1964	10.89	37.46%	29.08
1965	11.34	36.43%	31.14
1966	11.91	35.49%	33.56
1967	12.38	34.78%	35.59
1968	13.23	33.97%	38.96
1969	13.93	32.48%	42.89
1970	14.48	30.94%	46.81

9.3. U.S. and World Petroleum Consumption, 1949–96 *(continued)*.

Year	U.S. Petroleum Consumption Million barrels per day	Percent Consumed by U.S.	World Petroleum Consumption Million barrels per day
1971	14.99	30.34%	49.42
1972	16.17	30.47%	53.09
1973	17.06	29.80%	57.24
1974	16.41	28.95%	56.68
1975	16.09	28.62%	56.20
1976	17.26	28.92%	59.67
1977	18.16	29.38%	61.83
1978	18.57	28.95%	64.16
1979	18.24	27.97%	65.22
1980	16.85	26.72%	63.07
1981	15.82	25.99%	60.90
1982	15.07	25.33%	59.50
1983	15.01	25.55%	58.74
1984	15.54	25.97%	59.84
1985	15.50	25.79%	60.10
1986	16.04	25.98%	61.76
1987	16.42	26.07%	63.00
1988	17.08	26.35%	64.82
1989	17.07	25.90%	65.92
1990	16.74	25.37%	65.99
1991	16.47	24.74%	66.58
1992	16.83	25.22%	66.74
1993	16.99	25.34%	67.04
1994	17.46	25.56%	68.31
1995	17.47	25.12%	69.55
1996	18.02	NA	NA

Note:
Percentages calculated by the author.

Source: Energy Information Administration, U.S. Department of Energy. Washington, DC. Data appeared at http://www.eia.doe.gov on April 30, 1998.

9.4. Biomass Energy Consumption by Sector and Census Region, United States, 1992–97 (trillion Btu).

Energy Source	1992	1993	1994	1995	1996	1997
Wood Energy [1]	2,249	2,228	2,266	R 2,250	R 2,335	2,103
Sector						
Residential	645	548	537	596	595	433
Commercial	[2]	44	45	45	49	42
Industrial	1,593	1,625	1,673	R 1,598	R 1,679	1,617
Electric Utility	11	11	11	11	12	11
Census Region						
Northwest	264	277	278	R 343	R 348	328
Midwest	286	222	223	R 269	R 269	226
South	1,234	1,405	1,437	R 1,024	R 1,074	957
West	466	324	328	R 615	R 644	592
Waste Energy [3]	460	468	475	492	R 529	523
Source						
Municipal Solid Waste	383	390	394	408	R 447	449
Combustion	311	318	323	333	R 359	359
Landfill Gas	72	72	71	75	88	90
Manufacturing	77	78	81	81	82	74

9.4. Biomass Energy Consumption by Sector and Census Region, United States, 1992–97 (trillion Btu) *(continued)*.

Energy Source	1992	1993	1994	1995	1996	1997
Census Region						
Northwest	148	151	171	173	R 188	191
Midwest	84	85	76	88	R 80	88
South	128	130	134	134	R 158	151
West	100	102	95	96	R 103	93
Alcohol Fuels (Ethanol)	79	88	97	104	74	97
Census Region						
Northwest	*	*	*	3	R 7	9
Midwest	55	61	68	74	R 43	56
South	13	14	16	10	R 8	11
West	10	11	12	17	R 16	21
Total Biofuel Energy Consumption	2,788	2,784	2,838	R 2,846	R 2,938	2,723

Notes:
[1] Assuming an average energy yield of 17 million Btu per ton.
[2] Commercial wood energy use for 1990–92 is not included because there are no accurate data sources to provide reliable estimates. However, from the "1986 Nonresidential Energy Consumption Survey" conducted by the Energy Information Administration, it is estimated that commercial sector use is about 20 to 40 trillion Btu.
[3] Municipal solid waste, manufacturing waste, refuse-derived fuel, and methane recovered from landfills.
* = Less than 0.5 trillion Btu.

Sources: U.S. Department of Energy, Energy Information Administration. *Annual Energy Review 1996.* DOE/EIA-0384(96). Washington, DC: U.S. Department of Energy, Energy Information Administration, July 1997. Also U.S. Department of Energy, Energy Information Administration. Renewable *Energy Annual 1997, Volume 1.* Washington, DC: U.S. Department of Energy, Energy Information Administration, 1997. Table 8. U.S. Department of Energy, Energy Information Administration. *Renewable Energy Annual 1998 with Data for 1997.* Washington, DC: U.S. Department of Energy, Energy Information Administration, 1998. Table 8.

9.5. Renewable Energy Consumption by Sector and Energy Source, United States, 1992–97 (quadrillion Btu).

Sector and Source	1992	1993	1994	1995	1996	1997
Residential/Commercial						
Biomass	0.645	0.592	0.582	0.641	0.644	0.475
Solar	0.060	R 0.062	R 0.064	R 0.065	0.066	0.065
Geothermal [a]	NA	0.010	0.010	0.011	R 0.012	0.013
Total	NA	R 0.664	R 0.656	R 0.717	R 0.722	0.553
Industrial [b]						
Biomass	2.042	2.084	R 2.138	R 2.084	R 2.200	2.132
Geothermal [a]	0.179	R 0.206	R 0.214	R 0.210	R 0.217	0.194
Conventional Hydroelectric [c]	0.097	R 0.119	0.136	0.152	R 0.171	0.185
Solar	0.008	0.009	0.008	0.008	R 0.009	0.009
Wind	0.030	0.031	0.036	0.033	R 0.035	0.035
Total	2.357	R 2.449	R 2.533	R 2.487	R 2.633	2.555
Transportation						
Biomass [d]	0.079	0.088	R 0.097	0.104	0.074	0.097

9.5. Renewable Energy Consumption by Sector and Energy Source, United States, 1992–97 (quadrillion Btu) *(continued)*.

Sector and Source	1992	1993	1994	1995	1996	1997
Electric Utility						
Biomass	0.022	0.020	0.020	0.017	R 0.020	0.019
Geothermal	0.169	0.158	0.145	0.099	R 0.110	0.115
Conventional Hydroelectric [c]	2.511	R 2.774	R 2.538	R 3.054	R 3.422	3.528
Solar and Wind	*	*	*	*	*	*
Net Renewable Energy Imports /e	0.263	R 0.272	0.309	0.284	R 0.334	0.219
Total	2.965	R 3.225	R 3.012	R 3.454	R 3.886	3.881
Total Renewable Energy Consumption	6.106	R 6.426	R 6.309	R 6.763	R 7.315	7.086

Notes:
* Less than .0005 quadrillion Btu.
[a] Includes geothermal heat pump and direct use energy. The Industrial and Electric Utility sectors also include grid-connected electricity.
[b] Includes generation of electricity by cogenerators, independent power producers, and small power producers.
[c] Hydroelectricity generated by pumped storage is not included in renewable energy.
[d] Ethanol blended into gasoline.
[e] Includes only net imports of electricity known to be from renewable resources (geothermal and hydroelectric).
R = Revised data.
 Totals may not equal sum of components due to independent rounding.

Sources: 1992–96 data from Energy Information Administration (EIA), *Annual Energy Review 1996*, DOE/EIA-0384(96) (Washington, DC, July 1997), Table 10.1; 1996 data from Electricity Consumption—EIA, *Electric Power Monthly March 1997*, DOE-EIA-0348(97/03) (Washington, DC, March 1997). Nonelectricity Consumption (except imports)—Based on analysis by the Office of Coal, Nuclear, Electric and Alternate Fuels; Net Renewable Energy Imports, 1992–96 data from based on analysis by the Office of Coal, Nuclear, Electric and Alternate Fuels. Cited in U.S. Department of Energy, Energy Information Administration. *Renewable Energy Annual 1997, Volume 1.* Washington, DC: U.S. Department of Energy, Energy Information Administration, 1997. Table 2. and U.S. Department of Energy, Energy Information Administration. *Renewable Energy Annual 1998 With Data for 1997.* Washington, DC: U.S. Department of Energy, Energy Information Administration, 1998. Table 2.

9.6. Housing Units Heated by Various Types of Energy, United States, 1950–95.

Year	Occupied Housing Units Heated by Coal (number)	Share of Occupied Housing Units Heated by Coal (percent)	Occupied Housing Units Heated by Distillate Fuel and Kerosene (number)	Share of Occupied Housing Units Heated by Distillate Fuel and Kerosene (percent)	Occupied Housing Units Heated by Electricity (number)	Share of Occupied Housing Units Heated by Electricity (percent)	Occupied Housing Units Heated by Kerosene (number)	Share of Occupied Housing Units Heated by Kerosene (percent)	Occupied Housing Units Heated by LPG* (number)	Share of Occupied Housing Units Heated by LPG* (percent)	Occupied Housing Units Heated by Natural Gas (number)	Share of Occupied Housing Units Heated by Natural Gas (percent)	Occupied Housing Units with No Heating (number)	Share of Occupied Housing Units with No Heating (percent)	Occupied Housing Units (number)	Occupied Housing Units Heated by Other Energy Sources (number)	Occupied Housing Units Heated by Other Energy Sources (percent)	Solar Heated Housing Units (number)	Share of Solar Heated Housing Units / Occupied Solar Heated Housing Units (percent)	Occupied Housing Units Heated by Wood (number)	Housing Units Heated by Wood (percent)
1950	—	33.82%	9,460,560	22.09%	276,240	0.65%	0	0	975,435	2.28%	11,120,000	25.97%	1,567,686	3.66%	42,830,000	769,390	1.80%	NA	NA	4,171,690	9.74%
1960	6,455,565	12.18%	17,160,000	32.36%	933,023	1.76%	0	0	2,685,770	5.07%	22,850,000	43.10%	480,019	0.91%	53,020,000	223,015	0.42%	NA	NA	2,236,866	4.22%
1970	1,821,000	2.87%	16,470,000	25.96%	4,876,000	7.69%	0	0	3,807,000	6%	35,010,000	55.19%	395,000	0.62%	63,450,000	266,000	0.42%	NA	NA	794,000	1.25%
1973	800,000	1.15%	17,230,000	24.86%	7,213,000	10.40%	0	0	4,422,000	6.38%	38,460,000	55.47%	452,000	0.65%	69,340,000	151,000	0.22%	NA	NA	604,000	0.87%
1974	741,000	1.05%	16,840,000	23.77%	8,407,000	11.87%	0	0	4,143,000	5.85%	39,470,000	55.73%	484,000	0.68%	70,830,000	90,000	0.13%	NA	NA	658,000	0.93%
1975	573,000	0.79%	16,300,000	22.48%	9,173,000	12.65%	0	0	4,146,000	5.72%	40,930,000	56.44%	468,000	0.65%	72,520,000	78,000	0.11%	NA	NA	852,000	1.18%
1976	484,000	0.65%	16,450,000	22.23%	10,150,000	13.72%	0	0	4,239,000	5.73%	41,220,000	55.70%	463,000	0.63%	74,000,000	86,000	0.12%	NA	NA	912,000	1.23%
1977	454,000	0.60%	15,620,000	20.75%	11,150,000	14.81%	442,000	0.59%	4,182,000	5.56%	41,540,000	55.18%	507,000	0.67%	75,280,000	150,000	0.20%	NA	NA	1,239,000	1.65%
1978	402,000	0.52%	15,650,000	20.28%	12,260,000	15.89%	423,000	0.55%	4,130,000	5.35%	42,520,000	55.10%	597,000	0.77%	77,170,000	121,000	0.16%	NA	NA	1,066,000	1.38%
1979	358,000	0.46%	15,300,000	19.47%	13,240,000	16.86%	413,000	0.53%	4,134,000	5.26%	43,320,000	55.14%	570,000	0.72%	78,570,000	100,000	0.13%	NA	NA	1,140,000	1.45%
1980	329,000	0.41%	14,500,000	18.11%	14,210,000	17.74%	369,000	0.46%	4,170,000	5.21%	44,400,000	55.45%	613,000	0.77%	80,070,000	110,000	0.14%	NA	NA	1,380,000	1.72%
1981	361,000	0.43%	14,130,000	16.98%	15,490,000	18.62%	368,000	0.44%	4,165,000	5.01%	46,080,000	55.41%	589,000	0.71%	83,170,000	100,000	0.12%	NA	NA	1,890,000	2.28%
1983	432,000	0.51%	12,590,000	14.87%	15,680,000	18.53%	446,000	0.53%	3,869,000	4.57%	46,700,000	55.18%	677,000	0.80%	84,640,000	160,000	0.19%	NA	NA	4,090,000	4.83%
1985	447,000	0.51%	12,440,000	14.07%	18,360,000	20.76%	1,064,000	1.20%	3,584,000	4.05%	45,330,000	51.26%	530,000	0.60%	88,420,000	360,000	0.41%	50,000	0.06%	6,250,000	7.07%
1987	407,000	0.45%	12,740,000	14.02%	20,610,000	22.67%	1,075,000	1.18%	3,659,000	4.03%	45,960,000	50.57%	660,000	0.73%	90,890,000	280,000	0.31%	40,000	0.05%	5,450,000	6%
1989	335,000	0.36%	12,470,000	13.31%	23,060,000	24.62%	1,073,000	1.15%	3,664,000	3.91%	47,400,000	50.59%	663,000	0.71%	93,680,000	400,000	0.42%	40,000	0.04%	4,590,000	4.89%
1991	319,000	0.34%	11,470,000	12.32%	23,710,000	25.46%	989,000	1.06%	3,882,000	4.17%	47,020,000	50.48%	861,000	0.92%	93,150,000	410,000	0.44%	30,000	0.04%	4,440,000	4.77%
1993	297,000	0.31%	11,170,000	11.79%	25,110,000	26.51%	1,021,000	1.08%	3,922,000	4.14%	47,670,000	50.32%	911,000	0.96%	94,720,000	500,000	0.52%	30,000	0.03%	4,100,000	4.33%
1995	210,000	0.22%	10,980,000	11.23%	26,770,000	27.40%	1,055,000	1.08%	4,251,000	4.35%	49,200,000	50.37%	1,044,000	1.07%	97,690,000	640,000	0.65%	20,000	0.02%	3,530,000	3.62%

Notes:
LPG = liquefied petroleum gas.

Source: Energy Information Administration, U.S. Department of Energy. Washington, DC. From the Department of Energy's query system at http://tonto.eia.doe.gov/aer/, April 1998.

9.7. U.S. Electric Generating Capacity from Renewable Sources, 1992–97 (megawatts).

Source	1992	1993	1994	1995	1996	1997
Hydroelectric a	74,580	R 77,405	R 78,042	78,563	R 76,437	79,795
Geothermal	2,910	2,978	3,006	2,968	R 2,893	2,854
Biomass	9,701	10,045	10,465	R 10,263	R 10,531	10,702
Solar/Photovoltaic	339	340	333	333	333	334
Wind	1,823	1,813	1,745	1,731 b	R 1,677	1,620
Total Renewables	89,353	R 92,582	R 93,591	R 93,857	R 91,868	95,303
Nonrenewables c	656,563	R 662,373	R 670,423	R 675,660	R 684,004	683,210
Total	745,916	R 754,955	R 764,014	R 769,517	R 775,872	R 778,513

Notes:
a Excludes pumped storage, which is included in "Nonrenewables."
b Excludes 6.6 megawatts of utility capacity and 35 megawatts of nonutility capacity that were not captured by EIA sources.
c Includes hydrogen, sulfur, batteries, chemicals, spent sulfite liquore, and hydroelectric pumped storage.
Capacity ratings for nonrenewables have been revised to reflect estimated net summer capability rather than nameplate capacity.

Sources: Energy Information Administration, Form EIA-860, "Annual Electric Generator Report," and Form EIA-867, "Annual Nonutility Power Producer Report." Cited in U.S. Department of Energy, Energy Information Administration *Renewable Energy Annual 1997, Volume 1.* Washington, DC: U.S. Department of Energy, Energy Information Administration, 1997. Table 7. U.S. Department of Energy, Energy Information Administration. *Renewable Energy Annual 1998 With Data for 1997.* Washington, DC: U.S. Department of Energy, Energy Information Administration, 1998. Table 7.

9.8. Electricity Generation from Renewable Energy by Energy Source, United States, 1992–97 (thousand kilowatt hours).

Source	1992	1993	1994	1995	1996	1997
Industrial Sector (Gross Generation) [1]						
Biomass	R 53,606,891	R 55,745,781	57,391,594	R 57,513,666	R 57,937,058	55,886,586
Geothermal	R 8,577,891	9,748,634	10,122,228	9,911,659	R 10,197,514	9,110,297
Hydroelectric	R 9,446,439	11,510,786	13,226,934	14,773,801	R 16,555,389	17,904,653
Solar	R 746,277	896,796	823,973	824,193	R 902,830	892,892
Wind	R 2,916,379	3,052,416	3,481,616	3,185,006	R 399,642	3,384,576
Total	R 75,293,877	R 80,954,413	85,046,345	R 86,208,325	R 88,992,433	87,179,004
Electric Utility Sector (Net Generation) [2]						
Biomass	2,092,945	R 1,986,535	R 1,985,463	R 1,647,247	R 1,912,472	1,861,532
Geothermal	8,103,809	7,570,999	6,940,637	4,744,804	5,233,927	5,469,110
Conventional Hydroelectric	243,736,029	269,098,329	247,070,938	R 296,377,840	R 331,058,055	341,273,443
Solar	3,169	3,802	3,472	3,909	3,169	3,481
Wind	308	243	309	11,097	10,123	5,977
Total	253,936,260	R 278,659,908	R 256,000,819	R 302,784,897	R 338,217,746	348,613,543
Imports and Exports						
Geothermal (Imports)	889,864	877,058	1,172,117	884,950	649,514	10,313
Conventional Hydroelectric (Imports)	26,948,408	28,558,134	30,478,863	28,823,244	33,359,983	27,990,905
Conventional Hydroelectric (Exports)	3,254,289	3,938,973	2,806,712	3,059,261	2,335,340	6,790,778
Total Net Imports	24,583,983	25,496,219	28,844,268	26,648,933	31,573,157	21,210,440
Total Renewable Electricity Generation	R 353,814,120	R 385,110,540	R 369,891,432	R 415,642,155	R 458,883,336	457,002,987

Notes:
[1] Includes generation of electricity by cogenerators, independent power producers, and small power producers.
[2] Excludes imports.
R = Revised data.
Totals may not equal sum of components due to independent rounding.

Sources: U.S. Department of Energy, Energy Information Administration, Form EIA-759, "Monthly Power Plant Report; Form EIA-867, "Annual Nonutility Power Producer Report," and *Electric Power Monthly*, DOE/EIA-0226(75/03)(Washington, DC, March 1997). Personal communication with Dave Walker of Natural Resources Canada (Ottawa, Canada, March 1997). U.S. Department of Energy, Office of Fossil Energy, Form FE-781R, "Annual Report of International Electricity Export/Import Data." Cited in U.S. Department of Energy, Energy Information Administration. *Renewable Energy Annual 1997.* Volume 1. Washington, DC: U.S. Department of Energy, Energy Information Administration, 1997. Table 4. and U.S. Department of Energy, Energy Information Administration. *Renewable Energy Annual 1998 With Data for 1997.* Washington, DC: U.S. Department of Energy, Energy Information Administration, 1998. Table 4.

9.9. Nuclear Power Plants, United States, Various Statistics, 1957–97.

Year	Nuclear Electric Power Plants Capacity Factor (percent)	Net Generation of Electricity from Nuclear Electric Power at Electric Utilities (billion killowatt hours)	Share of U.S. Net Generation of Electricity from Nuclear Electric Power (percent)	Nuclear Reactor Units, Operable Number
1957	0	0.0097	0.002%	1
1958	0	0.1647	0.026%	1
1959	0	0.1881	0.026%	1
1960	0	0.5182	0.069%	3
1961	0	1.6921	0.213%	3
1962	0	2.2697	0.266%	5
1963	0	3.2118	0.350%	6
1964	0	3.3427	0.340%	6
1965	0	3.6567	0.347%	6
1966	0	5.5199	0.482%	8
1967	0	7.6552	0.630%	10
1968	0	12.5284	0.942%	11
1969	0	13.9278	0.966%	14
1970	0	21.8044	1.423%	18
1971	0	38.1045	2.363%	21
1972	0	54.0911	3.092%	29
1973	53.5%	83.4795	4.486%	39
1974	47.8%	113.9757	6.104%	48
1975	55.9%	172.5051	8.996%	54
1976	54.7%	191.1035	9.378%	61
1977	63.3%	250.8833	11.810%	65
1978	64.5%	276.4031	12.528%	70
1979	58.4%	255.1546	11.353%	68
1980	56.3%	251.1156	10.983%	70
1981	58.2%	272.6735	11.882%	74
1982	56.6%	282.7732	12.617%	77
1983	54.4%	293.6771	12.712%	80
1984	56.3%	327.6335	13.559%	86
1985	58.0%	383.6907	15.535%	95
1986	56.9%	414.0381	16.646%	100
1987	57.4%	455.2704	17.700%	107
1988	63.5%	526.9730	19.487%	108
1989	62.2%	529.3547	19.012%	110
1990	66.0%	576.8617	20.542%	111
1991	70.2%	612.5651	21.684%	111
1992	70.9%	618.7763	22.121%	109
1993	70.5%	610.2912	21.172%	109
1994	73.8%	640.4398	22.003%	109
1995	77.4%	673.4021	22.488%	109
1996	76.4%	674.7786	21.923%	110
1997	70.8%	629.4200	20.140%	—

Source: Energy Information Administration, U.S. Department of Energy. Washington, DC. From the Department of Energy's query system at http://tonto.eia.doe.gov/aer/, April 1998.

9.10. Motor Gasoline, All Types, Retail Price (Real), United States, 1978–97 (cents per gallon).

Year	Cents per Gallon
1978	128.1
1979	159.8
1980	202.5
1981	205.0
1982	182.5
1983	167.3
1984	157.8
1985	152.4
1986	115.5
1987	115.2
1988	111.8
1989	118.2
1990	130.0
1991	122.9
1992	119.0
1993	114.3
1994	111.7
1995	111.8
1996	116.9
1997	114.9

Source: Energy Information Administration, U.S. Department of Energy. Washington, DC. From the Department of Energy's query system at http://tonto.eia.doe.gov/aer/, April 1998.

9.11. Motor Vehicle Fuel Consumption Rate, United States, 1960–96.

Year	Gallons per Vehicle Consumed	Motor Vehicle Annual Fuel Rate in Miles per Gallon	Passeger Car Fuel Rate in Miles per Gallon	Miles per Vehicle
1960	784	12.42	14.28	9,732
1961	781	12.44	14.38	9,708
1962	779	12.43	14.37	9,687
1963	780	12.48	14.26	9,737
1964	787	12.47	14.25	9,805
1965	787	12.48	14.27	9,826
1966	780	12.40	14.11	9,675
1967	786	12.40	14.10	9,751
1968	805	12.25	13.87	9,864
1969	821	12.05	13.62	9,885
1970	830	12.02	13.52	9,976
1971	839	12.08	13.60	10,133
1972	857	11.99	13.50	10,279
1973	850	11.89	13.40	10,099
1974	788	12.05	13.60	9,493
1975	790	12.18	14.00	9,627
1976	806	12.12	13.80	9,774
1977	814	12.26	14.10	9,978
1978	816	12.35	14.04	10,077
1979	776	12.52	14.41	9,722
1980	712	13.29	15.46	9,458
1981	697	13.57	15.94	9,477
1982	686	14.07	16.65	9,644
1983	686	14.24	17.14	9,760
1984	691	14.49	17.83	10,017
1985	685	14.62	17.50	10,020
1986	692	14.66	17.40	10,143
1987	694	15.07	18.00	10,453
1988	688	15.58	18.80	10,721
1989	688	15.90	19.00	10,932

9.11. Motor Vehicle Fuel Consumption Rate, United States, 1960–96 *(continued)*.

Year	Gallons per Vehicle Consumed	Motor Vehicle Annual Fuel Rate in Miles per Gallon	Passeger Car Fuel Rate in Miles per Gallon	Miles per Vehicle
1990	677	16.40	20.30	11,107
1991	669	16.90	21.20	11,294
1992	683	16.91	21.00	11,558
1993	693	16.73	20.60	11,595
1994	698	16.74	20.80	11,683
1995	700	16.91	21.10	11,793
1996p	698	16.90	21.30	11,807

Notes:
p = preliminary.

Source: 1960–94, Federal Highway Administration, *Highway Statistics Summary to 1995*. 1995 foreword, Federal Highway Administration, *Highway Statistics*, annual. Cited in U.S. Department of Energy, Energy Information Administration. *Annual Energy Review 1997*. Washington, DC: Energy Information Administration, 1997. Based on Table 2.9. http://www.eia.doe.gov/emeu/aer/.

9.12. U.S. Energy Prices, by Type of Fuel, 1970–94 (current dollars per million Btu).

Year	Consumer Price Estimates for Energy—Biofuels (Dollars per) (million Btu)	Consumer Price Estimates for Energy—Coal (Dollars per) (million Btu)	Consumer Price Estimates for Energy—Distillate Fuel (Dollars per) (million Btu)	Consumer Price Estimates for Energy Electricity Purchased by End Users (Dollars per) (million Btu)	Consumer Price Esitmates for Energy—Jet Fuel (Dollars per) (million Btu)	Consumer Price Estimates for Energy Liquid Petroleum Gas (Dollars per) (million Btu)	Consumer Price Estimates for Energy—Motor Gasoline (Dollars per) (million Btu)	Consumer Price Estimates for Energy—Natural Gas (Dollars per) (million Btu)	Consumer Price Estimates for Energy—Nuclear Fuel (Dollars per) (million Btu)	Consumer Price Estimates for Energy—Total Petroleum (Dollars per) (million Btu)
1970	$0.65	$0.37	$1.16	$4.99	$0.73	$1.46	$2.85	$0.59	$0.18	$1.71
1971	$0.69	$0.42	$1.22	$5.30	$0.77	$1.49	$2.90	$0.63	$0.18	$1.78
1972	$0.72	$0.45	$1.22	$5.54	$0.79	$1.52	$2.88	$0.68	$0.18	$1.77
1973	$0.77	$0.48	$1.46	$5.86	$0.92	$2.02	$3.10	$0.73	$0.19	$1.96
1974	$0.84	$0.88	$2.44	$7.42	$1.58	$2.81	$4.32	$0.89	$0.20	$3.04
1975	$0.92	$1.03	$2.60	$8.61	$2.05	$2.97	$4.65	$1.18	$0.24	$3.33
1976	$0.98	$1.04	$2.77	$9.13	$2.25	$3.21	$4.84	$1.46	$0.25	$3.45
1977	$1.04	$1.11	$3.11	$10.11	$2.59	$3.65	$5.13	$1.76	$0.27	$3.71
1978	$1.12	$1.28	$3.26	$10.92	$2.87	$3.60	$5.24	$1.95	$0.30	$3.82
1979	$1.56	$1.36	$4.69	$11.78	$3.90	$4.50	$7.11	$2.31	$0.34	$5.20
1980	$1.74	$1.47	$6.70	$13.95	$6.36	$5.64	$9.84	$2.86	$0.43	$7.35
1981	$1.24	$1.65	$8.03	$16.14	$7.57	$6.18	$10.94	$3.43	$0.48	$8.68
1982	$1.28	$1.73	$7.78	$18.16	$7.23	$6.66	$10.39	$4.23	$0.54	$8.40
1983	$1.12	$1.71	$7.32	$18.62	$6.53	$7.17	$9.12	$4.72	$0.58	$7.78
1984	$1.28	$1.71	$7.36	$18.50	$6.25	$6.93	$8.89	$4.75	$0.67	$7.68
1985	$0.79	$1.70	$7.18	$19.05	$5.91	$6.33	$9.01	$4.61	$0.71	$7.61
1986	$0.32	$1.62	$5.66	$19.06	$3.92	$6.21	$6.79	$4.07	$0.70	$5.72
1987	$0.95	$1.54	$5.94	$18.74	$4.03	$5.85	$7.22	$3.77	$0.71	$6.01
1988	$0.87	$1.50	$5.80	$18.68	$3.80	$5.65	$7.32	$3.78	$0.73	$5.89
1989	$0.67	$1.49	$6.45	$18.98	$4.39	$5.35	$8.01	$3.85	$0.70	$6.42
1990	$1.94	$1.49	$7.70	$19.33	$5.68	$6.51	$9.12	$3.85	$0.67	$7.46
1991	$2.00	$1.49	$7.28	$19.85	$4.83	$6.54	$8.93	$3.78	$0.63	$7.18
1992	$2.01	$1.46	$7.11	$20.06	$4.52	$5.95	$8.96	$3.89	$0.59	$7.07
1993	$1.91	$1.43	$7.10	$20.38	$4.29	$5.97	$8.82	$4.16	$0.56	$7.00
1994	$1.88	$1.40	$7.03	$20.34	$3.95	$6.22	$8.91	$4.15	$0.56	$7.02

Notes:
There is a discontinuity between 1989 and 1990 due to expanded coverage of nonelectricity utility use of biofuels beginning in 1990.

Source: U.S. Department of Energy, Energy Information Administration. *Annual Energy Review 1997*. Washington, DC: U.S. Department of Energy, Energy Information Administration, 1997. Table 3–3. http://www.eia.doe.gov/emeu/aer/

9.13. Shipments of Solar Collectors by Market Sector, End Use, and Type, United States, 1995–97 (Detail is for 1997)   (thousand square feet).

Type	Low-Temperature Liquid/Air		Medium-Temperature Air			High-Temperature Liquid		1995 Total	1996 Total	1997 Total
	Metallic	Nonmetallic	ICS/ Thermosiphon	Flat-Plate (Pumped)	Evacuated Tube	Concentrator	Parabolic Dish/Trough			
Market Sector										
Residential	6,791	53	24	491	1	0	0	6,966	6,873	7,360
Commercial	726	1	9	25	0	0	7	604	682	768
Industrial	7	0	0	0	0	0	0	82	54	7
Utility	0	0	0	0	0	0	0	9	*	0
Other /a	0	1	0	0	1	0	0	6	7	2
Total	7,524	55	33	516	2	0	7	7,666	7,616	8,137
End Use										
Pool Heating	7,517	0	0	4	0	0	0	6,763	6,787	7,528
Hot Water	0	52	7	509	1	0	7	755	765	595
Space Heating	7	2	26	0	0	0	0	132	57	10
Space Cooling	0	0	0	0	0	0	0	1	0	0
Combined Space and Water Heating	0	0	0	3	0	0	0	2	3	3
Process Heating	0	0	0	0	0	0	0	*	4	0
Electricity Generation	0	0	0	0	1	0	0	10	*	0
Other /b	0	0	0	0	0	0	0	2	0	1
Total	7,524	54	33	516	2	0	7	7,666	7,616	8,137

Notes:

*Less than 500 square feet

/a = Other market sectors include shipments of solar thermal collectors to other sectors such as government, including the military but excluding space applications.

/b = Other end use includes shipments of solar thermal collectors for other uses such as cooking, water pumping, water purification, desalinization, distilling, etc.

ICS = Integral Collector Storage.

Totals may not equal sum of components due to independent rounding.

Source: Energy Information Administration, Form EIA-63A, "Annual Solar Thermal Collector Manufacturers Survey." 1995 and 1996 data cited in U.S. Department of Energy, Energy Information Administration. *Renewable Energy Annual 1997, Volume 1.* Washington, DC. Table 16. 1997 data comes from selected preliminary renewable energy tables at the Energy Information Administration Web site, ftp://ftp.eia.doe.gov/ pub/solar.renewables/solar.txt

9.14. Shipments of Photovoltaic Modules and Cells by Market Sector, End Use, and Type, United States, 1995–97 (peak kilowatts).

Sector and End Use	Crystalline Silicon[a]	Thin-Film Silicon	Concentrator Silicon	Other	1995 Total	1996 Total	1997 Total
Market Sector							
Industrial	11,244	504	0	0	7,198	8,300	11,748
Residential	10,691	300	2	0	6,272	8,475	10,993
Commercial	7,621	340	150	0	8,100	5,176	8,111
Transportation	3,378	196	0	0	2,383	3,995	3,574
Utility	5,331	320	0	0	3,759	4,753	5,651
Government[b]	3,772	135	2	0	2,000	3,126	3,909
Other[c]	2,276	91	0	0	1,347	1,639	2,367
Total	44,314	1,886	154	0	31,059	35,464	46,354
End Use							
Electricity Generation:							
Grid Interactive	7,402	871	0	0	4,585	4,844	8,273
Remote	8,433	195	2	0	8,233	10,844	8,630
Communications	7,289	94	0	0	5,154	6,041	7,383
Consumer Goods	72	275	0	0	1,025	1,063	347
Transportation	6,645	60	0	0	4,203	5,196	6,705
Water Pumping	3,748	35	0	0	2,727	3,261	3,783
Cells/Modules to OEM	4,984	261	0	0	3,188	2,410	5,245
Health	1,267	36	0	0	776	977	1,303
Other[d]	4,473	59	152	0	1,170	789	4,684
Total	44,314	1,886	154	0	31,059	35,464	46,354

Notes:
[a] = Includes single-crystal and cast and ribbon types.
[b] = Includes federal, state, and local governments, excluding military.
[c] = Other includes shipments that are manufactured for private contractors for research and development projects.
[d] = Other uses include shipments of photovoltaic modules and cells for other uses, such as cooking food, desalinitation, distilling, etc.
OEM = Original equipment manufacturers.

Source: Energy Information Administration, Form EIA-63B, "Annual Photovoltaic Module/Cell Manufacturers Survey." Cited in U.S. Department of Energy, Energy Information Administration. *Renewable Energy Annual 1997, Volume 1.* Washington, DC: Energy Information Administration, 1997. Table 29. And U.S. Department of Energy, Energy Information Administration. *Renewable Energy Annual 1998 with Data for 1997.* Washington, DC: Energy Information Administration, 1998. Table 30.

9.15. Annual Photovoltaic and Solar Thermal Shipments, 1977–97.

	Domestic Shipments [1]	
Year	Photovoltaic Modules and Cells (Peak Kilowatts)	Solar Thermal Collectors (Thousand Square Feet)
1977	NA	10,312
1978	NA	10,020
1979	NA	13,396
1980	NA	18,283
1981	NA	19,362
1982	6,897	18,166
1983	10,717	16,669
1984	7,759	16,843
1985	4,099	19,166 [2]
1986	3,224	9,136
1987	3,029	7,087
1988	4,318	8,016
1989	5,462	11,021
1990	6,293	11,164

9.15. Annual Photovoltaic and Solar Thermal Shipments, 1977–97 *(continued)*.

	Domestic Shipments [1]	
Year	Photovoltaic Modules and Cells (Peak Kilowatts)	Solar Thermal Collectors (Thousand Square Feet)
1991	6,035	6,242
1992	5,760	6,770
1993	6,137	6,557
1994	8,363	7,222
1995	11,188	7,136
1996	13,016	7,162
1997	12,561	7,759
Total	114,858	227,177

Notes:
[1] Total shipments minus export shipments.
[2] Estimated data
NA = Not available.

Sources: 1977: Federal Energy Administration telephone survey. 1978–84: Energy Information Administration, Form EIA-63, "Annual Solar Thermal Collector and Photovoltaic Module Manufacturers Survey." 1985–97: Energy Information Administration, Form EIA-63A, "Annual Solar Thermal Collector Manufacturers Survey," and Form EIA-63B, "Annual Photovoltaic Module/Cell Manufacturers Survey." Cited in U.S. Department of Energy, Energy Information Administration. *Renewable Energy Annual 1997, Volume 1,* and *Renewable Energy Annual 1998 with Data for 1997.* Washington, DC: Energy Information Administration, 1997 and 1998. Tables 10 and 11, respectively.

9.16. Number of Lights by Bulb Type by Room (Residential), United States, 1993.

		Incandescent				Fluorescent			Other	
Room	Total	Low (10–40 watts)	Medium (41–149 watts)	High (150 watts or more)	Unknown	Short	Long	Compact	Halogen	Other/ Unknown
Total	4,196	431	2,811	409	14	159	173	34	24	141
Bathroom	621	44	417	81	4	35	14	2	1	23
Bedroom	1,121	119	868	66	6	8	5	9	8	32
Dining Room	218	28	119	51	2	0	1	1	2	14
Den/Family/ Rec Room	279	28	171	42	1	7	12	2	3	13
Hallway/Stairs	193	42	136	6	0	1	3	0	0	5
Kitchen	820	80	440	70	1	85	104	12	1	27
Living Room	711	65	511	80	0	11	8	8	8	20
Laundry Room/ Other	233	25	149	13	0	12	26	0	1	7

Notes:
These data are from the 474 households included in the Lighting Supplement to the Residential Energy Consumption Survey. The supplement was not designed to weight the data to the population level.

Source: U.S. Department of Energy, Energy Information Administration, Office of Energy Markets and End Use. *Residential Lighting Use and Potential Savings.* Washington, DC, 1996. Table 4.20.

ENVIRONMENT

9.17. Number and Types of National Priorities List (NPL) Cleanup Sites in Construction Completion Stage as of September 30, 1996, United States.

Type	Number
Industrial Waste	124
Landfill	123
Manufacturing Plant	87
Ground Water	70
Inorganic Waste	36
Chemical Plants	35
Wells	23
Lagoons	20
Waterways/Creeks/Rivers	14
Housing Area/Farm	12
Mines/Tailings	9
Military-related	7
Radioactive Site	3
Other	45
Total	410

Notes:
There are 410 sites total, but many sites need more than one type of technology.

Source: U.S. Environmental Protection Agency. Washington, DC: From the Environmental Protection Agency's Web site at http://www.epa.gov/superfund/accomp/400/stats.htm, May 1998.

9.18. Cleanup Technologies Used at U.S. Superfund's 410 Construction Completion Sites as of September 30, 1996.

Site Cleanup Methods*	Sites
Excavation and Removal	188
Surface Capping/Soil Cover	161
Surface Drainage Control	51
Backfilling	61
Solidification/Stabilization & Immobilization	30
Treatment	
Ground Water Pump & Treat	142
Air Stripping	47
Incineration	
On Site	16
Off Site	20
Innovative Technologies	
Soil Vapor Extraction	33
Bioremediation	12
Thermal Desorption	4
Dechlorination	3
In-Situ Flushing	3
Soil Washing	2
Other Actions	
Ground Water Monitoring/Wells	293
Institutional Controls	153
Alternate Water supplies	56

Notes:
* = More than one technology may be associated with a construction completion site.
   At construction completion sites, EPA employed more than two dozen different types of cleanup approaches that are tailored both to the types of contaminants and the natural resources that are polluted (such as soil and ground water). For contaminated soils, "excavation and removal" was the most common method used. This method commonly removes polluted soil and debris by trucking it from a site and treating it at a licensed hazardous waste facility. The technology most often used at contaminated ground water sites was "pump and treat." This method pumps water out of the ground through a series of wells, cleans it by treating the contaminants, and either reinjects it back into the ground, discharges it into surface water, or sends it to a municipal water treatment plant. Most Superfund sites contain more than one type of chemical, and often both water and soil are affected. When this is the case, EPA may use a combination of solutions including one or more of the following: 1. Containing the contaminants (surface drainage control, soil capping, solidification. 2. Separating harmful chemicals from the soil or water (soil vapor extraction, air stripping, carbon adsorption, soil flushing, thermal desorption). 3. Rendering the material less toxic (bioremediation, incineration). Innovative technologies recently added include: 1. Flushing chemicals from soils while the soils remain in place. 2. Heating soil to vaporize contaminants and capture them. 3. Introducing microorganisms such as bacteria and fungi to break down hazardous chemicals into less-harmful substances.

Source: U.S. Environmental Protection Agency. Construction *Completion Statistics, Site Leads and Technologies Used as of September 30, 1996.* Washington, DC: From the Environmental Protection Agency's Web site *at* http://www.epa.gov/superfund/accomp/400/stats.htm, May 1998.

9.19. Quantity of RCRA [1] Hazardous Waste Generated, and Number of Hazardous Waste Generators, by State, 1993, 1995, and 1997.

State	1993 Rank	1993 Tons Generated	1993 Percentage	1993 Rank	1993 Number of Large-Quantity Generators	1993 Percentage	1995 Rank	1995 Tons Generated	1995 Percentage	1995 Rank	1995 Number of Large-Quantity Generators	1995 Percentage	1997 Rank	1997 Tons Generated	1997 Percentage	1997 Rank	1997 Number of Large-Quantity Generators	1997 Percentage
Alabama	26	779,645	0.3%	26	295	1.2%	17	1,409,582	0.7%	24	279	1.3%	16	457,492	1.1%	24	267	1.4%
Alaska	50	5,534	0	43	75	0.3%	51	3,432	0	43	64	0.3%	46	4,547	0.0%	43	50	0.3%
Arizona	41	46,913	0	27	233	1.0%	41	66,865	0	29	199	1.0%	56	107	0.0%	53	4	0.0%
Arkansas	25	794,801	0.3%	32	162	0.7%	20	992,794	0.5%	28	204	1.0%	10	972,059	2.4%	26	206	1.1%
California	7	14,055,553	5.4%	3	1,872	7.7%	6	11,109,924	5.2%	2	1,640	7.9%	14	494,914	1.2%	6	1,007	5.4%
Colorado	23	1,079,332	0.4%	35	146	0.6%	36	169,554	0.1%	32	156	0.7%	22	263,548	0.6%	31	160	0.9%
Connecticut	21	1,169,205	0.5%	17	441	1.8%	32	295,928	0.1%	18	395	1.9%	34	64,714	0.2%	14	404	2.2%
Delaware	42	22,173	0	44	71	0.3%	44	22,263	0	43	64	0.3%	39	19,369	0.0%	41	66	0.4%
District of Columbia	55	628	0	52	15	0.1%	54	764	0	49	18	0.1%	53	499	0.0%	49	20	0.1%
Florida	34	213,888	0.1%	18	438	1.8%	31	368,904	0.2%	17	418	2.0%	19	398,535	1.0%	17	378	2.0%
Georgia	24	921,076	0.4%	18 (sic)	438 (sic)	1.8%	28	459,543	0.2%	16	430	2.1%	21	276,342	0.7%	14	404	2.2%
Guam	51	2,453	0	53	14	0.1%	55	299	0	53	13	0.1%	54	190	0.0%	54	3	0.0%
Hawaii	53	1,774	0	48	44	0.2%	24	592,900	0.3%	45	53	0.3%	45	7,918	0.0%	46	42	0.2%
Idaho	20	1,255,865	0.5%	47	57	0.2%	18	1,209,841	0.6%	46	52	0.2%	8	1,014,825	2.5%	44	48	0.3%
Illinois	8	12,494,369	4.8%	6	1,238	5.1%	5	12,756,271	6.0%	6	1,156	5.5%	3	1,914,354	4.6%	4	1,047	5.6%
Indiana	14	1,751,572	0.7%	10	683	2.8%	14	1,733,026	0.8%	10	609	2.9%	7	1,077,849	2.6%	8	628	3.4%
Iowa	37	158,908	0.1%	28	196	0.8%	42	39,329	0	30	170	0.8%	18	434,895	1.1%	29	182	1.0%
Kansas	12	3,144,665	1.2%	25	297	1.2%	15	1,722,380	0.8%	27	210	1.0%	6	1,363,140	3.3%	25	214	1.1%
Kentucky	31	397,488	0.2%	16	472	1.9%	19	1,149,881	0.5%	15	440	2.1%	24	192,318	0.5%	20	348	1.9%
Louisiana	3	31,715,905	12.3%	23	347	1.4%	3	17,460,601	8.2%	21	359	1.7%	2	4,510,658	10.9%	18	363	1.9%
Maine	47	8,651	0	34	148	0.6%	45	19,459	0	34	144	0.7%	13	585,338	1.4%	34	136	0.7%
Maryland	33	308,621	0.1%	14	566	2.3%	30	448,707	0.2%	25	221	1.1%	36	34,150	0.1%	30	173	0.9%
Massachusetts	36	163,037	0.1%	13	569	2.3%	23	610,135	0.3%	13	476	2.3%	27	127,462	0.3%	12	455	2.4%
Michigan	4	21,014,255	8.1%	8	789	3.2%	4	13,446,389	6.3%	9	718	3.4%	9	994,047	2.4%	7	681	3.6%
Minnesota	11	5,993,221	2.3%	24	300	1.2%	39	77,720	0	23	284	1.4%	12	588,656	1.4%	23	272	1.5%
Mississippi	13	1,882,053	0.7%	31	163	0.7%	16	1,579,260	0.7%	33	152	0.7%	5	1,654,338	4.0%	28	193	1.0%
Missouri	28	528,922	0.2%	20	415	1.7%	27	508,963	0.2%	22	354	1.7%	29	116,796	0.3%	19	357	1.9%
Montana	44	11,282	0	45	60	0.2%	50	7,668	0	46	52	0.2%	41	12,266	0.0%	45	47	0.3%
Navajo Nation	56	245	0	54	9	0	56	195	0	54	11	0.1%	55	150	0.0%	52	6	0.0%
Nebraska	40	90,471	0	40	96	0.4%	37	99,702	0	40	86	0.4%	38	23,727	0.1%	40	68	0.4%
Nevada	45	10,773	0	41	82	0.3%	48	11,354	0	41	80	0.4%	42	11,820	0.0%	37	90	0.5%
New Hampshire	43	17,249	0	33	158	0.6%	46	15,169	0	35	130	0.6%	44	9,751	0.0%	32	152	0.8%
New Jersey	5	17,977,002	7.0%	1	3,120	12.8%	7	10,342,432	4.8%	5	1,178	5.6%	23	243,743	0.6%	10	507	2.7%
New Mexico	35	176,409	0.1%	45	60	0.2%	35	204,494	0.1%	48	44	0.2%	30	99,474	0.2%	47	39	0.2%
New York	16	1,498,421	0.6%	2	2,036	8.4%	11	2,306,232	1.1%	1	2,144	10.3%	17	452,373	1.1%	1	2,772	14.8%
North Carolina	30	447,718	0.2%	11	623	2.6%	33	286,339	0.1%	11	587	2.8%	33	66,501	0.2%	11	505	2.7%
North Dakota	27	594,815	0.2%	51	16	0.1%	25	520,226	0.2%	52	16	0.1%	49	2,686	0.0%	50	16	0.1%
Ohio	15	1,739,928	0.7%	4	1,524	6.3%	13	1,823,547	0.9%	3	1,373	6.6%	4	1,656,470	4.0%	2	1,258	6.7%
Oklahoma	22	1,145,732	0.4%	29	193	0.8%	26	511,918	0.2%	31	168	0.8%	20	313,355	0.8%	33	143	0.8%
Oregon	17	1,392,152	0.5%	30	184	0.8%	40	68,187	0	26	220	1.1%	35	49,877	0.1%	27	203	1.1%
Pennsylvania	9	9,441,256	3.7%	7	1,215	5.0%	9	6,446,730	3.0%	7	1,134	5.4%	15	491,547	1.2%	5	1,011	5.4%
Puerto Rico	18	1,373,639	0.5%	36	109	0.4%	21	900,567	0.4%	39	88	0.4%	37	29,850	0.1%	39	77	0.4%
Rhode Island	46	10,169	0	39	102	0.4%	43	25,428	0	37	112	0.5%	40	16,362	0.0%	36	107	0.6%
South Carolina	32	310,399	0.1%	21	388	1.6%	34	261,015	0.1%	19	371	1.8%	43	10,793	0.0%	21	341	1.8%

9.19. Quantity of RCRA [1] Hazardous Waste Generated, and Number of Hazardous Waste Generators, by State, 1993, 1995, and 1997 (continued).

State	1993						1995						1997					
	Rank	Tons Generated	Percentage	Rank	Number of Large-Quantity Generators	Percentage	Rank	Tons Generated	Percentage	Rank	Number of Large-Quantity Generators	Percentage	Rank	Tons Generated	Percentage	Rank	Number of Large-Quantity Generators	Percentage
South Dakota	54	767	0	50	24	0.1%	53	1,119	0	50	17	0.1%	52	948	0.0%	48	21	0.1%
Tennessee	2	33,937,638	13.1%	15	518	2.1%	2	38,686,622	18.1%	14	467	2.2%	11	690,519	1.7%	13	420	2.2%
Texas	1	63,435,688	24.6%	5	1,286	5.3%	1	68,513,285	32.0%	4	1,329	6.4%	1	18,973,406	45.9%	3	1,219	6.5%
Trust Territories	49	6,045	0	55	3	0	47	15,134	0	55	3	0	51	1,101	0.0%	54	3	0.0%
Utah	38	104,623	0	37	106	0.4%	29	456,847	0.2%	38	101	0.5%	31	78,555	0.2%	38	89	0.5%
Vermont	48	8,337	0	41	82	0.3%	49	10,497	0	42	66	0.3%	47	4,064	0.0%	42	65	0.3%
Virgin Islands	52	2,049	0	56	2	0	52	3,329	0	56	1	0	48	2,811	0.0%	56	2	0.0%
Virginia	39	96,850	0	22	379	1.6%	38	98,678	0	19	371	1.8%	32	74,340	0.2%	22	327	1.7%
Washington	6	14,397,985	5.6%	9	766	3.1%	10	3,088,487	1.4%	8	748	3.6%	28	126,601	0.3%	9	595	3.2%
West Virginia	10	8,471,643	3.3%	37	106	0.4%	8	8,489,828	4%	36	117	0.6%	26	147,213	0.4%	35	118	0.6%
Wisconsin	29	522,523	0.2%	12	605	2.5%	22	664,609	0.3%	12	558	2.7%	25	148,403	0.4%	16	400	2.1%
Wyoming	19	1,316,689	0.5%	49	26	0.1%	12	1,972,177	0.9%	50	17	0.1%	50	1,478	0.0%	51	15	0.1%
CBI Data [2]	N/A	N/A	N/A	N/A	N/A	N/A	N/A	5,977	N/A	N/A	6	N/A	N/A	N/A	N/A	N/A	N/A	N/A
Total		258,449,001	100%		24,362	100%		214,092,505	100%		20,873	100%		41,309,241	100%		18,724	100%

Notes:

[1] RCRA = Resource Conservation and Recovery Act of 1976.

[2] Some respondents from Georgia have submitted Confidential Business Information (CBI) pursuant to 40 CFR 260.2(b) (a section of the Code of Federal Regulations). CBI has been incorporated into this report wherever possible.

Columns may not add up due to rounding.

Percentages do not include CBI data.

Changes to the 1997 biennial reporting requirements will make comparisons of the 1997 National Biennial Report to previous years' National Biennial Reports misleading. However, comparisons between the underlying Biennial Reporting System data can be done with sufficient care.

Source: U.S. Environmental Protection Agency. *National Biennial RCRA Hazardous Waste Reports.* Various years. And U.S. Environmental Protection Agency. *Preliminary National Biennial RCRA Hazardous Waste Report:* Washington, DC: U.S. Environmental Protection Agency, RCRA Information Center. Based on 1997 Data. http://www.epa.gov/epaoswer/hazwaste/data/index.htm

9.20. 1997 CERCLA (Comprehensive Environmental Response, Compensation, and Liability Act) Priority List of Hazardous Substances.

1997 Rank	Name	Total Points*	1995 Rank	CAS Number**
1	Arsenic	1627.32	2	007440-38-2
2	Lead	1522.67	1	007439-92-1
3	Mercury	1492.21	3	007439-97-6
4	Vinyl chloride	1397.67	4	000075-01-4
5	Benzene	1372.88	5	000071-43-2
6	Polychlorinated biphenyls	1340.07	6	001336-36-3
7	Cadmium	1315.14	7	007440-43-9
8	Benzo(a)pyrene	1285.71	8	000050-32-8
9	Benzo(b)fluoranthene	1257.85	10	000205-99-2
10	Polycyclic aromatic hydrocarbons	1251.50	New	130498-29-2
11	Chloroform	1249.95	9	000067-66-3
12	Aroclor 1254	1186.77	14	011097-69-1
13	DDT, P,P'-	1183.50	11	000050-29-3
14	Aroclor 1260	1182.97	12	011096-82-5
15	Trichloroethylene	1160.83	13	000079-01-6
16	Chromium, hexavalent	1152.07	15	018540-29-9
17	Dibenzo(a,h)anthracene	1141.98	17	000053-70-3
18	Dieldrin	1130.30	20	000060-57-1
19	Hexachlorobutadiene	1129.40	18	000087-68-3
20	Chlordane	1121.25	16	000057-74-9
21	Creosote	1121.05	23	008001-58-9
22	DDE, P,P'-	1117.44	22	000072-55-9
23	Benzidine	1116.19	26	000092-87-5
24	Cyanide	1113.09	24	000057-12-5
25	Aldrin	1108.37	29	000309-00-2
26	DDD, P,P'-	1104.19	19	000072-54-8
27	Aroclor 1248	1098.88	27	012672-29-6
28	Phosphorus	1094.63	30	007723-14-0
29	Aroclor 1242	1091.22	25	053469-21-9
30	Heptachlor	1087.81	39	000076-44-8
31	Toxaphene	1084.63	32	008001-35-2
32	Hexachlorocyclohexane, gamma-	1084.48	21	000058-89-9
33	Tetrachloroethylene	1082.21	28	000127-18-4
34	Aroclor 1221	1070.41	34	011104-28-2
35	Hexachlorocyclohexane, beta-	1056.50	36	000319-85-7
36	1,2-Dibromoethane	1054.20	31	000106-93-4
37	Disulfoton	1052.88	35	000298-04-4
38	Benzo(a)anthracene	1047.78	37	000056-55-3
39	Endrin	1040.58	38	000072-20-8
40	Aroclor 1016	1031.03	43	012674-11-2
41	Hexachlorocyclohexane, delta-	1029.75	44	000319-86-8
42	Di-n-butyl phthalate	1028.68	40	000084-74-2
43	1,2-Dibromo-3-chloropropane	1027.45	33	000096-12-8
44	Beryllium	1023.27	46	007440-41-7
45	Pentachlorophenol	1022.95	45	000087-86-5
46	Carbon tetrachloride	1018.10	42	000056-23-5
47	Cobalt	1011.99	148	007440-48-4
48	Endosulfan, alpha	1010.08	52	000959-98-8
49	Nickel	1004.94	49	007440-02-0
50	Endosulfan sulfate	1004.02	50	001031-07-8
51	3,3'-Dichlorobenzidine	998.02	53	000091-94-1
52	Heptachlor epoxide	992.17	54	001024-57-3
53	Endosulfan	990.07	58	000115-29-7
54	Dibromochloropropane	987.92	59	067708-83-2
55	Aroclor	986.15	63	012767-79-2

9.20. 1997 CERCLA (Comprehensive Environmental Response, Compensation, and Liability Act) Priority List of Hazardous Substances *(continued)*.

1997 Rank	Name	Total Points*	1995 Rank	CAS Number**
56	Endosulfan, beta	981.77	61	033213-65-9
57	Endrin ketone	978.82	56	053494-70-5
58	Aroclor 1232	974.72	60	011141-16-5
59	Benzo(k)fluoranthene	972.84	New	000207-08-9
60	2-Hexanone	970.35	65	000591-78-6
61	Toluene	969.44	55	000108-88-3
62	Cis-chlordane	959.77	66	005013-71-9
63	Methane	956.21	67	000074-82-8
64	Trans-chlordane	951.80	62	005103-74-2
65	Zinc	932.34	68	007440-66-6
66	2,3,7,8-Tetrachlorodibenzo-p-dioxin	928.58	69	001746-01-6
67	Di(2-ethylhexyl)phthalate	923.15	70	000117-81-7
68	Methoxychlor	916.15	51	000072-43-5
69	Chromium	908.93	72	007440-47-3
70	Methylene chloride	908.12	71	00075-09-2
71	Benzofluoranthene	907.82	48	056832-73-6
72	Naphthalene	902.82	73	000091-20-3
73	1,1-Dichloroethene	894.46	74	000075-35-4
74	1,2-Dichloroethane	871.62	77	000107-06-2
75	Cyclotrimethylenetrinitramine (rdx)	871.35	230	000121-82-4
76	Bis(2-chloroethyl) ether	869.02	78	000111-44-4
77	2,4-Dinitrophenol	867.30	79	000051-28-5
78	2,4,6-Trinitrotoluene	866.63	82	000118-96-7
79	1,1,1-Trichloroethane	858.26	76	000071-55-6
80	1,1,2,2-Tetrachloroethane	851.93	85	000079-34-5
81	Ethyl benzene	847.51	80	000100-41-4
82	2,3,6-Trichlorophenol	845.69	86	000088-06-2
83	Total xylenes	844.63	83	001330-20-7
84	Endrin aldehyde	842.70	64	007421-93-4
85	Thiocyanate	842.29	New	000302-04-5
86	Asbestos	841.33	89	001332-21-4
87	4,6-Dinitro-0-cresol	837.64	90	000534-52-1
88	Uranium	831.94	92	007440-61-1
89	Chlorobenzene	829.90	91	000108-90-7
90	Radium	829.28	97	007440-14-4
91	Radium-226	829.10	93	013982-63-3
92	Ethion	827.76	98	000563-12-2
93	Hexachlorobenzene	827.25	94	000118-74-1
94	Dimethylarsinic acid	826.85	87	000075-60-5
95	Thorium	824.99	96	007440-29-1
96	Fluoranthene	817.60	101	000206-44-0
97	Radon	817.53	100	010043-92-2
98	Barium	816.92	95	007440-39-3
99	2,4-Dinitrotoluene	816.26	105	000121-14-2
100	Diazinon	814.01	104	000333-41-5

Notes:
Full list includes 275 substances.
* = Points are based on a toxicological profile algorithm that considers frequency of occurrence, toxicity, and potential for human exposure to the substances found at National Priorities List sites.
** = CAS Number. Chemical Abstracts Service registry number that identifies chemical substances.
The Comprehensive Environmental Response, Compensation, and Liability Act (CERCLA) section 104 (I) requires ATSDR (Agency for Toxic Substances and Disease Registry) and EPA to prepare a list, in order of priority, of substances that are most commonly found at facilities on the National Priorities List (NPL) and that are determined to pose the most significant potential threat to human health. The CERCLA priority list is revised and published on a 2-year basis. The list is based on an algorithm that utilizes the following three components: frequency of occurrence, toxicity, and potential for human exposure of the substances found at NPL sites. This algorithm utilizes data from ATSDR's HazDat database, which contains information from ATSDR's public health assessments and health consultations.

Source: U.S. Department of Health and Human Services, Agency for Toxic Substances and Disease Registry and Environmental Protection Agency. From the Agency's Web site at http://atsdr1.atsdr.cdc.gov, July 1999.

9.21. Shipments for Selected Industrial Air Pollution Control Equipment, 1992–97 (Quantity in number of units. Value in thousands of dollars).

Product	Product Description	1992 Number of Companies	1992 Shipments Quantity	1992 Value	1993 Number of Companies	1993 Shipments Quantity	1993 Value	1994 Number of Companies	1994 Shipments Quantity	1994 Value	1995 Number of Companies	1995 Shipments Quantity	1995 Value	1996 Number of Companies	1996 Shipments Quantity	1996 Value	1997 Number of Companies	1997 Shipments Quantity	1997 Value
	Industrial air pollution control equipment	100	N/A	$713,429	110	65,736	$762,324	110	67,927	$711,390	113	S	$774,004	112	83,553	$709,650	108	88,677	$722,485
	Particulate emissions collectors	N/A	N/A	$463,288	87	54,257	$466,429	90	54,744	$453,261	91	S	$506,298	90	74,154	$518,469	88	76,199	$497,447
35646 51	Electrostatic precipitators	19	236	$152,741	18	r/ 469	$171,771	19	552	$136,763	17	S	$152,022	15	D	$118,512	16	D	$110,978
35646 54	Fabric filters	51	42,059	$228,143	54	47,287	$205,391	59	47,213	$211,340	57	52,978	r/$ 242,123	53	55,510	$286,828	52	57,662	$266,870
35646 55	Mechanical collectors	35	3,597	$43,192	r/ 40	5,225	$48,083	39	5,692	$68,028	37	6,273	$63,343	38	D	$63,500	36	D	$62,014
35646 58	Wet scrubbers	32	1,101	$39,212	31	1,276	$41,184	31	1,287	$37,130	31	1,249	$48,777	29	1,240	$46,629	27	1,249	$57,585
	Gaseous emissions control devices	N/A	N/A	$196,899	r/37	r/1,097	$227,980	38	1,169	$206,546	39	1,245	$223,545	39	1,020	$166,624	41	978	$194,020
35646 70	Catalytic oxidation systems	7	104	$18,722	r/12	r/ 146	$20,549	10	S	$28,267	12	249	$34,832	11	189	$19,943	12	185	$34,373
35646 71	Nitric oxide (NO) control systems	N/A	N/A	N/A	2	D	D	3	D	D	4	D	$4,402	3	D	D	3	D	D
35646 72	Thermal and direct oxidation systems	11	258	$28,897	r/16	302	$33,339	14	272	$30,878	15	311	$71,402	r/17	246	$67,065	17	235	$40,890
35646 73	Scrubbers (gas absorber non-FGD [1])	12	597	$19,782	r/12	554	$11,360	11	658	$13,997	12	r/610	$26,527	10	525	$26,661	14	470	$23,140
35646 74	Wet flue gas desulfurization systems	4	N/A	N/A	5	r/22	$109,687	5	S	$105,821	4	D	$72,559	5	D	D	4	D	D
35646 75	Dry flue gas desulfurization systems	5	28	$120,290	4	D	D	4	14	$11,533	r/6	r/16	r/ $ 10,736	4	9	$4,884	4	D	$4,884
35646 76	Gas adsorbers	8	37	$9,208	8	D	D	9	D	D	5	38	$3,087	8	43	$24,487	7	35	$9,900
35646 79	Other types of industrial air pollution control equipment	19	N/A	53,242	22	10,382	67,915	19	12,014	51,583	r/19	10,111	44,161	18	8,375	24,557	15	11,500	31,018

Notes:
D = Data withheld to avoid disclosing data for individual companies.
N/A = Not available.
r/ = Revised by 5 percent or more from previously published data.
S = Does not meet publication standards.
[1] FGD = Flue gas desulfurization.

Source: Bureau of the Census, Current Industrial Reports, series MA35J. *Industrial Air Pollution Control Equipment.* Washington, DC: U.S. Government Printing Office, Various years. Table 2. http://www.census.gov/cir/www/ma35j.html

9.22. Value of Shipments of Selected Air Pollution Control Equipment by End Use, United States, 1992–97 (thousands of dollars).

	Particulate Emissions Collectors					Gaseous Emissions Control Devices					Industrial Air Pollution Control Devices				
	1992	1993	1995	1996	1997	1992	1993	1995	1996	1997	1992	1993	1995	1996	1997
Shipments	$477,910	$466,801	$506,298	$518,469	$497,447	$195,417	$226,530	$223,545	$166,624	$194,020	$54,701	$67,915	$44,161	$24,557	$31,018
Steam power plants—electric utility[11]	$96,376	$88,701	$46,823	$26,542	$28,801	r/$117,682	$98,466	$73,137	$21,172	D	D	[6]D	D	D	D
Steam power plants—industrial[1][12]	$17,766	$11,680	r/$8,596	$3,366	D	[5]r/	[7]$54,183	D	D	D	D	D	D	D	D
Coal mining and cleaning[1][12]	r/$651	$1,670	$1,320	$888	D	[8]r/$12,383	[7]	D	D	D	0	0	r/S	S	S
Petroleum refining[1][11][13]	$32,358	$24,795	$13,265	D	D	[8]r/	[7]	$14,99	D	S	D	0	r/S	S	D
Foundries[2][3]	$4,228	$5,005	$6,136	$2,694	$2,970	D	D	r/D	D	D	[9]r/190	0	r/S	D	D
Iron and steel mills[2]	r/$5,929	$16,380	$20,675	D	$9,993	r/$747	[10]$1,140				[9]r/	0	r/S	D	D
Primary nonferrous metal smelting plants[2][3]	r/$6,341	$3,614	$11,962	$5,064	$4,011	r/$704	[10]	r/$1,929	$2,485	$15,623	[9]r/	0	r/S	D	
Chemical and fertilizer production	r/$51,847	$43,719	$44,139	$38,950	$51,599	r/$7,392	$4,643	$10,441	$11,562	S	D	D	D	D	D
Cement manufacturing	r/$13,260	$10,744	$4,144	$2,594	$6,216	0	0	r/S	S	S	D	D	D	D	D
Grain milling and handling[4]	r/$3,299	$4,081	$3,339	$2,899	$2,915	0	0				D	D	D	D	D
Pulp and paper mill operations[4]	$35,284	$53,480	$40,090	$31,151	$32,485	[8]r/	D	$1,131	D	D	D	$924	$313	D	D
Municipal waste combusters[13]+A52	$2,499	$2,995	$12,483	$1,932	D	D	D	r/$4,186	$7,148	D	r/D	D	r/S	S	S
Other end users	r/$188,197	r/$176,409	r/$180,261	r/$172,992	$175,074	r/D	D	r/$1,732	r/$57,705	$54,205	r/D	D	$20,524	$12,590	$12,590
End use not specified	$19,875	$23,528	r/$113,065	r/$176,148	$155,958	$7,125	$4,509	r/$1,732	$60,009	$35,088	$2,615	$4,699	r/$84,992	$5,587	$14,473

Notes:

D = Withheld to avoid disclosing data for individual companies.
r/ = Revised by 5 percent or more from previously published data.
S = Does not meet publication standards.

1 "Industrial steam power plants," "Coal mining and cleaning," and "Petroleum refining" are combined for 1995 to avoid disclosing data for individual companies.
2 "Foundries," "Iron and steel mills," and "Primary nonferrous metal smelting plants" are combined for "Gaseous emissions control devices" for 1996 to avoid disclosing data for individual companies.
3 "Foundries" and "Primary nonferrous metal smelting plants" are combined for "Gaseous emissions control devices" for 1995 to avoid disclosing data for individual companies.
4 "Petroleum refining," "Grain milling and handling," and "Pulp and paper mill operations" are combined for "Other types of industrial air pollution control devices" for 1995 to avoid disclosing data for individual companies.
5 Data for the two categories of steam power plants are combined for gaseous emissions control devices for 1992 to avoid disclosing data for individual companies.
6 Data for "Steam power plants—electric utility" and "Chemical and fertilizer production" are combined for other types of industrial air pollution control devices for 1993 to avoid disclosing data for individual companies.
7 Data for the "Steam power plants—industrial," "Coal mining and cleaning," and "Petroleum refining" are combined for gaseous emissions control devices for 1993 to avoid disclosing data for individual companies.
8 Data for "Coal mining and cleaning," "Petroleum refining," and "Paper and Pulp mill operations" are combined for gaseous emissions control devices for 1992 to avoid disclosing data for individual companies.
9 Data for "Foundries," "Iron and steel mills," and "Primary nonferrous metal smelting plants" are combined for other types of industrial air pollution control devices for 1992 to avoid disclosing data for individual companies.
10 Data for iron and steel mills and primary nonferrous metal smelting plants are combined for gaseous emissions control devices for 1993 to avoid disclosing data for individual companies.
11 "Steam power plants—electric utility" and "Petroleum refining" total 61,372 for Gaseous Emissions Control Devices for 1997.
12 "Steam power plants—industrial" and "Coal mining and cleaning" total 2,468 for Particulate Emissions Collectors for 1997.
13 "Petroleum refining" and "Municipal waste combusters" total 24,957 for Particulate Emissions Collectors for 1997.

Source: U.S. Bureau of the Census. Current Industrial Reports, Series MA35J. *Selected Air Pollution Control Equipment,*1993, 1995, and 1997. Washington, DC. U.S. Government Printing Office, 1994, 1996, and 1998. Table 2. http://www.census.gov/cir/www/ma35j.html

9.23. Approximate Quantities of Wood and Wood-Fiber Materials Recovered for Recycling in the United States, 1994.

Use category	Approximate quantity (metric tons)
Paper and paperboard	28,100,000
Recovered paper for export	7,000,000
Insulation and related products	500,000
Molded-pulp products	300,000
Fiberboard products	275,000
Wooden pallets	250,000
Animal bedding	100,000
Mulch	100,000
Particleboard, hardboard [1]	50,000
Reclaimed lumber [1]	50,000
Roof systems, siding [1]	50,000
"Plastic lumber" and other [1]	10,000

Notes:
[1] Estimated, not based on actual survey data.

Source: USDA, Forest Service, Forest Products Laboratory. Cited in U.S. Department of Agriculture, Economic Research Service, Commercial Agriculture Division. *Industrial Uses of Agricultural Materials.* IUS-6. Washington, DC, August 1996.

9.24. Estimated Emissions of Greenhouse Gases, 1985–96 (million metric tons).

Year	Carbon Dioxide [1]	Methane	Nitrous Oxide	Halocarbons and Minor Gases				Criteria Pollutants		
				CFC-11 CFC-12 CRC-113	HCFC-22	HFC-23 and PFCs	Methyl Chloroform	Carbon Monoxide	Nitrogen Oxides	Nonmethane VOCs
1985	4,667.7 /r	30.1	0.4	0.3	0.1	(s)	0.3	104.9 /r	21.3 /r	22.0 /r
1986	4,666.3 /r	29.5	0.4	0.3	0.1	(s)	0.3	100.2 /r	21.2 /r	21.3 /r
1987	4,819.9 /r	30.3	0.4	0.3	0.1	(s)	0.3	98.8 /r	20.7 /r	21.0 /r
1988	5,044.7 /r	30.8	0.4	0.3	0.1	(s)	0.3	106.3 /r	22.2 /r	21.9 /r
1989	5,091.8 /r	31.3 /r	0.4	0.3	0.1	(s)	0.3	94.8 /r	21.8 /r	20.3 /r
1990	5,037.1 /r	31.6 /r	0.4	0.2	0.1	(s)	0.3	87.6 /r	21.6 /r	19.0 /r
1991	4,987.3 /r	31.6 /r	0.5	0.2	0.1	(s)	0.2	89.3 /r	21.6 /r	19.1 /r
1992	5,059.8 /r	31.7 /r	0.5	0.2	0.1	(s)	0.2	86.3 /r	21.9 /r	18.8 /r
1993	5,175.9 /r	30.8 /r	0.5	0.1 /r	0.1	(s)	0.1	86.4 /r	22.2 /r	19.0 /r
1994	5,256.1 /r	31.4 /r	0.5	0.1	0.1	(s)	0.1	90.4 /r	22.6 /r	19.5 /r
1995	5,296.9 /r	30.9 /r	0.5	0.1	0.1	(s)	(s)	81.4 /r	21.7 /r	18.7 /r
1996 /p	5,484.9	30.9	0.4	0.1	0.1	(s)	(s)	80.6	21.2	17.3

Notes:
/r = Revised.
/p = Preliminary.
(s) = Less than 0.05 million metric tons.
CFC = Chlorofluorocarbons.
HCFC = Chlorodifluoromethane.
HFC = Hydrofluorocarbons.
PFC = Perfluorocarbons.
VOC = Volatile organic compound.
[1] Carbon dioxide gas can be converted to units of carbon by dividing by 3.667. One ton of carbon = 3.667 tons of carbon dioxide gas. Emissions are from anthropogenic sources. Anthropogenic means produced as the result of human activities, including emissions from agricultural activity and domestic livestock.
Emissions from natural sources, such as wetlands and wild animals, are not included.

Sources: Carbon dioxide, methane, nitrous oxide, and halocarbons and minor gases: 1985–88, Energy Information Administration, Office of Integrated Analysis and Forecasting. 1989 forward: Energy Information Administration, *Emissions of Greenhouse Gases in the United States 1996,* October 1997. Criteria pollutants: 1985 forward: Environmental Protection Agency, *National Air Pollutant Emission Trends 1990-96,* December, 1997. Cited in U.S. Department of Energy, Energy Information Administration. *Annual Energy Review 1997.* Washington, DC: Energy Information Administration, 1997. Table 12.1. http://www.eia.doe.gov/emeu/aer/.

Chapter 10. The Military and Law Enforcement

It often seems as though the history of the world is a history of armed conflict. Attempting to control others through force remains such a central human pursuit, and the technologies to do so are so influential, that we hunger for information about the devices used to threaten, kill, and capture even as we recoil from it. Nations tax their citizens, whole industries flourish, and careers are made in the name of "defense."

At the same time, we fight smaller-scale wars on our streets as we struggle to eradicate crime. The lawman stands tall in our history and culture—Wyatt Earp, Elliott Ness, the Lone Ranger—but today's officers operate in a more complex environment and need sophisticated tools like forensic labs, computers, and video cameras to bring bad guys to justice.

Advances in military and law enforcement technologies spawn new devices and gadgets that eventually become available to the rest of us. Global positioning satellites, night vision/infrared imaging devices, and Primaloft TM—an insulating material now used for cold-weather clothing—were originally classified technologies confined to military and law enforcement uses. Medical products and techniques developed by the military, such as sterile all-type blood substitutes, will help save lives in the general population. And let's not forget that the Internet was born in the Department of Defense's nursery (ARPAnet). Unfortunately, some of these technologies end up being used unscrupulously and dangerously—for eavesdropping, urban warfare, and terrorism.

The Defense Technology Area Plan divides military technologies into the following useful groups:

- Air platforms: vehicles, engines, propulsion, and fuels
- Ground and sea vehicles: ships, submarines, unmanned undersea vehicles, and tanks
- Space platforms: launch vehicles, space vehicles, and space propulsion
- Chemical, biological, and nuclear weapons: detection, protection, decontamination, and survivability

- Weapons: guns, mines, undersea weapons, microwave weapons, missiles, guidance and control systems, lasers, threat warning systems, self-protection, ordnance, and mission support
- Materials and processes: manufacturing technology, environmental quality, materials for survivability and life extension, and civil engineering
- Sensors, electronics, and battlespace environment: microelectronics, radar sensors, ocean battlespace environments, and lower and upper atmosphere environments
- Biomedical: infectious diseases, combat casualty care, chemical and biological defense, dentistry, and operational medicine
- Information systems: modeling and simulation, decision making, information management and distribution, and communications
- Human systems: training, and warrior protection and sustainment

Some other interesting and important military technologies are

- Stealth technology, which renders an aircraft virtually invisible to radar. (See Tables 10.1, B-2; 10.3, B-2; 10.14, B-1 and B-2 for numbers of planes and costs. These aircraft cost billions of dollars each!)
- Endurance unmanned aerial vehicles (UAV's), which can be as small as half a foot. UAV's are used for reconnaissance. (See Table 10.1 almost at the end.)
- Aided/Automatic Target Recognition (ATR). ATR is a class of image/data analysis algorithms that find targets in noise backgrounds in real time. ATR has the potential to help surgeons pinpoint tumors while operating and to enhance medical diagnostic imaging capabilities. (See Table 10.2, Sensors, Electronics,and Battlespace Environment.)
- Precision weapons like The Tomahawk Cruise Missile and the Joint Stand-Off Weapon (JSOW), designed to be self-guiding and to improve accu-

racy, operate well in adverse weather, and increase lethality. (See Table 10.1.)

Some emerging technologies in law enforcement include

- Flexible, lightweight body armor
- Vehicle disabling systems to curtail high-speed pursuits
- Squad car computers that allow for video capture, GPS integration, and online references
- Less-than-lethal devices: shotguns that fire nets, thermal guns that incapacitate suspects by raising their body temperatures, blinding strobe lights, and drug-tipped darts
- Noninvasive in-field drug testing systems
- Technologies for detecting weapons, plastic explosives, and drugs

This chapter examines some of these new and old technologies and presents information on chemical and nuclear weapons. Much of the chapter details defense technologies and their budgets. Table 10.1 lists Department of Defense aircraft, missiles, vessels, and space programs at a general level. Here you can see how much the United States spends per year to acquire—not maintain—B-2 stealth bombers, Tomahawk cruise missiles, global positioning satellites, airborne lasers, and other technologies. Remember that these figures are per year, not total lifetime for each program. Table 10.2 presents data that is both more general and more comprehensive. Here you will find funding for technologies by function rather than specific weapon or system, but for much more than just Table 10.1's aircraft, missiles, vessels, and space programs. Table 10.2 includes chemical, biological, and nuclear defenses; biomedical technologies; materials and processes; sensors and electronics; human systems; and information systems, in addition to the types of systems covered in Table 10.1. Table 10.2 includes both summary and detail data for fiscal years 1997 through 2003. Of course, since Congress has to approve Defense Department appropriations, data for later years represent requested funds rather than those earmarked.

Table 10.3 illustrates current and historical nuclear weapons data at a glance. The number of nuclear weapons is supposed to decline from 40,000 to 21,000 by 2003. Almost all counted nuclear weapons belong to the United States and Russia and will continue to do so, according to the Center for Defense Information. The amount spent by the United States to prepare for, fight, and prevent nuclear war is staggering: $3.5 trillion for preparation between 1940 and 1995, with $29.2 billion, most for preparation, in 1995 alone.

U.S. chemical weapons stores are declining as well. In 1985, Congress directed the army to destroy all chemical weapons. Table 10.4 shows the amounts remaining and already destroyed.

Tables 10.5 through 10.12 highlight some specific Defense Department technologies under development in the areas of biomedical defense, ground and sea vehicles, materials and processes, air platforms, chemical/biological/nuclear, sensors/electronics/battlespace environment, weapons, and information superiority. See Table 10.5 for some biomedical technologies that could benefit everyone, like doubling the storage life of blood and blood products and evaluating products to keep hemorrhages from becoming life-threatening. Table 10.7 shows a laser eye protection program that could also benefit nonmilitary personnel and a hazardous and toxic waste treatment/destruction program for treating military, space-related, and industrial wastes. Here you will also find a program for materials and processes related to space vehicle reentry into earth's atmosphere, during which vehicles must be shielded from extreme friction-generated heat. Table 10.9 extends the latter with a program to develop new chemical and biological weapon decontaminants. See Table 10.11 for more about laser-based weapons.

The last few tables in the chapter deal with law enforcement technologies. Table 10.15 shows how many police forces are using semiautomatic weapons, nonlethal weapons, and body armor. Table 10.16 does the same for 911 systems, computers, automated fingerprint identification systems, and specialized vehicles like armored vehicles and aircraft. Table 10.17 details the number of DNA samples analzyed by forensic laboratories.

MILITARY

10.1. U.S. Department of Defense FY 1999 Budget, Program Acquisition Costs with History for 1997 and 1998: Aircraft, Missiles, Vessels, Space Programs, and Other Programs (millions of dollars).

	Aircraft	FY 1997	FY 1998	FY 1999
Army				
OH-58D	Kiowa Warrior	$199.6	$57.1	$40.4
AH-64D	Longbow Apache	$421.3	$512.9	$633.7
RAH-66	Comanche Helicopter	$325.3	$272.2	$367.8
UH-60	Blackhawk Helicopter	$286.9	$292.8	$220.7
Navy				
AV-8B	Harrier	$374.9	$329.8	$377.9
CH-60	Helicopter	$6.9	$61.5	$152.8
EA-6B	Prowler	$255.5	$114.4	$141.1
E-2C	Hawkeye	$357.4	$380.2	$457.1
F/A-18E/F	Hornet	$2,449.8	$2,455.7	$3,275.3
T-45TS	Goshawk	$309.7	$297.9	$363.9
Air Force				
B-2	Stealth Bomber	$684.2	$698.8	$376.3
C-17	Airlift Aircraft	$2,232.7	$2,312.3	$3,206.9
C-130J	Airlift Aircraft	$68.8	$28.2	$125.8
CAP	Civil Air Patrol	$2.6	$3.0	$2.6
E-8C	Joint Surveillance Target Attack Radar System (Joint STARS)	$759.4	$536.3	$654.4
F-15E	Eagle Multi-Mission Fighter	$416.0	$361.8	$104.2
F-16	Falcon Multi-Mission Fighter	$279.8	$176.0	$125.1
F-22	Advanced Tactical Fighter (ATF)	$1,827.4	$2,032.1	$2,393.1
C-37A/C-32A	Small/Large VCX	$102.6	$191.6	$160.9
ABL	Airborne Laser	$56.0	$151.4	$292.2
Dept. of Defense-wide/Joint				
JPATS	Joint Primary Aircraft Training System	$109.9	$133.7	$152.1
JSF	Joint Strike Fighter	$565.2	$905.0	$919.5
V-22	Osprey	$1,322.4	$1,206.1	$1,069.8
	Missiles			
Army				
ATACMS	Army Tactical Missile System	$208.9	$183.1	$142.4
BAT	Brilliant Anti-Armor Submunition	$94.8	$140.8	$183.5
JAVELIN	AAWS-M	$200.7	$146.9	$330.0
HELLFIRE II	Laser Hellfire Missile	$110.8	$9.5	$14.3
LONGBOW	Longbow Hellfire Missile	$249.3	$232.7	$346.3
MLRS	Multiple Launch Rocket System	$210.6	$175.2	$129.0
AVENGER	Missile System	$62.4	$7.2	$35.3
Navy				
RAM	Rolling Airframe Missile	$66.7	$56.8	$51.3
STANDARD	Missile (Air Defense)	$224.0	$178.1	$237.6
TOMAHAWK	Cruise Missile	$245.8	$140.2	$199.7
TRIDENT II	Submarine Launched Ballistic Missile	$346.3	$308.5	$385.6
Marine Corps				
JAVELIN	AAWS-M	$38.6	$58.0	$83.5
Dept. of Defense-wide/Joint				
AMRAAM	Advanced Medium Range Air-to-Air Missile	$178.9	$206.2	$234.4
JASSM	Joint Air-to-Surface Standoff Missile	$160.7	$129.3	$135.0
JSOW	Joint Standoff Weapon	$193.3	$181.6	$265.9
AIM-9X	Sidewinder	$74.5	$109.1	$118.9

10.1. U.S. Department of Defense FY 1999 Budget, Program Acquisition Costs with History for 1997 and 1998: Aircraft, Missiles, Vessels, Space Programs, and Other Programs (millions of dollars) *(continued).*

	Aircraft	FY 1997	FY 1998	FY 1999
	Vessels			
Navy				
DDG-51	AEGIS Destroyer	$3,723.7	$3,659.3	$2,904.3
NSSN	New Attack Submarine	$1,229.9	$2,909.1	$2,302.5
SSN-21	Seawolf Attack Submarine	$757.0	$224.8	$70.2
	Tracked Combat Vehicles			
Army				
M1A2	Abrams Tank Upgrade	$540.4	$634.4	$691.8
M2A3	Bradley Base Sustainment	$333.6	$300.6	$361.0
Crusader	Artillery Systems	$234.1	$320.2	$310.9
	Space Programs			
Army				
DSCS	Defense Satellite Communications System (Ground Systems)	$112.4	$107.3	$126.8
Air Force				
DSP	Defense Support Program	$85.1	$125.5	$105.2
MLV	Medium Launch Vehicles	$165.7	$206.9	$195.8
MILSTAR	Satellite Communications	$659.7	$628.0	$550.9
NAVSTAR GPS	NAVSTAR Global Positioning System	$274.3	$249.6	$258.6
Titan	Heavy Launch Vehicles	$395.6	$521.4	$666.0
EELV	Evolved Expendable Launch Vehicle	$44.3	$87.0	$283.6
SBIRS-H	Space Based Infrared System—High	$193.0	$316.5	$538.4
SBIRS-L	Space Based Infrared System—Low	$252.5	$202.4	$193.6
	Other Programs			
Army				
FHTV	Family of Heavy Tactical Vehicles	$246.3	$119.1	$189.6
FMTV	Family of Medium Tactical Vehicles	$242.1	$207.6	$336.3
HMMWV	High Mobility Multipurpose Wheeled Vehicle	$164.4	$128.0	$12.1
SADARM	Sense and Destroy Armor Munition	$103.3	$77.2	$77.3
WAM	Wide Area Munition	$35.4	$34.2	$32.8
Air Force				
SFW	Sensor Fuzed Weapon	$168.2	$166.6	$129.6
WCMD	Wind Corrected Munitions Dispenser	$46.1	$29.5	$21.5
Dept. of Defense-wide/Joint				
BMD	Ballistic Missile Defense	$3,703.8	$3,844.6	$4,001.6
JDAM	Joint Direct Attack Munition	$85.9	$117.4	$118.5
UAV	Unmanned Aerial Vehicles	$537.4	$651.5	$619.8
U.S. Special Operations Forces				
MK V	Special Operations Craft	$50.3	$54.0	—

Source: U.S. Department of Defense. *Program Acquisition Costs by Weapon System: Department of Defense Budget for Fiscal Year 1999.* February 1998. http://www.dtic.mil/comptroller/99budget/weaponsb.pdf

10.2. U.S. Defense Technology Area Plan, Resource Funding, 1997–2003.

Summary Data
(in thousands of dollars)

TECHNOLOGY AREA	FY97	FY98	FY99	FY00	FY01	FY02	FY03
Air Platforms	$487,328	$625,236	$622,804	$637,996	$685,231	$712,005	$743,504
Chemical/Biological Defense & Nuclear	$219,368	$218,034	$216,489	$219,825	$222,614	$236,896	$242,538
Information Systems & Technology	$1,133,756	$1,210,080	$1,300,500	$1,373,995	$1,426,951	$1,518,655	$1,543,153
Ground and Sea Vehicles	$335,269	$308,716	$349,235	$375,992	$429,576	$430,496	$465,644
Materials and Processes	$585,561	$430,523	$426,605	$423,697	$418,068	$485,194	$510,706
Biomedical	$340,983	$295,618	$324,020	$317,215	$314,688	$312,554	$314,769
Sensors, Electronics, and Battlespace Environment	$1,340,953	$1,390,717	$1,487,011	$1,565,426	$1,533,331	$1,489,350	$1,527,867
Space Platforms	$238,324	$163,343	$180,992	$200,685	$214,154	$221,250	$217,103
Human Systems	$271,300	$253,933	$248,128	$244,138	$232,062	$236,468	$241,293
Weapons	$1,015,150	$914,967	$942,836	$1,012,222	$1,004,250	$1,030,973	$1,058,870
Total	$5,967,991	$5,811,166	$6,098,619	$6,371,191	$6,480,925	$6,673,841	$6,865,446

DETAILED DATA—DEFENSE TECHNOLOGY SUBAREA
FUNDING (IN THOUSANDS OF DOLLARS)

Subarea	FY97	FY98	FY99	FY00	FY01	FY02	FY03
Fixed-wing Vehicles	$283,795	$420,938	$401,861	$398,295	$433,405	$449,018	$464,941
Rotary-wing Vehicles	$31,991	$23,593	$32,929	$49,322	$61,940	$68,630	$79,379
IHPTET*	$123,461	$132,239	$136,686	$140,900	$145,086	$148,036	$151,717
Aircraft Power	$20,977	$24,692	$28,920	$27,605	$22,007	$22,715	$23,392
High-speed Propulsion and Fuels	$27,104	$23,774	$22,408	$21,874	$22,793	$23,606	$24,075
Air Platforms Total	$487,328	$625,236	$622,804	$637,996	$685,231	$712,005	$743,504

CHEMICAL/BIOLOGICAL (CB) DEFENSE & NUCLEAR

Subarea	FY97	FY98	FY99	FY00	FY01	FY02	FY03
	$56,501	$40,919	$37,529	$34,008	$33,112	$38,913	$40,786
CB Detection	$13,626	$10,103	$8,673	$9,788	$10,225	$11,677	$11,200
CB Protection	$10,604	$4,458	$4,251	$4,220	$6,555	$6,873	$7,609
CB Decontamination	$5,923	$4,673	$4,987	$6,496	$6,382	$8,008	$8,116
CB Studies, Analysis, & Simulation	$29,838	$36,463	$37,659	$40,929	$42,918	$46,712	$47,661
Systems Effects & Survivability (Nuclear)	$36,603	$45,205	$44,306	$43,465	$41,197	$41,575	$41,955
Test & Simulation Technology (Nuclear)	$14,563	$17,530	$18,008	$17,604	$17,390	$17,390	$17,390
Scientific & Operational Computing (Nuclear)	$51,710	$58,683	$61,076	$63,315	$64,835	$65,748	$67,821
Chemical/Biological Nuclear & Defense Total	$219,368	$218,034	$216,489	$219,825	$222,614	$236,896	$242,538

INFORMATION SYSTEMS & TECHNOLOGY

Subarea	FY97	FY98	FY99	FY00	FY01	FY02	FY03
Decision Making	$164,414	$207,641	$215,937	$196,094	$201,522	$167,243	$168,273
Modeling & Simulation Technology	$194,605	$168,789	$182,260	$219,311	$235,701	$254,644	$280,735
Information Management & Distribution	$351,861	$381,251	$419,739	$466,986	$584,512	$673,814	$657,230
Seamless Communications	$120,948	$114,802	$125,684	$119,515	$76,943	$84,512	$84,616
Computers & Software Technology	$301,928	$337,597	$356,880	$372,090	$328,273	$338,442	$352,299
Information Systems & Technology Total	$1,133,756	$1,210,080	$1,300,500	$1,373,995	$1,426,951	$1,518,655	$1,543,153

GROUND AND SEA VEHICLES

Subarea	FY97	FY98	FY99	FY00	FY01	FY02	FY03
Ground Vehicles	$125,227	$107,949	$143,803	$144,523	$170,313	$191,076	$212,791
Surface Ship Combatants	$143,847	$128,535	$125,927	$150,542	$179,158	$157,938	$169,632
Submarines	$52,595	$60,077	$66,671	$67,727	$66,549	$67,926	$69,295
Unmanned Undersea Vehicles	$13,600	$12,155	$12,834	$13,200	$13,556	$13,556	$13,926
Ground and Sea Vehicles Total	$335,269	$308,716	$349,235	$375,992	$429,576	$430,496	$465,644

MATERIALS AND PROCESSES

Subarea	FY97	FY98	FY99	FY00	FY01	FY02	FY03
M&P for Survivability, Life Extension, & Affordability	$224,166	$198,204	$200,330	$202,950	$222,852	$286,475	$307,192
Manufacturing Technology	$157,108	$97,661	$84,592	$74,080	$47,714	$41,037	$43,114
Civil Engineering	$54,380	$26,888	$31,288	$34,736	$36,276	$37,544	$35,942
Environmental Quality	$149,907	$107,770	$110,395	$111,931	$111,226	$120,138	$124,458
Materials and Processes Total	$585,561	$430,523	$426,605	$423,697	$418,068	$485,194	$510,706

BIOMEDICAL

Subarea	FY97	FY98	FY99	FY00	FY01	FY02	FY03
Infectious Diseases of Military Importance	$58,246	$65,298	$63,895	$54,683	$55,762	$56,414	$57,522

10.2. U.S. Defense Technology Area Plan, Resource Funding, 1997–2003 *(continued)*.

Summary Data
(in thousands of dollars)

Technology Area	FY97	FY98	FY99	FY00	FY01	FY02	FY03
Combat Casualty Care	$97,345	$123,608	$146,478	$147,726	$141,466	$135,584	$133,401
Medical Biological Defense	$21,057	$25,334	$26,783	$26,664	$27,148	$27,728	$28,378
Medical Chemical Defense	$46,558	$23,089	$24,508	$24,330	$24,805	$25,337	$25,929
Military Operational Medicine	$106,205	$46,200	$49,931	$51,265	$52,800	$54,509	$56,256
Military Dentistry	$311	$324	$424	$309	$277	$288	$290
Medical Radiological Defense	$11,261	$11,765	$12,001	$12,238	$12,430	$12,694	$12,993
Biomedical Total	$340,983	$295,618	$324,020	$317,215	$314,688	$312,554	$314,769

SENSORS, ELECTRONICS, AND BATTLESPACE ENVIRONMENT

Subarea	FY97	FY98	FY99	FY00	FY01	FY02	FY03
Radar Sensors	$99,315	$112,273	$123,058	$114,100	$105,035	$94,122	$96,040
Electro-Optic Sensors	$127,821	$120,268	$103,151	$129,597	$122,161	$109,969	$115,391
Acoustic Sensors	$84,653	$116,693	$142,517	$156,500	$120,150	$122,727	$125,628
Automatic Target Recognition	$128,330	$169,601	$192,668	$181,256	$182,144	$187,142	$199,022
Integrated Platform Electronics	$78,639	$87,884	$87,854	$108,578	$116,973	$119,100	$116,145
Radio Frequency Components	$106,755	$88,073	$81,882	$70,482	$74,544	$79,589	$81,334
Electro-Optic Technology	$206,943	$155,519	$164,683	$139,965	$144,589	$104,489	$106,056
Microelectronics	$187,051	$206,523	$232,098	$273,427	$273,661	$277,019	$275,690
Electronic Materials	$75,866	$64,672	$78,825	$80,705	$84,091	$78,950	$84,476
Electronics Integration Technology	$90,387	$88,144	$114,960	$156,695	$163,336	$167,766	$174,087
Terrestrial Environments	$9,613	$10,426	$11,007	$11,023	$11,111	$11,296	$12,906
Ocean Battlespace Environment	$50,117	$34,864	$46,303	$49,009	$40,492	$41,084	$42,089
Lower Atmosphere Environment	$54,491	$38,583	$42,610	$41,260	$42,830	$43,742	$45,018
Space/Upper Atmosphere Environment	$40,972	$97,192	$65,394	$52,827	$52,215	$52,356	$53,984
Sensors, Electronics, and Battlespace Environment Total	$1,340,953	$1,390,717	$1,487,011	$1,565,426	$1,533,331	$1,489,350	$1,527,867

SPACE PLATFORMS

Subarea	FY97	FY98	FY99	FY00	FY01	FY02	FY03
Space & Launch Vehicles	$178,035	$117,590	$122,449	$139,871	$151,823	$157,827	$153,733
Propulsion (IHPRPT)**	$60,289	$45,753	$58,543	$60,814	$62,331	$63,423	$63,370
Space Platforms Total	$238,324	$163,343	$180,992	$200,685	$214,154	$221,250	$217,103

HUMAN SYSTEMS

Subarea	FY97	FY98	FY99	FY00	FY01	FY02	FY03
Personnel Performance & Training	$84,996	$83,753	$88,592	$88,789	$92,430	$93,218	$95,293
Info Display & Performance Enhancement	$40,779	$38,076	$37,683	$36,681	$34,281	$35,253	$36,467
Warrior Protection & Sustainment	$75,571	$69,825	$66,541	$66,306	$50,859	$53,033	$52,713
Design Integration & Supportability	$69,954	$62,279	$55,312	$52,362	$54,492	$54,964	$56,820
Human Systems Total	$271,300	$253,933	$248,128	$244,138	$232,062	$236,468	$241,293

WEAPONS

Subarea	FY97	FY98	FY99	FY00	FY01	FY02	FY03
Conventional Weapons							
Countermine/Mines	$146,041	$125,182	$133,888	$145,867	$133,208	$135,839	$139,008
Guidance & Control	$317,053	$324,792	$297,772	$263,484	$244,872	$256,232	$264,423
Guns	$89,556	$79,445	$95,819	$98,874	$113,887	$119,128	$120,232
Missiles	$18,165	$11,012	$13,056	$9,303	$9,137	$11,758	$10,384
Ordnance	$64,689	$58,276	$56,807	$91,601	$99,689	$103,000	$107,963
Undersea Weapons	$32,314	$32,063	$39,002	$41,122	$36,671	$37,637	$38,376
Weapons L/V	$28,784	$36,015	$38,639	$38,779	$31,260	$29,052	$29,914
Directed-Energy Weapons							
Lasers	$141,311	$62,752	$62,413	$61,455	$59,741	$61,033	$62,739
High-Power Microwave	$38,927	$33,445	$35,089	$37,383	$38,521	$39,463	$40,519
Electronic Warfare							
Threat Warning	$23,958	$23,808	$24,085	$24,654	$25,270	$24,654	$25,129
Self-Protection	$62,656	$57,392	$61,594	$66,992	$70,536	$67,483	$69,655
Mission Support	$51,696	$70,785	$84,672	$132,708	$141,459	$145,694	$150,528
Weapons Total	$1,015,150	$914,967	$942,836	$1,012,222	$1,004,250	$1,030,973	$1,058,870

Notes:
*IHPTEP = integrated high-performance turbine engine technology.
**IHPRPT = integrated high-performance rocket propulsion technology.

Source: U.S. Department of Defense, Office of the Director, Defense Research and Engineering, Office of Secretary of Defense, Joint Staff, Military Services, and Defense Agencies. *1997 Defense Technology Area Plan.* Ft. Belvoir, VA: Defense Technical Information Center, 1997. Appendix.

10.3. Nuclear Facts and Figures.

Approximate number of U.S. nuclear weapons: 15,000

Approximate number of Russian nuclear weapons: 24,000

Approximate number of U.S. nuclear weapons, year 2003, under START II**: 10,000

Approximate number of Russian nuclear weapons, year 2003, under START II: 10,000

Approximate number of nuclear weapons worldwide: 40,000

Approximate number of nuclear weapons worldwide, year 2003: 21,000

U.S. spent roughly $3.5 trillion from 1940 to 1995 to prepare to fight a nuclear war.

U.S. spent roughly $27 billion in 1995 to prepare to fight a nuclear war.

U.S. spent roughly $2.2 billion in 1995 to prevent nuclear war.

Each of the first 20 B-2 bombers authorized by Congress costs roughly $2.2 billion (RDTE*, procurement).

The lifecycle cost of each B-2 (RDTE*, procurement, operations, maintenance, and support) is $2.5 billion.

Buying 20 more B-2 bombers will cost taxpayers an estimated $31 billion over the airplanes' 20-year

expected lifespan. That's a lifecycle cost of $1.5 billion per plane.

Notes:
* RDTE = Research, development, test, and evaluation.
**START II = The second Strategic Arms Reduction Treaty between the U.S and Russia, signed January 3, 1993.

Source: Center for Defense Information. Washington, DC. From the Center's Web site at http://www.cdi.org, July 1999.

10.4. U.S. Chemical Weapons Storage and Destruction.

Originally 31,495 tons of chemical weapons were stored by the United States at nine facilities. Eight of those facilities fall within the continental United States. One is located on Johnston Atoll in the Pacific Ocean. Johnston Atoll is also one of the two facilities dedicated to the disposal of chemical weapons, the other being Tooele Chemical Agent Disposal Facility in Tooele, Utah. Destruction began at Johnston Atoll in 1990, and at Tooele in August of 1996. In 1985, Congress directed the Army to destroy all chemical weapons. Complete destruction should be accomplished by 2004.

As of August 30, 1998, 27,871 tons of chemical weapons remained, with 3,624 tons (11.51%) having been destroyed.

Storage Facilities:

Anniston Chemical Activity, 50 miles east of Birmingham, Alabama, stores 2,254 tons (7.4% of the original U.S. chemical weapons stockpile).

Agent	Item	Quantity	Pounds
HT-Blister	4.2-inch Cartridges	183,552	1,064,600
HT-Blister	4.2-inch Cartridges	75,360	452,160
HT-Blister	105mm Cartridges	23,064	68,500
HT-Blister	155mm Projectiles	17,643	206,420
HT-Blister	Ton Containers	108	185,080
GB-Nerve	105mm Cartridges	74,014	120,640
GB-Nerve	105mm Projectiles	26	40
GB-Nerve	155mm Projectiles	9,600	62,400
GB-Nerve	8-inch Projectiles	16,026	232,380
GB-Nerve	M55 Rockets	42,738	457,300
GB-Nerve	M56 Rocket Warheads	24	260
VX-Nerve	155mm Projectiles	139,581	837,480
VX-Nerve	Mines	44,131	463,380
VX-Nerve	M55 Rockets	35,636	356,360
VX-Nerve	M56 Rocket Warheads	26	260

10.4. U.S. Chemical Weapons Storage and Destruction *(continued).*

Blue Grass Chemical Activity is situated in central Kentucky.
Blue Grass stores 523 tons (1.7% of the original U.S. chemical weapons stockpile).

Agent	Item	Quantity	Pounds
H-Blister	155mm Projectiles	15,492	181,260
GB-Nerve	8-inch Projectiles	3,977	57,660
GB-Nerve	M55 Rockets	51,716	553,360
GB-Nerve	M56 Rocket Warheads	24	260
VX-Nerve	155mm Projectiles	12,816	76,900
VX-Nerve	M55 Rockets	17,733	177,340

Deseret Chemical Depot lies 20 miles south of Tooele, Utah. It has 13,616 tons (44.5% of the original U.S. chemical weapons stockpile) stored there.

Agent	Item	Quantity	Pounds
HT-Blister	4.2-inch Cartridges	62,590	363,020
HD-Blister	4.2-inch Cartridges	976	5,860
HD-Blister	Ton Containers	6,398	11,383,420
H-Blister	155mm Cartridges	54,663	699,540
Lewisite	Ton Containers	10	25,920
GA-Nerve	Ton Containers	2	2,820
TGA-Nerve	Ton Containers	2	1,280
TGB-Nerve	Ton Containers	7	6,960
GB-Nerve	105mm Cartridges	119,400	194,620
GB-Nerve	105mm Projectiles	679,303	1,107,260
GB-Nerve	155mm Projectiles	89,141	579,420
GB-Nerve	M55 Rockets	28,945	309,720
GB-Nerve	M56 Rocket Warheads	1,056	11,300
GB-Nerve	WETEYE	888	308,140
GB-Nerve	750-lb. Bombs	4,463	981,860
GB-Nerve	Ton Containers	5,709	8,598,200
VX-Nerve	155mm Projectiles	53,216	319,300
VX-Nerve	Mines	22,690	238,240
VX-Nerve	8-inch Projectiles	1	20
VX-Nerve	M55 Rockets	3,966	39,660
VX-Nerve	Spray Tanks	862	1,168,880
VX-Nerve	Ton Containers	640	910,960

Note:
Quantity and pounds as of May 1997. However, the Army's Chemical Stockpile Disposal Program began destroying the stockpile at the Tooele Chemical Disposal Facility in August 1996, and these numbers do not reflect chemical weapons destroyed under that program.

Edgewood Chemical Activity is a tenant of Aberdeen Proving Ground, 21 miles northeast of Baltimore. It houses 1,625 tons of bulk mustard agent (5.3% of U.S. chemical weapons stockpile).

Agent	Item	Quantity	Pounds
HD-Mustard	Ton Containers	1,818	3,249,740

Note:
Pounds comprise total weight of agent plus storage containers. Ton containers actually weigh 1600 pounds.

Newport Chemical Depot is located in western Illinois. It houses 1,269 tons (4.1% of the U.S. chemical weapons stockpile).

Agent	Item	Quantity	Pounds
VX-Nerve	Ton Containers	1,690	2,538,660

Pine Bluff Chemical Activity lies 35 miles southeast of Little Rock, Arkansas. It houses 3,850 tons (12.5% of U.S. chemical weapons).

Agent	Item	Quantity	Pounds
HT-Blister	Ton Containers	3,591	6,249,100
HD-Blister	Ton Containers	107	188,400

10.4. U.S. Chemical Weapons Storage and Destruction *(continued).*

Agent	Item	Quantity	Pounds
GB-Nerve	M55 Rockets	90,231	965,480
GB-Nerve	M56 Rocket Warheads	178	1,900
VX-Nerve	M55 Rockets	19,582	195,820
VX-Nerve	M56 Rocket Warheads	26	260
VX-Nerve	Mines	9,378	98,460

Pueblo Chemical Depot is located 14 miles east of Pueblo, Colorado. It houses 2,611 tons of chemical weapons (8.5% of the U.S. stockpile).

Agent	Item	Quantity	Pounds
HT-Blister	4.2-inch Cartridges	20,384	118,220
HD-Blister	4.2-inch Cartridges	76,722	460,340
HD-Blister	105mm Cartridges	383,418	1,138,760
HD-Blister	155mm Projectiles	299,554	3,504,780

Umatilla Chemical Depot, 7 miles west of Hermiston, Oregon, holds 3,717 tons (12.1% of the U.S. chemical weapons stockpile).

Agent	Item	Quantity	Pounds
H-Blister	Ton Containers	2,635	4,679,040
GB-Nerve	155mm Projectiles	47,406	308,140
GB-Nerve	8-inch Projectiles	14,246	206,560
GB-Nerve	M55 Rockets	91,375	977,720
GB-Nerve	M56 Rocket Warheads	67	720
GB-Nerve	500-lb Bombs	27	2,920
GB-Nerve	750-lb Bombs	2,418	531,960
VX-Nerve	155mm Projectiles	32,313	193,880
VX-Nerve	8-inch Projectiles	3,752	54,400
VX-Nerve	Mines	11,685	122,700
VX-Nerve	M55 Rockets	14,513	145,140
VX-Nerve	M56 Rocket Warheads	6	60
VX-Nerve	Spray Tanks	156	211,540

Source: U.S. Soldier and Biological Chemical Command. From the Command's Web site at http://www.sbccom.army.mil/programs/stockpile.htm

Destruction Facilities

Johnston Atoll Chemical Agent Disposal System
Status as of June 27, 1999

Original tonnage:	2,031 tons
Remaining tonnage:	358 tons
Destroyed to date:	1,673 tons
Percent of original tonnage destroyed:	82.3%
Percent of original munitions destroyed:	82.6%

Munitions and bulk containers destroyed:

4,763 HD blister gaent-filled projectiles (155mm)
36 HD blister agent-filled projectiles (155mm-Solomon Island)
43,660 HD blister agent-filled mortars (4.2-inch)
13,020 GB nerve agent–filled projectiles (8 inch)
49,360 GB nerve agent–filled projectiles (105mm)
107,197 GB nerve agent–filled projectiles (155mm)
2,570 MK-94 (500-pound) bombs filled with GB nerve agent
3,047 MC-1 (750-pound) bombs filled with GB nerve agent
72,242 M-55 GB & VX nerve agent-filled rockets/warheads
45,154 blister agent–filled projectiles (105mm)
68 blister agent–filled ton containers
66 GB nerve agent–filled ton containers

10.4. U.S. Chemical Weapons Storage and Destruction *(continued)*.

Tooele Chemical Agent Disposal Facility
Status as of June 27, 1999

Original tonnage:	13,616 tons
Remaining tonnage:	10,736 tons
Destroyed to date:	2,880 tons (based on nominal fill weight or 1506.08 lbs per ton container)
Percent of original tonnage destroyed:	21.1%
Percent of original munitions destroyed:	16.6%

Munitions and bulk containers destroyed:

162,162 GB nerve agent-filled projectiles (105 mm)
2,853 GB nerve agent–filled ton containers
4,463 NC-1 (750-pound) bombs filled with GB nerve agent
20,219 M-55 GB nerve agent–filled rockets

Source: U.S. Army, Program Manager for Chemical Demilitarization. From the Program Manager's Web site at http://www-pmcd.apgea.army.mil, July 1999. See "At a Glance" section.

10.5. U.S. Defense Technology: Programmed Funding for Various Technologies, 1997–2003. Biomedical Defense Technology Objectives.

Advanced Medical Technology—Advanced Field Medical Support in Forward Combat Areas. Focuses on development of biophysical and biochemical smart sensors and information fusion to monitor warfighter physiological status. Will help avoid casualties, provide physiologic status of combat casualties, and support rapid casualty care. Will develop computer-assisted monitoring, diagnosis, and medical decision assist support software and hardware, such as portable and handheld medical imaging devices. Will also involve development of virtual reality technology for combat-care training and triage.
Programmed Funding (in millions of dollars)

	FY97	FY98	FY99	FY00	FY01	FY02	FY03
Total	$19.5	$13.5	$8.6	$5.9	$6.3	$6.6	$6.7

Antiparasitic Drug Program (to prevent or treat 80% of infections caused by malaria or leishmania parasites).
Programmed Funding (in millions of dollars)

	FY97	FY98	FY99	FY00	FY01	FY02	FY03
Total	$4.3	$3.8	$3.9	$3.7	$4.0	$4.5	$4.3

Chemical Agent Prophylaxes (pre-treatments for nerve agents).
Programmed Funding (in millions of dollars)

	FY97	FY98	FY99	FY00	FY01	FY02	FY03
Total	$2.3	$2.3	$2.2	$-	$-	$-	$-

Far-Forward Assessment and Treatment for Blood Loss; Development of Blood Products and Resuscitation. Goals are to double storage life of liquid whole blood and blood products and to define optimum resuscitation perfusion pressures, volumes, and temperatures for early versus delayed field resuscitation of hemorrhage. Will evaluate products to keep 20% of hemorrhages from becoming life-threatening. (In pre-hospital setting of modern battlefields, 50% of combat deaths are due to hemorrhage!)

Programmed Funding (in millions of dollars)

	FY97	FY98	FY99	FY00	FY01	FY02	FY03
Total	$8.1	$8.2	$8.1	$8.5	$8.7	$-	$-

Far-Forward Assessment, Treatment, and Management of Combat Trauma and Severe Hemorrhage and Sequelae.
Programmed Funding (in millions of dollars)

	FY97	FY98	FY99	FY00	FY01	FY02	FY03
Total	$11.3	$10.9	$10.3	$10.5	$10.5	$10.9	$11.3

10.5. U.S. Defense Technology: Programmed Funding for Various Technologies, 1997–2003. Biomedical Defense Technology Objectives *(continued)*.

Laser Bioeffects Countermeasures (against eye damage and visual performance).
Programmed Funding (in millions of dollars)

	FY97	FY98	FY99	FY00	FY01	FY02	FY03
Total	$3.8	$4.3	$4.6	$4.8	$4.9	$5.2	$5.2

Medical Countermeasures for Botulinum Toxin (a biological weapon).
Programmed Funding (in millions of dollars)

	FY97	FY98	FY99	FY00	FY01	FY02	FY03
Total	$2.7	$1.8	$-	$-	$-	$-	$-

Medical Countermeasures for Encephalomyelitis Viruses (biological weapons).
Programmed Funding (in millions of dollars)

	FY97	FY98	FY99	FY00	FY01	FY02	FY03
Total	$1.2	$1.2	$0.8	$0.7	$-	$-	$-

Medical Countermeasures for Staphylococcal Enterotoxin B (a biological weapon).
Programmed Funding (in millions of dollars)

	FY97	FY98	FY99	FY00	FY01	FY02	FY03
Total	$3.3	$3.9	$4.1	$1.8	$-	$-	$-

Medical Countermeasures for Vesicant Agents (chemical agents causing blisters and burns).
Programmed Funding (in millions of dollars)

	FY97	FY98	FY99	FY00	FY01	FY02	FY03
Total	$8.6	$8.9	$8.8	$8.9	$-	$-	$-

Medical Countermeasures for Yersinia pestis (a biological weapon that causes plague).
Programmed Funding (in millions of dollars)

	FY97	FY98	FY99	FY00	FY01	FY02	FY03
Total	$1.8	$1.9	$-	$-	$-	$-	$-

Prevention of Diarrheal Diseases (vaccines).
Programmed Funding (in millions of dollars)

	FY97	FY98	FY99	FY00	FY01	FY02	FY03
Total	$2.8	$2.6	$2.9	$2.7	$3.1	$1.8	$2.0

Sustained Operations Enhancement Ensemble. Develops and adapts countermeasures for behavioral and physiological degradation caused by the demands of sustained operation. Focuses on developing drugs to counter fatigue and sleep loss, on monitoring alertness, and on understanding sleep physiology.
Programmed Funding (in millions of dollars)

	FY97	FY98	FY99	FY00	FY01	FY02	FY03
Total	$6.6	$7.4	$8.9	$8.7	$8.9	$9.2	$-

Toxic Hazards Evaluation Tools. Will develop tools to identify and analyze toxological exposures.
Programmed Funding (in millions of dollars)

	FY97	FY98	FY99	FY00	FY01	FY02	FY03
Total	$3.2	$3.4	$3.4	$3.4	$4.1	$4.2	$-

Vaccines for Prevention of Malaria.
Programmed Funding (in millions of dollars)

	FY97	FY98	FY99	FY00	FY01	FY02	FY03
Total	$4.0	$3.8	$3.9	$3.8	$4.1	$4.8	$4.8

Source: U.S. Department of Defense. *1997 Defense Technology Area Plan.* Ft. Belvoir, VA: Defense Technical Information Center, 1997. Unnumbered tables.

10.6. Defense Technology: Programmed Funding for Various Technologies, 1997–2003. Ground and Sea Vehicles Technology Objectives.

Future Combat System. Will develop a ground combat vehicle weapons platform with leap-ahead capabilities in the areas of mobility, lethality, survivability, deployability, fightability, and sustainability.
Programmed Funding (in millions of dollars)

	FY97	FY98	FY99	FY00	FY01	FY02	FY03
Total	$10.7	$24.0	$22.9	$15.2	$15.8	$23.2	$37.2

Ground Vehicle Electronic Systems. To reduce manning requirements to operate and support ground vehicles. Involves integration of advanced vehicle electronic architecture and crewstation.
Programmed Funding (in millions of dollars)

	FY97	FY98	FY99	FY00	FY01	FY02	FY03
Total	$2.9	$2.9	$7.8	$10.6	$11.5	$8.7	$8.8

Submarine Signature Control. Will control acoustic and nonacoustic signatures to assure a stealth advantage.
Programmed Funding (in millions of dollars)

	FY97	FY98	FY99	FY00	FY01	FY02	FY03
Total	$23.0	$8.0	$2.3	$2.3	$2.4	$2.5	$2.5

Surface Ship Integrated Topside Concepts. Focuses on reducing signatures, controlling electromagnetic emissions, and improving sensor performance.
Programmed Funding (in millions of dollars)

	FY97	FY98	FY99	FY00	FY01	FY02	FY03
Total	$17.5	$17.4	$14.0	$6.4	$2.4	$2.4	$2.4

Source: U.S. Department of Defense. *1997 Defense Technology Area Plan.* Ft. Belvoir, VA: Defense Technical Information Center, 1997, Unnumbered tables.

10.7. U.S. Defense Technology: Programmed Funding for Various Technologies, 1997–2003. Materials and Processes Objectives.

Hazardous and Toxic Waste Treatment/Destruction for DoD Operations. Technologies include advanced oxidation, reductive electrochemical processes, advanced chemical reactors, and biotechnology for treating wastes relating to propellants, explosives, pyrotechnics, and complex industrial wastes.
Programmed Funding (in millions of dollars)

	FY97	FY98	FY99	FY00	FY01	FY02	FY03
Total	$8.1	$5.5	$6.9	$6.3	$3.2	$4.8	$-

Laser Eye Protection. Will develop advanced materials (interference filters, holograms, dielectric stacks, rugates) to provide protection against low-energy-visible and near-infrared devices such as rangefinders, illuminators, and designators. Will also protect against high-energy laser weapons.
Programmed Funding (in millions of dollars)

	FY97	FY98	FY99	FY00	FY01	FY02	FY03
Total	$3.6	$6.8	$9.2	$9.4	$7.8	$6.8	$-

Materials and Processes for Reentry Vehicle Technology. Will develop advanced nosetip, heatshield, and antenna window materials and processes for intercontinental ballistic missiles and submarine-launched ballistic missiles.
Programmed Funding (in millions of dollars)

	FY97	FY98	FY99	FY00	FY01	FY02	FY03
Total	$6.8	$4.7	$5.1	$4.9	$5.1	$5.3	$-

Protective Materials for Combatant and Combat Systems against Conventional Weapons. Will develop ultra light materials and new armor principles.
Programmed Funding (in millions of dollars)

	FY97	FY98	FY99	FY00	FY01	FY02	FY03
Total	$2.1	$5.1	$6.3	$9.0	$9.0	$9.0	$5.0

Source: U.S. Department of Defense. *1997 Defense Technology Area Plan.* Ft. Belvoir, VA: Defense Technical Information Center, 1997. Unnumbered tables.

10.8. U.S. Defense Technology: Programmed Funding for Various Technologies, 1997–2003. Air Platform Technology Objectives.

Flight Control Technology for Affordable Global Reach/Power. Develops aircraft control systems that automatically adjust to and survive combat damage, systems to identify flight control component failures, etc.
  Programmed Funding (in millions of dollars)

	FY97	FY98	FY99	FY00	FY01	FY02	FY03
Total	$9.7	$15.2	$16.0	$16.7	$18.0	$17.7	$19.2

Source: U.S. Department of Defense. *1997 Defense Technology Area Plan.* Ft. Belvoir, VA: Defense Technical Information Center, 1997. Unnumbered tables.

10.9. U.S. Defense Technology: Programmed Funding for Various Technologies, 1997–2003. Chemical/Biological Defense and Nuclear Technology Objectives.

Enzymatic Decontamination. Will develop a new generation of chemical and biological warfare agent decontaminants that are nontoxic, noncorrosive, environmentally safe, and lightweight.
  Programmed Funding (in millions of dollars)

	FY97	FY98	FY99	FY00	FY01	FY02	FY03
Total	$0.8	$0.8	$0.8	$0.8	$0.8	$0.9	$-

Laser Standoff Chemical Detection Technology. Goal is to detect agents at a distance of 20 km.
Programmed Funding (in millions of dollars)

	FY97	FY98	FY99	FY00	FY01	FY02	FY03
Total	$0.5	$2.7	$4.1	$5.4	$-	$-	$-

Nuclear Hardness and Survivability Testing Technologies. Will develop technology to supplant nuclear testing.
Programmed Funding (in millions of dollars)

	FY97	FY98	FY99	FY00	FY01	FY02	FY03
Total	$54.9	$66.0	$63.6	$62.5	$59.4	$62.4	$63.3

Source: U.S. Department of Defense. *1997 Defense Technology Area Plan.* Ft. Belvoir, VA: Defense Technical Information Center, 1997. Unnumbered tables.

10.10. U.S. Defense Technology: Programmed Funding for Various Technologies, 1997–2003. Sensors, Electronics, and Battlespace Environment Objectives.

Advanced Infrared Search and Track Systems. Develops and demonstrates highly stabilized passive infrared search and track systems (IRSTs) sensing capabilities for over-the-horizon detection and precision tracking of TBMs and cruise missiles for ship self-defense, and sensors for detection and precision tracking of threat aircraft from ground combat vehicles.
  Programmed Funding (in millions of dollars)

	FY97	FY98	FY99	FY00	FY01	FY02	FY03
Total	$4.8	$2.8	$-	$-	$-	$-	$-

Advanced Pilotage. Develops and demonstrates advanced sensor technology for night/adverse weather pilotage/navigation requirements. Included will be all aspect viewing via fixed-mounted sensors providing full sphere coverage, large staring arrays, and multispectral image fusion.
  Programmed Funding (in millions of dollars)

	FY97	FY98	FY99	FY00	FY01	FY02	FY03
Total	$8.2	$7.4	$-	$-	$-	$-	$-

Foliage Penetration Detection Algorithm Demonstration. Will conduct experiments to yield statistics to support development of foliage penetration radar system technology.
  Programmed Funding (in millions of dollars)

	FY97	FY98	FY99	FY00	FY01	FY02	FY03
Total	$1.8	$0.6	$-	$-	$-	$-	$-

10.10. U.S. Defense Technology: Programmed Funding for Various Technologies, 1997–2003. Sensors, Electronics, and Battlespace Environment Objectives *(continued)*.

High-Density Radiation-Resistant Microelectronics.
Programmed Funding (in millions of dollars)

	FY97	FY98	FY99	FY00	FY01	FY02	FY03
Total	$8.5	$7.0	$3.5	$-	$-	$-	$-

Space Radiation Mitigation for Satellite Operations.
Programmed Funding (in millions of dollars)

	FY97	FY98	FY99	FY00	FY01	FY02	FY03
Total	$2.0	$3.8	$4.0	$4.4	$4.5	$-	$-

Source: U.S. Department of Defense. *1997 Defense Technology Area Plan.* Ft. Belvoir, VA: Defense Technical Information Center, 1997. Unnumbered tables.

10.11. U.S. Defense Technology: Programmed Funding for Various Technologies, 1997–2003. Weapons Objectives.

Airborne Lasers for Theater Missile Defense.
Programmed Funding (in millions of dollars)

	FY97	FY98	FY99	FY00	FY01	FY02	FY03
Total	$10.7	$12.3	$10.1	$9.8	$9.0	$9.1	$9.4

Ground-Based Laser Antisatellite Technology.
Programmed Funding (in millions of dollars)

	FY97	FY98	FY99	FY00	FY01	FY02	FY03
Total	$13.1	$13.4	$12.4	$10.6	$11.1	$11.3	$11.6

Multimission Space-Based Laser. Will support theater missile defense, national missile defense, ASAT, surveillance, target designation, and active and passive target discrimination.
Programmed Funding (in millions of dollars)

	FY97	FY98	FY99	FY00	FY01	FY02	FY03
Total	$98.5	$29.0	$28.7	$28.4	$27.8	$28.4	$28.0

Source: U.S. Department of Defense. *1997 Defense Technology Area Plan.* Ft. Belvoir, VA: Defense Technical Information Center, 1997. Unnumbered tables.

10.12. U.S. Defense Technology: Programmed Funding for Various Technologies, 1997–2003. Information Superiority Objectives.

Rapid Battlefield Visualization ACTD. Will develop the ability to rapidly collect source data and generate high-resolution terrain databases to support crisis response and force projection operations.
Programmed Funding (in millions of dollars)

	FY97	FY98	FY99	FY00	FY01	FY02	FY03
Total	$9.4	$11.7	$15.8	$14.8	$-	$-	$-

Source: U.S. Department of Defense. *1997 Joint Warfighting Science and Technology Plan.* Ft. Belvoir, VA: Defense Technical Information Center, 1997. Unnumbered tables.

10.13. U.S. Joint Warfighting Capability Objectives Funding, Fiscal Year 1998 (millions of dollars).

Objective	Funding
Information Superiority	$202.2
Precision Force	$213.8
Combat Identification	$19.3
Joint Theater Missile Defense	$54.7
Military Operations in Urban Terrain	$82.2
Joint Readiness & Logistics	$46.2
Joint Countermine	$47.4
Electronic Combat	$56.4
Chemical/Biological Warfare Defense & Protection	$27.5
Counter Weapons of Mass Destruction	$46.4

Source: U.S. Department of Defense. *1997 Joint Warfighting Science & Technology Plan.* Ft. Belvoir, VA: Defense Technical Information Center, 1997. Figure 11-2.

10.14. U.S. Department of Defense Strategic Forces Highlights, 1992–2000.

	FY 1992	FY 1993	FY 1994	FY 1995	FY 1996	FY 1997	FY 1998	FY 1999	FY 2000
Land-based ICBMs [2]									
Minuteman II (1 warhead each) plus Minuteman III (3 warheads each)	880	737	625	535	530	530	500	500	500
Peacekeeper (10 warheads each)	50	50	50	50	50	50	50	50	50
Heavy Bombers (PAI) [3]									
B-52	129	110	64	74	56	56	56	56	56
B-1	84	84	84	60	60	60	70	72	78
B-2	0	0	3	6	9	10	12	14	16
Submarine-launched Ballistic Missiles [2]									
Poseidon (C-3) and Trident (C-4) missiles on pre-Ohio-class subs	176	96	48	0	0	0	0	0	0
Trident (C-4 and D-5) missiles on Ohio-class submarines	288	312	336	360	384	408	432	432	432

Notes:

[1] Force levels shown are for the ends of the fiscal years in question. Inventory levels for future years reflect the force structures supported by the FY1998 budget. The actual force levels for FY1999 and FY2000 will depend on future decisions.

[2] Number of operational missiles. Not ones in maintenance or overhaul status.

[3] PAI = Primary Aircraft Inventory. PAI excludes backup and attrition reserve aircraft as well as aircraft in depot maintenance. Total inventory counts will be higher than the PAI figures given here.

Source: William H. Cohen, U.S. Secretary of Defense. *Annual Defense Report, 1998.* Washington, DC: U.S. Government Printing Office. http://www.dtic.mil/execsec/adr_intro.html.

LAW ENFORCEMENT

10.15. Selected Summary Data for State and Local Law Enforcement Agencies with 100 or More Officers, by Type of Agency, United States, 1990 and 1993: Weapons and Body Armor.

	1990					1993				
	County Police (33)	Municipal Police (411)	Sheriff (146)	Special Police (22)	State Police (49)	County Police (33)	Municipal Police (411)	Sheriff (146)	Special Police (22)	State Police (49)
Percent of agencies authorizing the use of semiautomatic sidearms:										
Any type	85%	92%	92%	59%	80%	88%	98%	94%	91%	96%
10mm	6%	7%	18%	6%	18%	15%	15%	29%	0	18%
9mm	76%	86%	88%	59%	74%	70%	82%	80%	86%	73%
.45	21%	38%	49%	12%	27%	27%	38%	58%	5%	35%
.380	6%	16%	17%	0	10%	12%	15%	20%	18%	18%
.357	0	4%	7%	0	4%	NA	NA	NA	NA	NA
Percent of agencies authorizing the use of nonlethal weapons:										
Baton, collapsible	NA	NA	NA	NA	NA	36%	48%	64%	23%	53%
Baton, PR-24	NA	NA	NA	NA	NA	48%	58%	68%	55%	47%
Baton, traditional	NA	NA	NA	NA	NA	79%	64%	62%	68%	55%
Capture net	NA	NA	NA	NA	NA	0	6%	1%	0	0
Carotid hold	NA	NA	NA	NA	NA	18%	20%	16%	18%	16%
Choke hold	NA	NA	NA	NA	NA	6%	4%	5%	0	10%
Flash/bang grenade	NA	NA	NA	NA	NA	61%	57%	63%	14%	41%
Pepper spray	NA	NA	NA	NA	NA	70%	69%	66%	36%	63%
Rubber bullet	NA	NA	NA	NA	NA	9%	9%	12%	0	10%
Soft projectile	NA	NA	NA	NA	NA	9%	7%	10%	0	6%
Stun gun	NA	NA	NA	NA	NA	6%	14%	28%	18%	4%
Tear gas—personal issue	NA	NA	NA	NA	NA	42%	31%	35%	41%	29%
Tear gas—large volume	NA	NA	NA	NA	NA	64%	42%	41%	18%	41%
Three-pole trip	NA	NA	NA	NA	NA	3%	—	0	5%	0
Chemical agents	82%	70%	69%	53%	61%	NA	NA	NA	NA	NA
Electrical devices	27%	22%	34%	24%	4%	NA	NA	NA	NA	NA
Impact devices	77%	77%	78%	82%	65%	NA	NA	NA	NA	NA
Restraining devices	29%	20%	26%	24%	25%	NA	NA	NA	NA	NA
Percent of agencies supplying or giving cash allowance for body armor to regular field officers:										
Armor supplied to all regular field officers	79%	65%	65%	71%	76%	85%	73%	76%	68%	80%
Armor supplied to some regular field officers	15%	8%	11%	0	8%	3%	5%	5%	9%	6%
Cash allowance given to all regular field officers	6%	10%	3%	6%	2%	6%	8%	3%	9%	6%

10.15. Selected Summary Data for State and Local Law Enforcement Agencies with 100 or More Officers, by Type of Agency, United States, 1990 and 1993: Weapons and Body Armor (continued).

	1990					1993				
	County Police (33)	Municipal Police (411)	Sheriff (146)	Special Police (22)	State Police (49)	County Police (33)	Municipal Police (411)	Sheriff (146)	Special Police (22)	State Police (49)
Percent of agencies authorizing the cash allowance given to some regular field officers	0	2%	3%	0	2%	0	1%	4%	0	0
Percent of agencies requiring that body armor be worn by:										
All regular field officers	18%	16%	26%	35%	12%	24%	29%	36%	55%	16%
Some regular field officers	6%	8%	7%	12%	6%	6%	8%	12%	9%	16%
All special operations officers	24%	21%	29%	35%	14%	18%	30%	38%	45%	29%
Some special operations officers	15%	15%	14%	18%	8%	21%	19%	22%	9%	18%

Source: U.S. Bureau of Justice Statistics. Law Enforcement Management and Administrative Statistics: *Data for Individual State and Local Agencies with 100 or More Officers.* 1990 and 1993 editions. Washington, DC: U.S. Department of Justice, Bureau of Justice Statistics.

10.16. Selected Summary Data for State and Local Law Enforcement Agencies with 100 or More Officers, by Type of Agency, United States, 1990 and 1993: 911 Systems, Computers and Information Systems, and Vehicles.

	1990					1993				
	County Police (33)	Municipal Police (411)	Sheriff (146)	Special Police (22)	State Police (49)	County Police (33)	Municipal Police (411)	Sheriff (146)	Special Police (22)	State Police (49)
Percent of agencies participating in a 911 emergency system:										
Expanded/enhanced 911 system	65%	57%	56%	12%	37%	73%	77%	79%	5%	33%
Basic 911 system	24%	27%	29%	24%	31%	21%	18%	15%	32%	39%
Percent of agencies operating vehicles other than cars:										
All-terrain vehicles	NA	NA	NA	NA	NA	27%	18%	41%	18%	29%
Armored vehicles	9%	8%	10%	12%	8%	9%	13%	16%	9%	20%
Boats	47%	23%	67%	12%	43%	39%	25%	71%	18%	33%
Fixed-wing aircraft	1 21%	1 6%	1 28%	1 0	1 86%	21%	6%	30%	0	84%
Helicopters	35%	13%	28%	0	59%	45%	12%	28%	0	53%
Bicycles	9%	20%	7%	0	4%	NA	NA	NA	NA	NA
Buses	41%	22%	47%	24%	49%	NA	NA	NA	NA	NA
Percent of agencies operating computers:										
Digital terminal, car-mounted	24%	21%	12%	18%	14%	39%	38%	19%	14%	29%
Digital terminal, handheld	0	7%	3%	6%	2%	12%	14%	4%	5%	6%
Laptop	35%	23%	25%	24%	53%	67%	50%	64%	41%	84%
Mainframe	88%	75%	88%	65%	92%	88%	80%	89%	64%	96%
Mini	53%	53%	44%	53%	55%	64%	56%	53%	23%	49%
Personal	100%	94%	92%	88%	98%	97%	98%	97%	95%	96%
Percent of agencies with Automated Fingerprint Identification System (AFIS) faciities										
Exclusive ownership of an AFIS	15%	6%	7%	6%	27%	21%	9%	14%	0	33%
Shared ownership of an AFIS	50%	21%	20%	12%	8%	55%	28%	28%	0	14%
Terminal with access to a remote AFIS site	44%	16%	20%	12%	22%	42%	26%	36%	0	33%
Percent of agencies using computers for:										
Budgeting	85%	69%	80%	71%	90%	91%	80%	88%	73%	86%
Crime analysis	94%	85%	69%	77%	53%	85%	90%	75%	73%	45%
Crime investigation	88%	82%	78%	82%	59%	88%	86%	84%	50%	63%
Dispatch	77%	77%	65%	41%	76%	76%	89%	76%	27%	67%
Fleet management	74%	47%	55%	35%	0	73%	53%	60%	55%	88%
Jail management	15%	17%	88%	0	0	6%	21%	78%	0	0
Manpower allocation	62%	54%	44%	82%	61%	64%	60%	51%	68%	59%
Record-keeping	94%	89%	94%	94%	98%	91%	94%	97%	91%	92%

10.16. Selected Summary Data for State and Local Law Enforcement Agencies with 100 or More Officers, by Type of Agency, United States, 1990 and 1993: 911 Systems, Computers and Information Systems, and Vehicles *(continued)*.

	1990					1993				
	County Police (33)	Municipal Police (411)	Sheriff (146)	Special Police (22)	State Police (49)	County Police (33)	Municipal Police (411)	Sheriff (146)	Special Police (22)	State Police (49)
Research	NA	NA	NA	NA	NA	67%	60%	40%	50%	63%
Percent of agencies maintaining computer files on:										
Arrests	82%	88%	88%	82%	67%	85%	93%	93%	91%	65%
Calls for service	74%	90%	77%	59%	39%	82%	94%	84%	59%	55%
Criminal histories	77%	76%	80%	59%	67%	70%	76%	77%	45%	63%
Driver's license information	NA	NA	NA	NA	NA	45%	42%	42%	36%	69%
Evidence	53%	65%	53%	24%	27%	82%	71%	66%	27%	39%
Fingerprints	NA	NA	NA	NA	NA	58%	38%	35%	0	43%
Inventory	NA	NA	NA	NA	NA	82%	62%	75%	55%	84%
Payroll	NA	NA	NA	NA	NA	79%	67%	82%	64%	84%
Personnel	NA	NA	NA	NA	NA	85%	82%	88%	77%	80%
Stolen property (other than vehicles)	[2] 68%	[2] 82%	[2] 77%	[2] 59%	[2] 57%	79%	80%	69%	45%	39%
Stolen vehicles	[2]	[2]	[2]	[2]	[2]	67%	79%	64%	41%	49%
Summonses	44%	37%	58%	24%	20%	42%	34%	55%	27%	24%
Traffic accidents	NA	NA	NA	NA	NA	76%	80%	45%	41%	78%
Traffic citations	59%	71%	63%	29%	78%	67%	73%	62%	50%	65%
Uniform Crime Reports—NIBRS	[3] 65%	[3] 82%	[3] 71%	[3] 53%	[3] 51%	42%	40%	46%	27%	41%
Uniform Crime Reports—Summary	[3]	[3]	[3]	[3]	[3]	79%	79%	71%	55%	57%
Vehicle registration	62%	40%	40%	41%	65%	45%	47%	47%	36%	65%
Warrants	85%	72%	91%	35%	51%	79%	75%	90%	41%	55%

Notes:
[1] For 1990, category described as simply "Airplanes"
[2] Stolen property other than vehicles and stolen vehicles combined for 1990
[3] Uniform Crime Reports—NIBRS and Uniform Crime Reports described simply as "Uniform Crime Reports" in 1990

Source: Bureau of Justice Statistics. Law Enforcement Management and Administrative Statistics: *Data for Individual State and Local Agencies with 100 or More Officers. 1990 and 1993 editions.* Washington, DC: U.S. Department of Justice, Bureau of Justice Statistics.

10.17. Crime Scene DNA Samples Received and Analyzed by U.S. Forensic Laboratories, 1995–98 and Projected for 1999 (number of samples).

Year	Samples Received	Samples Analyzed	Percent of Samples Received That Were Analyzed
1995	NA	12,400	NA
1996	19,000	14,200	75%
1997	24,586	16,545	67%
1998	32,359	25,055	77%
1999 projected	46,443	37,759	81%

Note:
As of July 1998, there were 132 laboratories in 49 states (of 161 surveyed) in the United States conducting DNA crime scene analysis.

Source: FBI Laboratory, Forensic Science Systems Unit. *1998 CODIS DNA Laboratory Survey.* Washington, DC: Federal Bureau of Investigation, 1999.

10.18. U.S. Law Enforcement Agency Use of Mobile Data Terminals and Laptop Computers, 1997 and Estimated for 1999 (percent of agencies).

	1997	1999 (projected)
Mobile Data Terminals	21%	31%
Laptop Computers	26%	56%
Both	9%	NA

Notes:
Methodology: A 10-page questionnaire was sent to all agencies that employ more than 100 sworn officers and to a stratified random sample of smaller agencies across the country. A total of 1,334 agencies responded, an overall response rate of 48%. Agencies were categorized by size (the number of sworn officers served as the basis for six size categories) and type (local police, sheriff's departments, special police, and State police) for analyses. A bias analysis was conducted, as were analyses based on weighted data to correct for under- or overrepresented groups. All data in this report are based on the respondent sample. The sample is broadly representative of the nation.
A mobile data terminal is a small, computer-like system usually installed in a patrol car. It allows the officer to receive and transmit a limited range of information from and to a communications center.

Source: Mary J. Taylor, Robert C. Epper, and Thomas K. Tolman. State and Local *Law Enforcement Wireless Communications and Interoperability: A Quantitative Analysis.* National Institute of Justice Research Report NCJ 168961. Washington, DC: National Institute of Justice, 1998. Page 34. http://www.ncjrs.org/pdffiles1/168961.pdf

10.19. U.S. Law Enforcement Use of Wireless Voice and/or Data Security, 1997 (percent of agencies).

Type of Security	State Agencies	Sheriff's Departments	Local Police	Special Police
Scrambling Devices	40%	27%	20%	7%
Digital Encryption	55%	19%	21%	14%
Digital Voice Processing	35%	NA	NA	NA

Notes:
Methodology: A 10-page questionnaire was sent to all agencies that employ more than 100 sworn officers and to a stratified random sample of smaller agencies across the country. A total of 1,334 agencies responded, an overall response rate of 48%. Agencies were categorized by size (the number of sworn officers served as the basis for six size categories) and type (local police, sheriff's departments, special police, and State police) for analyses. A bias analysis was conducted, as were analyses based on weighted data to correct for under- or overrepresented groups. All data in this report are based on the respondent sample. The sample is broadly representative of the nation.
Scrambling is a method of converting an input waveform to a digital representation, which is then encrypted and transmitted. The receiver decrypts the data and regenerates the original analog signal. Voice security in general refers to over-the-air audio that is unintelligible and/or inaccessible without appropriate means of decoding.

Source: Mary J. Taylor, Robert C. Epper, and Thomas K. Tolman. *State and Local Law Enforcement Wireless Communications and Interoperability: A Quantitative Analysis.* National Institute of Justice Research Report NCJ 168961. Washington, DC: National Institute of Justice, 1998. Page 37. http://www.ncjrs.org/pdffiles1/168961.pdf

10.20. U.S. Law Enforcement Use of Computerized Crime Mapping, 1997 and 1998* (percent of agencies that already use computerized mapping in some form).

Technology Used	Percent of Agencies
Personal or desktop computers	89%
Commercially available software	88%
Internet and other technically advanced resources	82%
Customized version of commercially available mapping application or custom-developed mapping program for internal use	38%
Global Positioning Systems (to assist in operations)	16%
Data Geocoded**	
Burglaries	95%
Arrest and incident data	91%
Motor vehicle thefts	87%
Robberies	86%
Maintain archive of geocoded crime data that begins between 1990 and 1998	76%
Rapes	71%
Homicides	69%
Larceny	69%
Calls-for-service data	65%
Aggravated assaults	62%
Vehicle recovery data	52%
Drug offenses	50%
Arson	40%
Uses for Computerized Crime Mapping	
Inform officers and investigators of crime incident locations	94%
Conduct crime cluster or hot spot analyses	77%
Make resource allocation decisions	56%
Evaluate interventions	49%
Inform residents about crime activity and changes in communities	47%
Identify repeat calls-for-service	44%

Notes:

* = Survey was mailed in March 1997 to a sample of law enforcement agencies in the United States. Departments that did not respond to the first mailing were sent a second survey. Surveys were accepted from the field until May 1, 1998. The total sample included 2,768 departments, with 2,004 agencies responding. However, only 261 (13%) of the departments surveyed currently use any computerized crime mapping.

** = Geocoding means to assign an x and y coordinate to an address so it can be placed on a map by using either street centerlines (every address within a block is encoded) or parcels (each piece of land that can be bought or sold is encoded). The majority of departments (77%) reported using street centerline reference files for geocoding and crime mapping. Many departments (25%) also reported using parcel database reference files for geocoding and crime mapping.

Source: Cynthia A. Mamalian, Nancy G. LaVigne, and the Staff of the Crime Mapping Research Center. *The Use of Computerized Crime Mapping by Law Enforcement: Survey Results.* Washington, DC: National Institute of Justice, 1999. http://www.ncjrs.org/pdffiles1/fs000237.pdf

Chapter 11. Medicine and Biotechnology

MEDICINE

Some medical technologies seem so bizarre that it is amazing that anyone ever thought to try them. How did the idea of inoculation come about? Whatever inspired someone to test for pregnancy by injecting women's urine into rabbits? Who first discovered acupuncture and what—besides sadism—possessed them to stick needles into another person? It all sounds kind of like a B movie.

But of course, many of these bizarre practices work, and now we are beginning to understand how and why. We have instruments for peering into hidden worlds on and in the body, and we know how to manipulate some of the structures and processes within them. We can predict and prevent and cure—within limits. We are unlocking the secrets of DNA, which tells the body how to look and function and sets a timetable for its doing so. And in many ways we are making patients more comfortable, though we still have light years to go on that score.

The statistics that follow may hold some surprises. Which drugs sell the most and are mentioned most, which procedures are most commonly performed, what is the dispensed prescription volume—these figures tell fascinating stories. For example, both Americans and the rest of the world suffer from poor stomachs. Tables 11.12, 11.13, 11.16, and 11.18 show that we spend more money on ulcer medicines like Prilosec/Losec and Zantac (see product class 28345 47 on the last) than any other type, including multivitamins, antidepressants, antihistamines, antivirals like Zovirax, and penicillin. (Note that the latest figures available do not reflect sales of Viagra, which burst onto the scene in 1998.) But antidepressants, hypertension drugs, and antibiotics are not far behind, representing an additional 14 out of the top 30 prescription drugs worldwide in 1996 (ulcer and stomach acid drugs accounted for another four of the 30). Nevertheless, Table 11.15, which illustrates prescriptions for classes of drugs, shows that the greatest growth in prescription volume in 1997 was for cholesterol reducers, followed by antihistamines. Prescriptions for the mighty ulcer medications,

which score the most in sales dollars, grew only 5% that same year. Could it be that ulcer medication is more expensive than the other highly prescribed types? By contrast, Table 11.11 shows that in 1995, the drugs mentioned most often in physician offices and outpatient and emergency departments were pain relievers and antimicrobials, with gastrointestinal agents ranked eighth out of 15 classes.

Major categories of medical technologies include

- Diagnostics
- Drugs
- Health care and patient information systems
- Medical devices and equipment
- Nonsurgical, nondrug treatments
- Nutrition
- Patient monitoring
- Surgical treatments
- Wellness and prevention

Hot topics in medical technology include

- Telemedicine (See Tables 11.1 and 11.2 which show number and percent of telemedicine facilities offering general and specialized services.)
- High-quality imaging equipment such as magnetic resonance imaging systems, and advanced imaging techniques, like image guidance of interventional procedures, elastography, and three-dimensional microscopy (See Table 11.23 for shipments of selected equipment.)
- Home diagnostic tests (See Table 11.3 for growth in home diagnostic revenues.)
- New drug delivery systems, such as implanted infusion pumps and drugs that are activated externally by the application of heat, ultrasound, light, or radiation
- DNA technology, including gene diagnostics and therapy
- Alternative medicine, like acupuncture and herbs (See Tables 11.9 and 11.20 for numbers of alternative practitioners and dollar sales of herbs.)

- Computerized patient records
- Cloning
- Noninvasive diagnostics, monitoring, and treatments

In addition to the statistics already mentioned, this chapter contains information about

- The types of procedures patients are undergoing (See Table 11.5 for a summary, Table 11.6 for surgical procedures, and Table 11.7 for nonsurgical procedures.)
- The growth in organ and tissue transplants (See Table 11.8. Cornea, bone marrow, and kidney transplants are the most often-performed transplants.)
- The size of the cancer treatment drug market (Table 11.14)
- The size of the pain management market (Table 11.21)
- Number of persons using assistive technology devices (Table 11.22—canes and hearing aids are used by more people than any other device)

BIOTECHNOLOGY

Many people are confused about what biotechnology is, and for good reason. Biotechnology is not a product, but a process. Biotechnologists use organisms or their cells or molecules to create products or modify the genetic and biochemical makeup of life forms, both animal and plant. But haven't people long been using microorganisms to make cheese and wine, breeding plants and animals, and creating vaccines?

What is new about today's biotechnology is the recent development of techniques that have allowed scientists to alter organisms faster and with laser-like precision. Genetic, protein, and tissue engineering (see Tables 2.11 and 2.12 in chapter 2 for some agricultural applications); DNA amplification; and monoclonal antibody technologies burst on the scene starting in the 1970s, inspiring new medical therapies, pharmaceuticals, and diagnostics; environmental cleanup methods; food

processing methodologies; and plant and animal breeding techniques.

Highly research intensive, the biotechnology industry includes the following applications:

- Human and veterinary diagnostics and therapeutics
- Breeding plants for resistance to pests, high nutrient content, better seed quality, and adjusting growth and ripening cycles
- Industrial and hazardous waste management, including converting waste into useful products
- Energy biomass
- Cosmetics
- Testing for toxicity and contamination in the environment
- Specialty and fine chemicals
- Industrial enzymes for processing pulp and paper and textiles

Notable products developed through biotechnology include

- Hepatitis A vaccine
- Serological detection of antibodies to HIV-1
- Alpha interferon (for hairy-cell leukemia, Kaposi's sarcoma, venereal warts, hepatitis C and B)
- Human growth hormone
- Alpha amylase (enzyme used in corn syrup and textile manufacturing)
- Environmental tests for detecting legionella bacteria in water
- OKT3 (monoclonal antibody used to treat kidney transplant rejection)

Tables 11.24 through 11.26 provide a glimpse at the biotechnology industry. Table 11.24 shows that the largest market area is therapeutics, with 315 (29.4%) companies participating. Diagnostics represents the second largest area (187 companies, 17.4%). Table 11.26 indicates that there were 500 biotechnology drugs in clinical trials in 1997, with a much lower number, 75, actually on the market. Of the 75, 6 generated worldwide sales of over $500 million in 1996.

MEDICAL SERVICES AND PROCEDURES

11.1. Telemedicine Facilities by Clinical Service, United States, as of July 1998.

Clinical Service	Number of Telemedicine Facilities	Percent of Telemedicine Facilities
Dental	22	10%
General medicine/primary care	67	30%
General surgery	55	24%
Mental health	63	28%
Nursing	18	8%
Nutrition	2	1%
Pharmacy	9	4%
Public health	15	7%
Specialty	196	87%

(225 of 525 facilities reporting)

Note:
Percentages calculated by the author.
Number of facilities totals more than 225 and percentages total more than 100% because a facility can offer more than one type of service.

Source: Joint Working Group on Telemedicine (JWGT). *Summary Report 7.* Found at http://www.tmgateway.org/gateway/, July 1998.

11.2. Top Ten Telemedicine Specialty Services, United States, as of July 1998.

Specialty Service	Number of Telemedicine Facilities	Percent of Telemedicine Facilities
Orthopedics	94	58%
Dermatology	91	56%
Radiology	86	53%
Cardiology	63	39%
Psychiatry	60	37%
Emergency medicine	50	31%
Gynecology	48	29%
Pediatrics	47	29%
Neurology	36	22%
Ophthalmology	35	21%

(163 of 525 facilities reporting)

Notes:
Percentages calculated by the author.
Number of facilities totals more than 163 and percentages total more than 100% because a facility can offer more than one service.

Source: Joint Working Group on Telemedicine (JWGT). *Summary Report 6.* Found at http://www.tmgateway.org/gateway/, July 1998.

11.3. Home Diagnostic Monitoring Revenues, 1994–97, and Projected for 1998–2000 (billions of dollars).

Year	Revenues
1994	$1.19
1995	$1.35
1996	$1.52
1997	$1.70
1998*	$1.89
1999*	$2.10
2000*	$2.34

Notes:
Includes test kits for pregnancy and ovulation, occult blood and feces, blood pressure, blood glucose, temperature, cholesterol, and HIV.
* = Projected.

Source: Frost & Sullivan. New York, NY. Unpublished. Cited in "Home Diagnostics Puts the Doctor Back in the House." *Drug Store News 19* (June 16, 1997): 142.

11.4. Medical Testing Chemicals, United States, 1994 and Forecast for 2000.

Demand (millions of dollars)	1994	2000F
U.S. total	$3,200	$5,120
In-vitro testing	$1,695	$2,735
In-vivo testing	$750	$1,195
Medical research testing	$613	$935
Home medical testing	$142	$255

Source: Freedonia Group. *Industry Study #760: Medical Testing Chemicals.* Cleveland, OH: Freedonia Group, 1996.

11.5. Number and Rate of Ambulatory and Inpatient Procedures by Selected Procedure Categories, United States, 1995.

Procedure category and ICD-9-CM code	Total		Ambulatory[1]		Inpatient[2]	
	Estimate	Standard error[3]	Estimate	Standard error[3]	Estimate	Standard error[3]
Number in thousands						
All procedures[4]	69,240	2,040	29,433	1,275	39,807	1,592
Endoscopy of small intestine with or without biopsy 45.11–45.14, 45.16	2,383	132	1,490	125	892	44
Arteriography and angiocardiography using contrast material 88.4–88.5	2,358	142	525	65	1,834	127
Extraction of lens 13.1–13.6	2,335	179	2,275	177	*	*
Endoscopy of large intestine with or without biopsy 45.21–45.25	2,321	139	1,809	135	512	32
Insertion of prosthetic lens (pseudophakos) 13.7	1,777	104	1,723	101	*	*
Injection or infusion of therapeutic or prophylactic substance 99.1–99.2	1,745	182	379	47	1,366	176
Episiotomy with or without forceps or vacuum extraction 72.1, 72.21, 72.31, 72.71, 73.6	1,411	63	*	*	1,410	63
Cardiac catheterization 37.21–37.23	1,389	85	321	43	1,068	74
Excision or destruction of lesion or tissue of skin and subcutaneous tissue 86.2–86.4	1,300	50	831	44	470	24
Diagnostic ultrasound 88.7	1,288	119	*	*	1,181	112
Respiratory therapy 93.9, 96.7	1,132	116	*	*	1,127	116
Computerized axial tomography 87.03, 87.41, 87.71, 88.01, 88.38	1,038	93	*	*	967	85
Rate per 10,000 population						
All procedures[4]	2,648.8	78.1	1,125.9	48.8	1,522.8	60.9
Endoscopy of small intestine with or without biopsy 45.11–45.14, 45.16	91.2	5.1	57.0	4.8	34.1	1.7
Arteriography and angiocardiography using contrast material 88.4–88.5	90.2	5.4	20.1	2.5	70.1	4.8
Extraction of lens 13.1–13.6	89.3	6.8	87.0	6.8	*	*
Endoscopy of large intestine with or without biopsy 45.21–45.25	88.8	5.3	69.2	5.2	19.6	1.2
Insertion of prosthetic lens (pseudophakos) 13.7	68.0	4.0	65.9	3.9	*	*
Injection or infusion of therapeutic or prophylactic substance 99.1–99.2	66.8	7.0	14.5	1.8	52.3	6.7
Episiotomy with or without forceps or vacuum extraction 72.1, 72.21, 72.31, 72.71, 73.6	54.0	2.4	*	*	53.9	2.4
Cardiac catheterization 37.21–37.23	53.1	3.3	12.3	1.6	40.9	2.8
Excision or destruction of lesion or tissue of skin and subcutaneous tissue 86.2–86.4	49.7	1.9	31.8	1.7	18.0	0.9
Diagnostic ultrasound 88.7	49.3	4.5	*	*	45.2	4.3
Respiratory therapy 93.9, 96.7	43.3	4.4	*	*	43.1	4.4
Computerized axial tomography 87.03, 87.41, 87.71, 88.01, 88.38	39.7	3.6	*	*	37.0	3.3

Notes:
* = Figure does not meet standard of reliability or precision.
[1] Data from the National Survey of Ambulatory Surgery. Excludes ambulatory surgery patients who became inpatients.
[2] Data from the National Hospital Discharge Survey. Excludes newborn infants.
[3] Standard error is a measure of the chance that data is incorrect and indicates the range within which correct answers probably fall. For example, in this table, the actual number of all procedures is most likely a value within plus or minus 2,040 from the estimated number of 69,240.
[4] All procedures estimated from the National Survey of Ambulatory Surgeries and the National Hospital Discharge Survey. Not all procedures are covered by these surveys. Procedure groupings and code numbers are based on the *International Classification of Diseases, 9th Revision, Clinical Modification* (ICD-0-CM).

Source: L. J. Kozak and M. F. Owings. "Ambulatory and Inpatient Procedures in the United States, 1995." National Center for Health Statistics. *Vital and Health Statistics* 13 (135). 1998. http://www.cdc.gov/ nchswww/data/sr13_135.pdf

11.6. Number and Rate of Ambulatory and Inpatient Surgical Procedures by Selected Procedure Categories, United States, 1995.

Procedure category and ICD-9-CM code	Total		Ambulatory [1]		Inpatient [2]	
	Estimate	Standard error**	Estimate	Standard error**	Estimate	Standard error**
Number in thousands						
All surgical procedures [3]	44,518	1,104	21,989	807	22,530	753
Extraction of lens 13.1–13.6	2,335	179	2,275	177	*	*
Insertion of prosthetic lens (pseudophakos) 13.7	1,777	104	1,723	101	*	*
Episiotomy with or without forceps or vacuum extraction 72.1, 72.21, 72.31, 72.71, 73.6	1,411	63	*	*	1,410	63
Cardiac catheterization 37.21–37.23	1,389	85	321	43	1,068	74
Excision or destruction of lesion or tissue of skin and subcutaneous tissue 86.2–86.4	1,300	50	831	44	470	24
Endoscopy of small intestine with biopsy 45.14, 45.16	1,197	74	754	69	443	28
Repair of current obstetric laceration 75.5–75.6	964	48	*	*	964	48
Operations on muscle, tendon, fascia, and bursa 82–83	881	39	636	36	245	17
Reduction of fracture 76.7,79.0–79.3	825	39	204	14	621	36
Caesarean section 74.0, 74.2, 74.4, 74.99	785	41	0	0	785	41
Artificial rupture of membranes 73.0	752	61	0	0	752	61
Cholecystectomy 73.0	711	35	241	23	470	27
Rate per 10,000 population						
All surgical procedures [3]	1703.0	42.2	841.2	30.9	861.9	28.8
Extraction of lens 13.1–13.6	89.3	6.8	87.0	6.8	*	*
Insertion of prosthetic lens (pseudophakos) 13.7	68.0	4.0	65.9	3.9	*	*
Episiotomy with or without forceps or vacuum extraction 72.1, 72.21, 72.31, 72.71, 73.6	54.0	2.4	*	*	53.9	2.4
Cardiac catheterization 37.21–37.23	53.1	3.3	12.3	1.6	40.9	2.8
Excision or destruction of lesion or tissue of skin and subcutaneous tissue 86.2–86.4	49.7	1.9	31.8	1.7	18.0	0.9
Endoscopy of small intestine with biopsy 45.14, 45.16	45.8	2.8	28.9	2.6	16.9	1.1
Repair of current obstetric laceration 75.5–75.6	36.9	1.8	*	*	36.9	1.8
Operations on muscle, tendon, fascia, and bursa 82–83	33.7	1.5	24.3	1.4	9.4	0.6
Reduction of fracture 76.7,79.0–79.3	31.5	1.5	7.8	0.5	23.7	1.4
Caesarean section 74.0, 74.2, 74.4, 74.99	30.0	1.6	0	0	30.0	1.6
Artificial rupture of membranes 73.0	28.8	2.3	0	0	28.8	2.3
Cholecystectomy 73.0	27.2	1.3	9.2	0.9	18.0	1.0

Notes:
* = Figure does not meet standard of reliability or precision.
** = Standard error is a measure of the chance that data is incorrect and indicates the range within which correct answers probably fall. For example, in this table, the actual number of all procedures is most likely a value within plus or minus 1,104 from the estimated number of 44,518.
[1] Data from the National Survey of Ambulatory Surgery. Excludes ambulatory surgery patients who became inpatients.
[2] Data from the National Hospital Discharge Survey. Excludes newborn infants.
[3] All surgical procedures estimated from the National Survey of Ambulatory Surgery and the National Hospital Discharge Survey. Not all procedures are covered by these surveys. Procedure groupings and code numbers are based on the *International Classification of Diseases, 9th Revision, Clinical Modification* (ICD-9-CM).

Source: L. J. Kozak and M.F. Owings. "Ambulatory and Inpatient Procedures in the United States, 1995." National Center for Health Statistics. *Vital and Health Statistics* 13(135). 1998. http://www.cdc.gov/nchswww/data/sr13_135.pdf

11.7. Number and Rate of Ambulatory and Inpatient Nonsurgical Procedures by Selected Procedure Categories, United States, 1995.

Procedure category and ICD-9-CM code	Total		Ambulatory[1]		Inpatient[2]	
	Estimate	Standard error**	Estimate	Standard error**	Estimate	Standard error**
	Number in thousands					
All nonsurgical procedures[3]	24,722	1,148	7,444	594	17,278	983
Arteriography and angiocardiography using contrast material 88.4–88.5	2,358	142	525	65	1,834	127
Injection or infusion of therapeutic or prophylactic substance 99.1–99.2	1,745	182	379	47	1,366	176
Endoscopy of large intestine without biopsy 45.21–45.24	1,678	110	1,343	108	335	20
Diagnostic ultrasound 88.7	1,288	119	*	*	1,181	112
Endoscopy of small intestine without biopsy 45.11–45.13	1,186	75	736	72	450	21
Respiratory therapy 93.9, 96.7	1,132	116	*	*	1,127	116
Computerized axial tomography (CAT) 87.03, 87.41, 87.71, 88.01, 88.38	1,038	93	*	*	967	85
Fetal EKG (scalp) and fetal monitoring, not otherwise specified 75.32, 75.34	935	102	*	*	935	102
Manually assisted delivery 73.5	869	59	0	0	869	59
Cystoscopy without biopsy 57.31–57.32	842	68	629	66	213	16
Arthroscopy of knee 80.26	680	47	636	47	44	6
	Rate per 10,000 population					
All nonsurgical procedures[3]	945.7	44.0	284.8	22.7	660.9	37.6
Arteriography and angiocardiography using contrast material 88.4–88.5	90.2	5.4	20.1	2.5	70.1	4.8
Injection or infusion of therapeutic or prophylactic substance 99.1–99.2	66.8	7.0	14.5	1.8	52.3	6.7
Endoscopy of large intestine without biopsy 45.21–45.24	64.2	4.2	51.4	4.1	12.8	0.8
Diagnostic ultrasound 88.7	49.3	4.5	*	*	45.2	4.3
Endoscopy of small intestine without biopsy 45.11–45.13	45.4	2.9	28.2	2.7	17.2	0.8
Respiratory therapy 93.9, 96.7	43.3	4.4	*	*	43.1	4.4
Computerized axial tomography (CAT) 87.03, 87.41, 87.71, 88.01, 88.38	39.7	3.6	*	*	37.0	3.3
Fetal EKG (scalp) and fetal monitoring, not otherwise specified 75.32, 75.34	35.8	3.9	*	*	35.8	3.9
Manually assisted delivery 73.5	33.2	2.2	0	0	33.2	2.2
Cystoscopy without biopsy 57.31–57.32	32.2	2.6	24.0	2.5	8.2	0.6
Arthroscopy of knee 80.26	26.0	1.8	24.3	1.8	1.7	0.2

Notes:

* = Figure does not meet standard of reliability or precision.

** = Standard error is a measure of the chance that data is incorrect and indicates the range within which correct answers probably fall. For example, in this table, the actual number of all procedures is most likely a value within plus or minus 1,148 from the estimated number of 24,722.

[1] Data from the National Survey of Ambulatory Surgery. Excludes ambulatory surgery patients who became inpatients.

[2] Data from the National Hospital Discharge Survey. Excludes newborn infants.

[3] All nonsurgical procedures estimated from the National Survey of Ambulatory Surgery and the National Hospital Discharge Survey. Not all procedures are covered by these surveys. Procedure groupings and code numbers are based on the *International Classification of Diseases, 9th Revision, Clinical Modification* (ICD-9-CM)

Source: L. J. Kozak and M. F. Owings. "Ambulatory and Inpatient Procedures in the United States, 1995." National Center for Health Statistics. *Vital and Health Statistics* 13 (135). 1998. http://www.cdc.gov/nchswww/data/sr13_135.pdf

11.8. Annual Number of Organ and Tissue Transplants in the United States for All Ages and Regardless of Citizenship of Recipient, 1989–97, and Projected for 1998 and 1999.

Year	Heart	Liver	Kidney	Kidney-Pancreas	Pancreas	Heart-Lung	Lung	Cornea	Bone Marrow
1989	1,705	2,201	8,655	334	83	67	93	38,464	4,563 P
1990	2,107	2,691	9,421	459	69	52	204	40,631	5,583 P
1991	2,126	2,953	9,674	452	78	51	405	41,393	6,755 P
1992	2,171	3,064	9,737	493	64	48	535	42,377	7,669 P
1993	2,297	3,440	10,359	661	113	60	667	42,469	8,464 P
1994	2,341	3,652	10,645	748	94	70	723	43,743	9,000 P1
1995	2,361	3,925	10,983	918	107	70	871	44,652	11,250 P1
1996	2,343	4,065	11,293	859	165	39	810	46,300	12,600 P1
1997	2,292	4,167	11,453	854	207	62	928	45,493	13,950 P1
Projected 1998	2,315	4,271	11,739	888	238	63	1,002	46,175	15,300
Projected 1999	2,338	4,356	12,032	897	262	64	1,082	46,868	16,700

Notes:

P = Based on Milliman & Robertson projections and sample data obtained from the IBMTR (International Bone Marrow Transplant Registry) and the ABMTR (Autologous Blood and Marrow Transplant Registry). The analysis has not been reviewed or approved by the Advisory Committee of the IBMTR and ABMTR.

P1 = Based on Milliman & Robertson adjustments to data obtained from the IBMTR and the ABMTR. The analysis has not been reviewed or approved by the Advisory Committee of the IBMTR and ABMTR.

Cornea transplants through 1997 are based on data from the Eye Bank Association of America.

Other organ transplants are based on data from UNOS (United Network for Organ Sharing).

Source: Richard H. Hauboldt, F.S.A. Milliman & Robertson, Inc. Brookfield, WI. Unpublished.

11.9. Supply of Alternative Nonphysician Clinicians, United States, 1990 to 2015.

	Naturopaths *	Chiropractors	Acupuncturists
1990	800	47,000	5,000
1997	1,900	52,000	11,000
2005 projected	3,500	95,200	21,000
2015 projected	6,600	145,000	40,000

Notes:

Data include the actual numbers of active clinicians in 1990 and 1997 and their projected supply in 2005 and 2015.

* = A naturopathic physician receives training in the same basic sciences as a conventional doctor, but also studies holistic and nontoxic therapies and emphasizes prevention and optimizing wellness. The naturopath also studies nutrition, acupuncture, homeopathic and botanical medicine, and psychology.

Source: Richard A. Cooper, Prakash Laud, and Craig L. Dietrich. "Current and Projected Workforce of Nonphysician Clinicians." *JAMA 280* (September 2, 1998): 788–94.

DRUGS AND NUTRITION

11.10. Generic Substances Most Frequently Used at Ambulatory Care Visits, United States, 1995.

Generic substance	Combined Settings Number of occurrences, in thousands [1]	Combined Settings Percent of all drug mentions [2]	Physician offices	Outpatient departments	Emergency departments
			Percent distribution		
All occurrences	1,395,873	—	79.5%	8.2%	12.3%
Acetaminophen	61,230	5.2%	59.5%	7.6%	32.9%
Amoxicillin	45,207	3.9%	82.3%	5.9%	11.9%
Ibuprofen	27,310	2.3%	58.6%	8.8%	32.7%
Albuterol	23,034	2.0%	70.5%	10.6%	18.8%
Aspirin	18,087	1.5%	82.0%	9.4%	8.6%
Hydrochlorothiazide	17,546	1.5%	89.0%	9.2%	1.8%
Furosemide	16,603	1.4%	81.4%	7.8%	10.9%
Estrogens	15,885	1.4%	89.5%	7.6%	2.9%
Hydrocodone	15,120	1.3%	67.8%	3.0%	29.3%
Fuaifenesin	14,709	1.3%	84.1%	7.0%	8.9%
Codeine	14,073	1.2%	67.8%	7.7%	24.5%
Erythromycin	13,951	1.2%	80.7%	5.1%	14.3%
Levothyroxine	13,034	1.1%	86.1%	10.6%	3.3%
Phenylephrine	13,006	1.1%	88.5%	6.0%	5.5%
Trimethoprim	12,918	1.1%	73.0%	10.8%	16.2%
Cephalexin	12,782	1.1%	73.7%	6.6%	19.6%
Prednisone	12,685	1.1%	79.6%	10.7%	9.7%
Sulfamethoxazole	12,077	1.0%	71.8%	11.2%	17.0%
Digoxin	11,839	1.0%	85.7%	7.1%	7.2%
Naproxen	11,455	1.0%	79.1%	7.8%	13.1%
Rantidine	11,400	1.0%	80.0%	10.1%	9.9%
Promethazine	11,152	1.0%	59.4%	3.3%	37.3%
Nifedipine	10,588	0.9%	77.9%	10.7%	11.4%
Phenylpropanolamine	10,189	0.9%	87.8%	5.6%	6.6%
Triamcinolone	10,023	0.9%	88.2%	7.7%	4.1%
Pseudoephedrine	9,752	0.8%	84.8%	7.8%	7.4%
Potassium replacement solutions	9,745	0.8%	82.8%	7.6%	9.6%
Beclomethasone	9,069	0.8%	87.6%	9.9%	2.5%
Enalapril	8,992	0.8%	84.0%	11.1%	4.9%
Insulin	8,965	0.8%	71.9%	18.0%	10.1%
Diltiazem	8,909	0.8%	82.2%	12.0%	5.8%

11.10. Generic Substances Most Frequently Used at Ambulatory Care Visits, United States, 1995 *(continued)*.

Generic substance	Combined Settings		Physician offices	Outpatient departments	Emergency departments
	Number of occurrences, in thousands [1]	Percent of all drug mentions [2]			
				Percent distribution	
Glyburide	8,850	0.8%	86.4%	9.9%	3.7%
Nitroglycerine	8,805	0.8%	67.6%	7.0%	25.4%
Propoxyphene	8,453	0.7%	75.4%	5.3%	19.4%
Hydrocortisone	8,398	0.7%	84.5%	8.3%	7.3%

Notes:
Percentages may not add to 100 due to rounding.
[1] Frequency of mention combines single-ingredient agents with mentions of the agent in a combination drug.
[2] Based on an estimated 1,167,162,000 drug mentions at physician office visits, hospital outpatient department visits, and hospital emergency department visits in 1995.

Source: S. M. Schappert. *Ambulatory Care Visits to Physician Offices, Hospital Outpatient Departments, and Emergency Departments: United States, 1995.* National Center for Health Statistics. *Vital and Health Statistics* 13 (129). 1997. http://www.cdc.gov/nchswww/data/sr13_129.pdf

11.11. Drug Mentions by Therapeutic Classification, According to Ambulatory Care Setting, United States, 1995.

Therapeutic classification [1]	Combined settings	Physician offices	Outpatient departments	Emergency departments	Physician offices	Outpatient departments	Emergency departments
	Number of drug mentions in thousands	Percent distribution			Number of mentions per 100 visits		
All drug mentions	1,167,162	79.3%	8.3%	12.3%	132.9	144.2	149.2
Drugs used for relief of pain	167,792	67.0%	7.6%	25.4%	16.1	18.9	44.2
Antimicrobial agents	163,062	78.4%	6.7%	14.9%	18.3	16.3	25.2
Cardiovascular-renal drugs	157,348	84.1%	8.9%	6.9%	19.0	20.9	11.3
Respiratory tract drugs	127,063	78.4%	8.1%	13.4%	14.3	15.4	17.7
Hormones and agents affecting hormonal mechanisms	101,539	83.9%	9.7%	6.4%	12.2	14.7	6.7
Central nervous system	78,952	83.5%	7.9%	8.7%	9.5	9.2	7.1
Skin/mucous membrane	62,052	85.0%	6.8%	8.1%	7.6	6.3	5.2
Gastrointestinal agents	55,336	76.9%	9.8%	13.3%	6.1	8.0	7.6
Metabolic and nutrient agents	54,852	82.0%	10.1%	7.8%	6.5	8.3	4.5
Ophthalmic drugs	40,936	92.3%	4.6%	3.1%	5.4	2.8	1.3
Immunologic agents	40,457	80.5%	9.8%	9.7%	4.7	5.9	4.1
Neurologic drugs	28,635	77.2%	8.2%	14.6%	3.2	3.5	4.3
Hematologic agents	19,403	77.4%	12.1%	10.5%	2.2	3.5	2.1
Radiopharmaceutical/contrast	8,470	92.7%	5.9%	*	1.1	0.7	*
Other and unclassifed media [2]	61,264	77.3%	10.6%	12.2%	6.8	9.6	7.7

Notes:
"Mentions" = all new or continued medications ordered, supplied, or administered at the visit. Up to six medications were coded per visit on the survey form.
* = Figure does not meet standard of reliability or precision.
[1] Based on the standard drug classification used in the *National Code Directory*, 1995 edition (NDC) (7).
[2] Includes anesthetics, antidotes, oncolytics, otologics, antiparasitics, and unclassified/miscellaneous drugs.
Percentages may not total 100 due to rounding.

Source: S. M. Schappert. *Ambulatory Care Visits to Physician Offices, Hospital Outpatient Departments, and Emergency Departments: United States, 1995.* National Center for Health Statistics. *Vital and Health Statistics* 13 (129). 1997. http://www.cdc.gov/nchswww/data/sr13_129.pdf

11.12. Drugs That Exceeded a Billion Dollars in Worldwide Sales in 1996.

Drug	Sales (in millions)
Prilosec/Losec	$3,675.4
Zantac	$3,012.4
Zocor	$2,800.0
Vasotec	$2,535.0
Prozac	$2,356.0
Norvasc	$1,795.0
Augmentin	$1,354.1
Zoloft	$1,337.0
Sandimmune	$1,300.0
Adalat	$1,290.0
Cipro	$1,290.0
Zovirax	$1,266.7
Mevacor	$1,255.0
Voltaren	$1,158.9
Biaxin	$1,150.0
Epogen	$1,150.0
Paxil	$1,101.4
Capoten	$1,091.0
Claritin	$1,075.0
Pravachol	$1,068.0
Premarin	$1,039.2
Pepcid	$1,030.0
Zestril	$1,022.5
Neupogen	$1,016.5
Procardia	$1,005.0

Source: Engel Publishing Partners. "Power Drugs: The Top 100 Prescription Drugs." *Med Ad News* 16 (December 1997): 59+.

11.13. The Top 100 Prescription Drugs by Worldwide Sales, 1995 and 1996.

Rank 1996	Rank 1995	U.S. Brand Name	Chemical Name	Indicated For	Worldwide Sales ($ millions)
1	2	Prilosec/Losec	Omeprazole	Ulcers	$3,675.4
2	1	Zantac	Ranitidine	Ulcers	$3,012.4
3	5	Zocor	Simvastatin	Cholesterol reduction	$2,800
4	3	Vasotec/Vaseretic	Enalapril	Hypertension, heart failure, left ventricular dysfunction	$2,535
5	4	Prozac	Fluoxetine	Depression	$2,356
6	10	Norvasc	Amlodipine	Hypertension, angina	$1,795
7	8	Augmentin/Augmentin BID	Amoxicillin	Infections	$1,354.1
8	17	Zoloft	Sertaline	Depression	$1,337
9	15	Sandimmune/Neoral	Cyclosporine	Organ rejection prevention	$1,300
10	12	Adalat/Adalat CC	Nifedipine	Hypertension, angina	$1,290
10	12	Cipro	Ciprofloxacin	Infections	$1,290
12	7	Zovirax	Acyclovir	Herpes	$1,266.7
13	11	Mevacor	Lovastatin	Cholesterol reduction	$1,255
14	9	Voltaren/Voltaren-XR	Diclofenac	Arthritis	$1,158.9
15	20	Biaxin	Clarithromycin	Infections	$1,150
15	23	Epogen	Epoetin alfa	Red blood cell enhancement	$1,150
17	30	Paxil	Paroxetine	Depression	$1,101.4
18	6	Capoten/Capozide	Captopril	Hypertension, heart failure, diabetic nephropathy	$1,091
19	29	Claritin line	Loratadine	Allergies	$1,075
20	32	Pravachol	Pravastatin	Cholesterol reduction	$1,068
21	21	Premarin line	Conjugated estrogens	Estrogen replacement	$1,039.2
22	22	Pepcid	Famotidine	Ulcers	$1,030
23	25	Zestril/Zestoretic	Lisinopril	Hypertension, congestive heart failure	$1,022.5
24	19	Neupogen	Filgrastim	Restoration of white blood cells	$1,016.5
25	16	Procardia/Procardia XL	Nifedipine	Hypertension, angina	$1,005
26	27	Procrit	Epoetin alfa	Red blood cell enhancement	$995
27	18	Cardizem line	Diltiazem	Hypertension, angina, atrial fibrillation	$945.3
28	14	Rocephin	Ceftriaxone	Infections	$933.3

11.13. The Top 100 Prescription Drugs by Worldwide Sales, 1995 and 1996 *(continued).*

Rank 1996	Rank 1995	U.S. Brand Name	Chemical Name	Indicated For	Worldwide Sales ($ millions)
29	31	Propulsid	Cisapride	Gastroesophageal reflux disease	$930
30	24	Diflucan	Fluconazole	Fungal infections	$910
31	28	Humulin	Insulin	Diabetes mellitus	$884
32	42	Imitrex	Sumatriptan	Migraine	$840.8
33	41	Taxol	Paclitaxel	Ovarian and breast cancers	$813
34	35	Lupron line	Leuprolide acetate	Prostate cancer and endometriosis	$810
35	—	Atrovent	Ipratropium bromide	Asthma	$746.7
36	26	Ventolin	Albuterol sulfate	Asthma	$734.8
37	—	Pulmicort Turbuhaler	Budesonide	Asthma	$706.4
38	48	Tenormin/Tenoretic	Atenolol	Hypertension, angina, myocardial infarction	$670.9
39	61	Risperdal	Risperidone	Psychotic disorders	$645
40	38	Ceftin	Cefuroxime axetil	Infections	$639.6
41	65	Zithromax	Azithromycin	Infections	$619
42	36	Beclovent	Beclomethasone	Asthma	$611.5
43	54	Diprivan	Propofol	Anesthesia	$589.8
44	47	Ortho-Novum	Norethindrone/mestranol	Prevention of pregnancy	$575
45	39	Zofran	Ondansetron	Nausea and vomiting	$574.1
46	37	Engerix-B	Recombinant hepatitis B	Prevention of hepatitis B	$567.8
47	74	Zoladex	Goserelin	Breast and prostate cancers	$562.8
48	40	Nolvadex	Tamoxifen	Breast cancer	$561.1
49	43	Primaxin	Imipenem-cilastatin	Infections	$555
50	59	Serevent	Salmeterol	Asthma	$544.4
51	34	Ceclor	Cefaclor	Infections	$542
52	51	Hytrin	Terazosin	Benign prostatic hypertrophy and hypertension	$540
53	62	Cardura	Doxazosin	Benign prostatic hypertrophy and hypertension	$533
54	44	Axid	Nizatidine	Ulcers	$531
55	57	Intron A	Interferon alpha-2b	Bone marrow transplantation	$524
56	53	Sporanox	Itraconazole	Fungal infections	$515
57	69	Prinivil/Prinzide	Lisinopril	Hypertension, congestive heart failure	$485
58	33	Seldane/Seldane-D	Terfenadine	Allergies	$478
59	46	Trental	Pentoxifylline	Circulation enhancement	$467.3
60	63	Relafen	Nabumetone	Arthritis	$457.1
61	85	Lamisil	Terbinafine	Fungal infections	$450
62	64	Proscar	Finasteride	Benign prostatic hypertrophy	$450
63	49	Fortaz	Ceftazidime	Infections	$449.3
64	81	Retrovir	Zidovudine	HIV infection	$441.5
65	72	Clozaril	Clozapine	Psychotic disorders	$435.5
66	82	Accupril	Quinapril	Hypertension	$406
67	58	Amoxil	Amoxicillin	Infections	$405.6
68	56	Timoptic/Timoptic-XE	Timolol maleate	Glaucoma	$405
69	66	Tegretol/Tegretol-XR	Carbamazepine	Epilepsy	$404
70	88	Lovenox	Enoxaparin sodium	Deep vein thrombosis	$401
71	67	Genotropin	Somatropin	Growth failure in children	$396.9
72	45	Versed	Midazolam	Anesthesia	$388.9
73	—	Toprol-XL	Metopolol succinate	Hypertension, angina	$388.4
74	52	Accutane	Isotretinoin	Acne	$385.2
75	87	Depakote	Divalproex	Epilepsy and migraine	$380
76	76	Nizoral	Ketoconazole	Fungal infections	$380
77	79	Paraplatin	Carboplatin	Ovarian cancer	$373
78	—	Lescol	Fluvastatin sodium	Cholesterol reduction	$371
79	83	BuSpar	Buspirone	Anxiety	$370
80	—	Isovue	Iopamidol	Imaging cardiovascular system, excretory urography, and head and body	$366.1
81	94	Lotensin	Benazepril	Hypertension	$365.3
82	71	Provera/Depo-Provera	Medroxyprogesterone	Secondary amenorrhea and endometrial renal cancer	$363.4
83	—	Betaseron	Interferon beta-1b	Multiple sclerosis	$353
84	80	Xanax	Alprazolam	Anxiety	$352.5
85	73	Estraderm	Estradiol	Estrogen replacement	$350.8
86	—	Iopamiron	Iopamidol	Imaging cardiovascular system, excretory urography, and head and body	$347
87	50	Tagamet	Cimetidine	Ulcers	$340.1
88	—	Cozaar/Hyzaar	Losartan	Hypertension	$339
89	70	Claforan	Cefotaxime	Infections	$335.3
90	68	Transderm-Nitro	Nitroglycerin	Angina	$333.1

11.13. The Top 100 Prescription Drugs by Worldwide Sales, 1995 and 1996 *(continued)*.

Rank 1996	Rank 1995	U.S. Brand Name	Chemical Name	Indicated For	Worldwide Sales ($ millions)
91	—	Glucophage	Metformin	Non-insulin-dependent diabetes	$332
92	78	Unasyn	Ampicillin, sulbactam	Bacterial infections	$326
93	—	Altace	Ramipril	Hypertension, congestive heart failure	$319.3
94	75	Klonopin	Clonazepam	Epilepsy	$319.3
95	—	Kytril	Granisetron	Nausea and vomiting	$316.7
96	60	Proventil	Albuterol	Asthma	$316
97	—	Synthroid	Levothyroxine	Hypothyroidism	$313.9
98	—	Zaditen	Ketotifen	Allergies, asthma	$313.7
99	—	Optiray	Ioversol	Tomography, urography, and angiography procedures	$313
100	96	Lodine/Lodine XL	Etodolac	Arthritis	$308.8

Source: Engel Publishing Partners. "Power Drugs: The Top 100 Prescription Drugs." *Med Ad News* 16 (December 1997): 59+.

11.14. Cancer and Cancer-related Treatments Worldwide, 1996 (sales, millions of dollars).

	Company	Value
Antineoplastics		$3,119.9
Taxoids		
Taxol	Bristol-Myers Squibb Co.	$813.0
Hormones		
Lupron line	TAP Pharmaceuticals Inc.	$810.0
Zoladex	Zeneca Inc.	$562.8
Antiestrogents		
Nolvadex	Zeneca Inc.	$561.1
Cytotoxic agents		
Paraplatin	Bristol-Myers Squibb Co.	$373.0
Biological Response Modifiers		$1,540.5
Neupogen	Amgen Inc.	$1,016.5
Intron A	Schering-Plough Corp.	$524.0
Total		$4,660.4

Source: Engel Publishing Partners. "Leading Therapeutic Categories." *Med Ad News* 16 (May 1997): 22.

11.15. Leading Therapeutic Classes—Dispensed Prescription Volume, United States, 1997.

Rank	Class	Sales Rxs* (thousands)	Growth/ Decline
1	Codeine & Combinations, Noninjectable	94,276	5%
2	Calcium Blockers	90,930	1%
3	ACE Inhibitors	77,737	8%
4	Systemic Antiarthritics	76,709	1%
5	Anti-ulcerants	73,934	5%
6	Antidepressants, Specific Neurotransmitter Modulators	65,709	17%
7	Amoxicillin	65,703	-7%
8	Oral Contraceptives	64,914	4%
9	Beta Blockers	64,520	8%
10	Cholesterol Reducers	61,228	28%
11	Benzodiazepines	60,925	0%

11.15. Leading Therapeutic Classes—Dispensed Prescription Volume, United States, 1997 *(continued)*.

Rank	Class	Sales Rxs* (thousands)	% Growth/ Decline
12	Oral estrogens	56,377	2%
13	Cephalosporins	55,468	-5%
14	Diabetes, Oral	54,376	20%
15	Thyroid, Synthetic	51,588	10%
16	Noninjectable Diuretics	45,334	3%
17	Seizure Disorders	39,150	12%
18	Beta Agonists, Aerosol	37,183	4%
19	Antihistamines	34,548	24%
20	Tri & Tetracyclics	32,730	-4%
	Top Twenty Total	1,203,339	
	Total Market	2,523,304	5%

Notes:
* = Represents total prescriptions in millions, dispensed through independent, chain, foodstore, long-term, and mail order pharmacies.

Source: IMS Health. National Prescription Audit Plus (TM) and Retail and Provider Perspective (TM) Audits. Cited on Web site at http://www.ims-america.com, September 1998.

11.16. Leading U.S. Rx Pharmaceutical Dollar Sales Volume Leaders, 1996–97 (millions of dollars).

Product	1996 Sales Dollars*	% Growth/ Decline
Zantac (Glaxo)	$1,761	-18%
Prilosec (Astra-Merck)	$1,742	46%
Prozac (Dista)	$1,685	14%
Epogen (Amgen)	$1,184	23%
Zoloft (Roerig)	$1,098	23%
Zocor (Merck)	$1,015	78%
Procardia XL (Pfizer)	$962	-13%
Vasotec (Merck)	$866	1%
Mevacor (Merck)	$792	-7%
Premarin (Wyeth-Ayerst)	$768	8%
Biaxin (Abbott)	$760	23%
Cardizem CD (Hoechst Marion Roussel)	$734	-3%
Lupron Depot (TAP)	$725	21%
Norvasc (Pfizer)	$723	45%
Paxil (SmithKline Beecham)	$696	43%
Cipro (Bayer)	$684	4%
Neupogen (Amgen)	$677	13%
Augmentin (SmithKline Beecham)	$672	11%
Claritin (Schering)	$657	48%
Pepcid (Merck)	$651	8%

Product	1997 Sales Dollars*	% Growth/ Decline
Prilosec	$2,282	33%
Prozac	$1,942	14%
Zocor	$1,379	38%
Epogen	$1,201	2%
Zoloft	$1,196	11%
Zantac	$1,097	-37%
Paxil	$949	39%

11.16. Leading U.S. Rx Pharmaceutical Dollar Sales Volume Leaders, 1996–97 (millions of dollars) (*continued*).

Product	1997 Sales Dollars*	% Growth/ Decline
Norvasc	$915	28%
Claritin	$908	40%
Vasotec	$843	-1%
Premarin	$807	7%
Augmentin	$805	21%
Imitrex	$792	38%
Procardia XL	$785	-17%
Pravachol	$768	38%
Biaxin	$744	-1%
Lupron Depot	$712	-1%
Cipro	$710	5%
Cardizem CD	$699	-3%
Pepcid	$698	9%
1997 Top Twenty Total	$20,231	
1997 Total Market	$93,974	11.4%

Notes:
Represents, in millions of dollars, pharmacy acquisition costs to independent, chain, foodstore, nonfederal and federal hospital, clinic, closed-wall HMO, and long-term care pharmacies.

Source: IMS Health. National Prescription Audit Plus (TM) and Retail and Provider Perspective (TM) Audits. Cited in "Business Watch." *Medical Marketing & Media* 32 (April 1997): 78+; plus information from Web site at http://www.ims-america.com, September 1998.

11.17. U.S. Rx Pharmaceutical Prescription Leaders, 1996–97.

Product	1996 Prescriptions Dispensed (millions)*	% Growth/ Decline
Premarin (Wyeth-Ayerst)	44.8	1%
Trimox (Apothecon)	35.4	16%
Synthroid (Knoll)	33.3	12%
Lanoxin (Allen & Hansbury)	25.7	-1%
Zantac (Glaxo)	23.1	-17%
Vasotec (Merck)	21.6	-1%
Prozac (Dista)	20.7	10%
Procardia XL (Pratt Pharm)	18.8	-13%
Hydrocodone w/APAP (Watson)	18.8	52%
Coumadin Sodium (DuPont)	17.5	17%
Zoloft (Roerig)	17.1	20%
Prilosec (Astra Merck)	16.1	43%
Norvasc (Pfizer)	15.6	42%
Cardizem CD (Hoechst Marion Roussel)	15.5	-4%
Albuterol (Warrick)	15.5	232%
Biaxin (Abbott)	14.7	14%
Amoxil (SmithKline Beecham)	14.1	-41%
Triamterene/HCTZ (Geneva)	13.6	10%
Zestril (Zeneca)	13.6	8%
Claritin (Schering)	13.5	35%

Product	1997 Prescriptions Dispensed (millions)	% Growth/ Decline
Premarin	45.1	1%
Synthroid	36.1	8%
Trimox	34.5	-2%

11.17. U.S. Rx Pharmaceutical Prescription Leaders, 1996–97 *(continued).*

Product	1997 Prescriptions Dispensed (millions)*	% Growth/ Decline
Lanoxin	25.3	-2%
Hydrocodone w/APAP	25.0	33%
Prozac	22.8	10%
Albuterol	21.8	41%
Prilosec	21.1	31%
Vasotec	20.2	-6%
Norvasc	19.7	26%
Coumadin Sodium	19.1	9%
Zoloft	18.6	8%
Claritin	18.3	36%
Zocor	17.5	31%
Paxil	15.9	32%
Procardia XL	15.3	-19%
Zantac	15.1	-35%
Zestril	15.0	11%
Furosemide	14.9	21%
Cardizem CD	14.7	-6%
Top Twenty Total	436.1	
Total Market	2,523.3	5%

Notes:
* = Total prescriptions, in millions, dispensed through independent, chain, foodstore, long-term care, and mail order pharmacies.

Source: IMS Health. National Prescription Audit Plus (TM) and Retail and Provider Perspective (TM) Audits. Cited in "Business Watch." *Medical Marketing & Media* 32 (April 1997): 78+; plus information from Web site at http://www.ims-america.com, September 1998.

11.18. Value of Shipments of Pharmaceutical Preparations, Except Biologicals, United States, 1995–97 (thousands of dollars).

Product Code	Product Description	1995	1996	1997
2834- —	Pharmaceutical preparations (except biologicals)	$48,864,310	$52,184,509	$62,593,374
28341 —	Pharmaceutical preparations affecting neoplasms, endocrine system, and metabolic diseases, for human use	$4,075,943	$4,775,056	$6,872,950
	Hormones and synthetic substitutes: Corticoids:			
28341 11	Systemic	$252,564	$213,457	$143,463
28341 15	Local and topical, including anti-infective combinations	$437,174	$424,457	$382,862
28341 17	Androgens	D	D	D
28341 19	Estrogens	$154,238	$173,558	$224,885
28341 21	Insulin and antidiabetic agents	D	D	$1,670,118
28341 25	ACTH (corticotropin)	D	$3,672	D
28341 27	Oral contraceptive preparations	$55,378	$58,223	D
28341 29	Progestogens (excludes premenstrual tension preps, see code 28345 85)	D	D	D
28341 31	Sex hormone combinations (except progestogen combinations)	D	D	D
28341 35	Thyroid and antithyroid preparations including iodides	D	D	$60,051
28341 37	Anabolic agents	D	D	D
28341 39	Other hormone preparations	$269,226	$284,118	$248,095
	Antineoplastic agents:			
28341 43	Radioactive isotopes for internal use	D	$18,333	$42,405

11.18. Value of Shipments of Pharmaceutical Preparations, Except Biologicals, United States, 1995–97 (thousands of dollars) *(continued).*

Product Code	Product Description	1995	1996	1997
28341 45	Specific antineoplastic agents	$645,145	$670,248	$2,154,269
28341 98	Other pharmaceutical preparations affecting neoplasms, the endocrine system, and metabolic diseases, for human use	$1,018,083	$1,488,202	$1,541,587
28342 —	Pharmaceutical preparations acting on the central nervous system and the sense organs, for human use	r/ $9,228,199	$10,207,294	$12,279,859
28342 11	Parasympathomimetic cholinergic drugs	$16,900	a/ $21,605	$22,940
28342 13	Skeletal muscle relaxants	$357,225	$313,009	$336,752
	Internal analgesics and antipyretics: Narcotic:			
28342 19	Opium and derivatives	$439,703	$398,655	a/$390,052
28342 20	Synthetic narcotics	$259,019	$319,226	$285,257
	Non-narcotic: Salicylates:			
28342 22	Aspirin (acetylsalicylic acid)	$213,783	$241,525	$196,320
28342 24	Other salicylates, such as sodium salicylate	1	1	1
28342 25	Aspirin combinations	$334,560	$276,490	$192,939
28342 26	Acetaminophen and combinations	$431,411	$445,727	$453,708
28342 27	Antiarthritics (nonhormonal)	r/ $877,323	$646,840	$731,302
28342 28	Other internal analgesics and antiparetics, including effervescent types and suppositories. Also includes other salicylates, such as sodium salicylate, code 28342 24.	r/ $263,880	$244,678	$187,054
28342 41	Anticonvulsants (except phenobarbitol)	r/ $241,232	$442,688	$585,521
	Psychotherapeutic agents:			
28342 51	Antidepressants	$1,708,726	$2,004,305	$2,454,149
	Tranquilizers:			
28342 55	Phenothiazine derivatives	$401,376	$382,824	$345,494
28342 57	Other tranquilizers	r/ $85,246	$73,789	$60,888
28342 59	Other psychotherapeutic agents	r/ $133,569	$253,997	$906,712
	Central nervous system (CNS) stimulants (respiratory and cerebral stimulants, including sympatho-mimetic agents employed mainly as CNS stimulants) (excludes nondrug dietaries for weight control):			
28342 61	Amphetamines	D	D	D
28342 63	Anorexiants, except amphetamines	$105,733	$276,717	$574,004
28342 69	Other CNS stimulants	$175,306	$256,468	D
	Sedatives and hypnotics: Prescription:	$44,978	$36,159	$38,789
28342 71	Barbiturates	$45,279	$49,163	D
28342 75	Nonbarbiturates			
	Non-prescription:			
28342 77	Sleep inducers	$52,846	$49,191	$50,395
28342 79	Calming agents	D	D	D
	Anesthetics, except urinary tract anesthetics and skin preparations used as antipruritics:			
28342 81	Local and topical:	$257,804	$254,784	$295,059
28342 85	General	$96,397	$111,490	$42,294

11.18. Value of Shipments of Pharmaceutical Preparations, Except Biologicals, United States, 1995–97 (thousands of dollars) *(continued).*

Product Code	Product Description	1995	1996	1997
	Eye and ear preparations, excluding anti-infectives, corticoids, and antibacterials and antiseptics:			
28342 91	Mydriatics and miotics	D	D	D
28342 93	Contact lens solutions	$613,652	$611,486	$643,573
28342 94	Other eye and ear preparations	$521,392	$446,349	$673,310
28342 98	Other pharmaceutical preparations acting on the central nervous system and the sense organs, for human use	$1,187,055	$1,666,808	$1,620,592
28343 —	Pharmaceutical preparations acting on the cardiovascular system, for human use	$5,987,883	$6,965,310	$8,261,537
28343 11	Anticoagulants	$763,122	$867,347	$1,664,026
28343 21	Hemostatics	$59,197	D	$26,298
28343 31	Digitalis preparations	D	D	D
	Hypotensives:			
28343 43	Rauwolfia-alkaloid preparations	$23,070	$14,876	$17,305
28343 47	Beta receptor blocking agents	$116,381	$100,423	$259,265
28343 48	Other hypotensives	$299,384	$226,437	$220,360
	Vasodilators:			
28343 51	Coronary	$422,384	$467,338	$631,380
28343 55	Peripheral	$119,417	$78,772	$72,699
28343 61	Anti-arrythmics, such as propanolol and quinidine	$228,222	$220,337	a/$269,931
28343 71	Calcium channel blockers such as isoptin, calan, procardia, cardizem, and adalat	r/ $622,747	$591,713	$697,931
28343 81	Ace inhibitors, such as vasotec, capoten, prinivil, and zestril	$2,074,735	$2,218,353	$2,538,520
28343 97	Other pharmaceutical preparations acting on the cardiovascular system including vasopressors, and antiheparin agents, for human use	r/ $953,566	$1,777,912	$1,533,363
28344 —	Pharmaceutical preparations acting on the respiratory system, for human use	$5,196,084	$5,178,997	$5,486,171
28344 11	Antihistamines (except cold preparations and antiemetics)	$570,151	$679,424	$1,077,668
28344 15	Bronchial dilators, including anti-asthmatics	$935,393	$923,745	$713,061
	Cough and cold preparations (prescription): Cough preparations and expectorants (containing antitussive or other ingredients intended primarily to treat cough only):			
28344 21	Narcotic	$104,171	$56,768	$58,044
28344 25	Non-narcotic	$112,798	$96,957	$101,450
	Cold preparations (containing combinations of the following ingredients [but not antitussive]: nasal decongestant, antihistamine, analgesic, bioflavanoid, or antibiotic):			
28344 31	Nasal decongestants	$14,096	$6,197	$4,555
28344 35	Antihistamine cold preparations	$18,499	$20,980	$4,208
28344 39	Other prescription cold preparations	$88,636	$60,810	$38,271
28344 41	Cough and cold combinations (prescription): Cough and cold preparations (nonprescription):	$138,023	$129,577	$90,790

11.18. Value of Shipments of Pharmaceutical Preparations, Except Biologicals, United States, 1995–97 (thousands of dollars) *(continued)*.

Product Code	Product Description	1995	1996	1997
	Decongestants:			
28344 51	Nasal sprays	$106,417	$103,044	$205,749
28344 55	Nose drops	452	D	D
28344 59	Other decongestants	$228,118	$222,353	$191,743
28344 61	Cough syrups	$422,935	$446,287	$467,303
28344 63	Capsules and tablets	$490,179	$452,747	$502,859
28344 65	Lozenges	$40,109	$33,217	$49,994
28344 67	Topical preparations	D	D	D
28344 71	Cough drops	$57,987	$35,517	D
28344 79	Other nonprescription cough and cold preparations	$183,014	$187,721	$71,790
28344 81	Beta agonists such as proventil and ventolin	$1,223,176	$1,167,772	$1,131,534
28344 98	Other pharmaceutical preparations acting on the respiratory system, for human use	$405,014	$461,774	$688,164
28345 —	Pharmaceutical preparations acting on the digestive or the genito-urinary systems, for human use	$8,593,410	$8,602,477	$10,099,604
28345 11	Enzymes	$70,467	$78,482	$78,249
28345 15	Antacids, including acid neutralizing products with coating functions, but excluding effervescent salicylate products classified as analgesics	$494,531	$450,858	$344,989
28345 19	Antidiarrheals	$223,963	$201,805	$263,271
	Laxatives:			
28345 21	Irritants	$83,675	$79,902	$88,318
28345 23	Bulk producing	$194,079	$199,159	$202,800
28345 25	Fecal softeners	$50,345	a/ $46,923	$41,619
28345 27	Emollients	$20,587	$21,223	D
28345 28	Saline	$14,321	$15,424	$6,802
28345 29	Enema specialties	$40,561	$37,767	D
28345 31	Digestants	D	D	D
28345 33	Bile therapy preparations, including bile products, choleretics, and cholagogues	D	D	D
28345 35	Antinauseants and motion sickness remedies (antiemetics), including antihistaminic antiemetic preparations	D	D	D
28345 37	Lipotropics and cholesterol reducers	D	D	D
28345 39	Diet aids containing local anesthetics such as benzocaine	D	D	D
	Antispasmodics and anticholinergics:			
28345 41	Synthetics	$16,870	$12,904	$23,491
28345 43	Ataractic combinations	D	D	D
28345 45	Belladonna and derivatives	D	D	D
28345 47	H_2 blocking agents, such as zantec, tagamet, carafate, and pepcid	$3,922,367	$3,772,091	$3,861,543
28345 49	Other antispasmodics and anticholinergics	D	D	D
28345 58	Other digestive system preparations, including emetics	$81,587	$96,666	$68,996
	Genito-urinary preparations:			
28345 61	Urinary antibacterials and antiseptics	$123,296	$112,857	$116,177
	Diuretics (excluding aminophylline, xanthine, and rauwolfia-diuretic combinations. See cardiovascular preparations.):			
28345 71	Thiazides and related agents	$62,081	$32,672	$28,245
28345 73	Other diuretics	$136,530	$105,186	$120,937

11.18. Value of Shipments of Pharmaceutical Preparations, Except Biologicals, United States, 1995–97 (thousands of dollars) *(continued)*.

Product Code	Product Description	1995	1996	1997
28345 81	Oxytocics	$18,510	D	D
28345 83	Contraceptive agents, except oral contraceptives (aerosols, gels, sponges, and creams)	$76,370	D	$35,524
28345 85	Premenstrual tension preparations	D	D	D
28345 87	Vaginal cleaners	$112,220	$89,899	D
28345 98	Other pharmaceutical preparations acting on the genito-urinary system, including urinary tract anesthetics, for human use	r/ $143,638	$152,158	$338,464
28346 —	Pharmaceutical preparations acting on the skin, for human use	$2,170,966	$2,074,398	$1,981,706
	Dermatological preparations:			
28346 11	Emollients and protectives, including burn remedies and ointment bases	$653,513	$774,907	$503,097
28346 13	Antipruritics and local anesthetic skin preparations	$147,538	$120,726	$79,910
28346 15	Coal tar, sulfur, and resorcinol preparations	$7,115	$29,645	D
28346 16	Antiacne preparations	$332,892	$208,982	$552,175
28346 18	Antidandruff and antiseborrheic preparations (except dandruff shampoos)	D	$12,405	$12,225
28346 19	Other dermatological preparations	$196,224	$191,188	$175,283
28346 21	Hemorrhoidal preparations	$127,621	$112,316	$116,850
	External analgesics and counterirritants:			
28346 31	Ointments, jellies, pastes, creams, cerates, and salves	$170,187	$222,190	$240,150
28346 35	Liquid, excluding rubbing alcohol but including liniments	$56,139	$69,109	$50,341
28346 37	Rubbing alcohol	$135,719	a/ $ 53,429	$15,949
28346 39	Other external analgesics and counterirritants	$74,378	$73,767	$30,114
28346 98	Other pharmaceutical preparations acting on the skin, for human use	r/ $208,259	$191,222	$132,582
28347 —	Vitamin, nutrient, and hematinic preparations, for human use	r/ $4,811,596	$5,412,698	$7,108,848
	Vitamins:			
28347 11	Multivitamins, plain and with minerals (except B complex vitamins and fish liver oils)	$1,154,404	$1,315,966	a/$2,262,869
28347 13	Pediatric vitamin preparations (drops, suspensions, and chewable tablets)	$91,775	$96,183	$95,695
28347 14	Prenatal vitamin preparations	$29,998	$30,264	$32,190
28347 15	B complex preparations	$104,388	$115,434	$111,042
28347 17	Fluoride preparations	$1,722	$1,397	$1,514
28347 19	All other vitamin preparations	$616,701	$791,380	$983,790
28347 21	Fish liver oils (cod, etc.)	$11,854	$13,746	$22,331
28347 31	Nutrients, excluding therapeutic dietary foods and infant formulas	r/ $261,181	$311,259	$364,926
28347 41	Tonics and alternatives	$19,918	$18,863	$14,017
	Hematinics:			
	With B^{12}:			
28347 51	Oral	$15,425	$18,806	$15,780
28347 53	Parenteral	D	D	D
	Other hematinics:			
28347 55	Oral	$24,585	$24,064	$15,583
28347 57	Parenteral	D	D	D
28347 61	Hospital solutions (includes dextran, etc., but excludes biologicals such as blood plasma)	$1,470,312	$1,484,077	$1,714,669

11.18. Value of Shipments of Pharmaceutical Preparations, Except Biologicals, United States, 1995–97 (thousands of dollars) *(continued)*.

Product Code	Product Description	1995	1996	1997
28347 98	Other vitamin, nutrient, and hematinic preparations, for human use	r/ $477,584	$596,928	a/$818,504
28348 —	Pharmaceutical preparations affecting parasitic and infective diseases, for human use	$7,195,493	$7,217,901	$8,689,724
	Anti-infective agents, excluding corticoid-anti-infective combinations (see code 28341 15):			
28348 11	Amebicides and trichomonicides	$91,408	$51,090	$58,343
28348 15	Anthelmintics	$8,735	$20,853	$21,587
	Systemic antibiotic preparations:			
	Broad and medium spectrum (single or in combinations with other antibiotics):			
28348 20	Tetracyclines, including chlortetracycline and congeners	$68,454	$71,554	$106,985
28348 22	Cephalosporins, such as cefamandole nafate and cephalexin	$1,721,354	$1,679,932	$2,102,079
28348 24	Erythromycins, such as erythromycin ethylsuccinate and erythromycin stearate	$138,315	$105,853	$89,657
	Penicillins (single):			
	Semisynthetic penicillins, such as amoxicillins and ampicillin:			
28348 26	Injectable	r/ $22,476	$21,514	$45,482
28348 28	Other forms	$1,871,298	$1,650,478	$1,984,685
	Other penicillins (except semisynthetic), such as penicillin V and penicillin G:			
28348 30	Injectable	D	D	D
28348 32	Other forms	$10,507	a/ $7,363	$15,349
28348 34	Other broad and medium spectrum antibiotics (except sulfa-antibiotic combinations)	$594,051	$557,088	$503,954
28348 36	Antibiotics in combination with sulfonamides	D	$17,030	$31,730
28348 38	Other systemic antibiotic preparations, including narrow-spectrum antibiotics and streptomycins	r/ $ 66,288	$87,026	$97,356
28348 41	Topical antibiotic preparations	$138,911	$138,195	$89,805
	Tuberculostatic agents:			
28348 51	Isoniazid (isonicotinic acid hydrazide) preparations	a/ $3,585	a/ $2,716	a/$3,359
28348 55	Other antituberculars	D	D	$27,861
28348 61	Antimalarials (plasmodicides)	$150,730	$72,227	D
28348 63	Sulfonamides (except antibiotic-sulfonamide combinations)	$62,819	$56,384	$55,118
28348 65	Antifungal preparations	$294,904	$393,335	$598,354
28348 67	Antivirals, systemic, such as zovirax and retrovir	D	$1,327,442	$1,869,578
28348 69	Other anti-infective agents	$346,995	$274,001	$47,514
	Antibacterials and antiseptics:			
28348 71	General	$67,415	$58,925	$91,098
28348 75	Mouth and throat preparations	D	D	D
28348 98	Other pharmaceutical preparations affecting parasitic and infective diseases, for human use	$122,913	$108,829	$185,557

Notes:
D = Withheld to avoid disclosing data for individual companies.
r/ = Revised by 5 percent.
X = Not applicable.
a/ 20 percent or more of this item is estimated.
[1] Code 28342 24 combined with code 28342 28.

Source: U.S. Bureau of the Census. Current Industrial Reports, Series MA28G: *Pharmaceutical Preparations, Except Biologicals, 1996 and 1997.* Washington, DC: U.S. Bureau of the Census, 1998 and 1999. http://www.census.gov/cir/www/ma28g.html

11.19. Value of Product Class Shipments of Pharmaceutical Preparations, Except Biologicals, United States, 1986–97 (in millions of dollars).

Product Code	Product Description	1986	1987	1988	1989	1990	1991	1992	1993	1994	1995	1996	1997
28341 —	Pharmaceutical preparations affecting neoplasms, endocrine system, and metabolic diseases, for human use	$2,140.6	$1,994.9	$2,071.2	$2,507.4	$2,743.2	$2,877.4	$3,317.8	$3,820.3	$4,119.8	$4,075.9	$4788.0	$6,873.0
28342 —	Pharmaceutical preparations acting on the central nervous system and sense organs, for human use	$5,946.0	$5,740.5	$6,172.7	$6,441.3	$7,218.5	$7,430.8	$8,319.3	$8,926.8	$8,989.9	$9,228.2	$10,123.1	$12,279.9
28343 —	Pharmaceutical preparations acting on the cardiovascular system, for human use	$3,570.3	$3,795.9	$4,449.9	$4,874.9	$4,814.6	$4,810.3	$4,908.7	$5,234.2	$5,673.9	$5,987.9	$6,911.9	$8,261.5
28344 —	Pharmaceutical preparations acting on the respiratory system, for human use	$2,492.2	$2,811.4	$3,224.0	$3,286.2	$3,724.1	$4,259.7	$5,276.8	$5,509.8	$5,566.4	$5,196.1	$4,993.9	$5,486.2
28345 —	Pharmaceutical preparations acting on the digestive or genito-urinary system, for human use	$1,864.5	$2,954.8	$3,860.8	$4,363.0	$4,840.2	$5,624.8	$7,110.9	$7,995.7	$8,479.2	$8,593.4	$8,494.4	$10,099.6
28346 —	Pharmaceutical preparations acting on the skin, for human use	$1,327.3	$1,288.3	$1,388.5	$1,451.7	$1,558.0	$1,578.5	$1,759.6	$1,979.7	$2,089.7	$2,171.0	$2,184.8	$1,981.7
28347 —	Vitamin, nutrient, and hematinic preparations, for human use	$2,665.9	$2,570.7	$2,538.5	$2,672.4	$2,587.9	$2,787.6	$2,923.7	$3,559.8	$4,402.1	$4,811.6	$5,280.8	$7,108.8
28348 —	Pharmaceutical preparations affecting parasitic and infective diseases, for human use	$3,821.0	$4,024.9	$4,593.6	$4,936.2	$5,411.0	$6,006.1	$7,017.6	$7,233.6	$7,257.4	$7,195.5	$7,304.1	$8,689.7
28349 —	Pharmaceutical preparations for veterinary use	$853.7	$997.1	$1,051.5	$1,071.1	$1,056.8	$1,462.2	$1,327.3	$1,352.5	$1,631.0	$1,604.7	$1,763.3	$1,813.0

Source: U.S. Bureau of the Census. Current Industrial Reports, Series MA28G: *Pharmaceutical Preparations, Except Biologicals.* Washington, DC: U.S. Bureau of the Census, 1996 and 1997. http://www.census.gov/cir/www/ma28g.html

11.20. Nutraceuticals. Market Size, 1996 (in millions of dollars).

Segment	Sales
Food supplements	$700
Slimming preparations	$200
Natural medicine (homeopathic)	$300
Sports nutritionals	$600
Herbs and natural botanicals	$600
Functional teas	$75
Isotonics	$1,200
Beverages (i.e., calcium orange juice)	$200
Lactose-free or -reduced dairy products	$100
Bran and high-fiber cereals	$1,500
Dietetic food and confections	$100
Vitamin and mineral supplements	$2,500
Total	**$8,075**

Source: Shear/Kershman Laboratories, Inc. St. Louis, MO. Cited in "Pharmaceuticals Spur New Product Growth." *Beverage Industry* 8 8(October 1997): 45.

11.21. Revenues in the U.S. Market for Pain Management, 1995 and 1996, and Forecast for 2000.

Revenues (millions of dollars)	1995	1996	2000F
Pharmaceuticals	$11,824	$12,970	$18,788
Devices	$766	$865	$1,411
Services	$774	$900	$1,626

Notes:

Class Definition: This market class includes pharmaceuticals (systemic and topical over-the-counter drugs, prescription drugs), devices (spinal cord stimulators, electrical stimulus devices, and patient-controlled analgesia infusion devices, PCA), and pain management services (home care, hospices, hospitals, pain management clinics, and others).

Source: Find/SVP. New York, NY. Cited in "Pain Management." *Medical & Healthcare Marketplace Guide* (January 1997).

MEDICAL EQUIPMENT

11.22. Number of Persons Using Assistive Technology Devices, by Age of Person and Type of Device, United States, 1994.

Assistive Device	All Ages	44 Years and Under	45–64 Years	65 Years and Over
Anatomical devices		Number in thousands		
Any anatomical device [1]	4,565	2,491	1,325	748
Back brace	1,688	795	614	279
Neck brace	168	76	78	13*
Hand brace	332	171	119	42
Arm brace	320	209	86	25*
Leg brace	596	266	138	192
Foot brace	282	191	59	31
Knee brace	989	694	199	96
Other brace	399	239	104	56
Any artificial limb	199	69	59	70
Artificial leg or foot	173	58	50	65
Artificial arm or hand	21*	9*	6*	6*
Mobility devices				
Any mobility device [1]	7,394	1,151	1,699	4,544
Crutch	575	227	188	160
Cane	4,762	434	1,116	3,212

11.22. Number of Persons Using Assistive Technology Devices, by Age of Person and Type of Device, United States, 1994 *(continued)*.

Assistive Device	All Ages	44 Years and Under	45–64 Years	65 Years and Over
Walker	1,799	109	295	1,395
Medical shoes	677	248	226	203
Wheelchair	1,564	335	365	863
Scooter	140	12	53	75
Hearing devices				
Any hearing device [1]	4,484	439	969	3,076
Hearing aid	4,156	370	849	2,938
Amplified telephone	675	73	175	427
TDD/TTY [2]	104	58	25*	21*
Closed caption television	141	66	32*	43
Listening device	106	26*	22*	58
Signaling device	95	37*	23*	35
Interpreter	57	27*	21*	9*
Other hearing technology	93	28*	24*	41
Vision devices				
Any vision device [1]	527	123	135	268
Telescopic lenses	158	40	49	70
Braille	59	28*	23*	8*
Readers	68	15*	14*	39
White cane	130	35*	48	47
Computer equipment	34*	19*	8*	7*
Other vision technology	277	51	76	151

Notes:

* = Figure does not meet standard of reliability or precision.

[1] Numbers do not add to these totals because categories are not mutally exclusive; a person could have used more than one device within a category.

[2] TDD/TTY = Telecommunication device for the deaf/Text telephone

Source: J. N. Russell, G. E. Hendershot, F. LeClere, et al. *Trends and Differential Use of Assistive Technology Devices: United States, 1994.* Advance data from Vital and Health Statistics, no. 292. National Center for Health Statistics. Hyattsville, MD, 1997.

11.23. Electromedical and Irradiation Equipment, Quantity and Value of Shipments, United States, 1996–97 (quantity in number of units; value in thousands of dollars).

Product Code	Product Description	Number of Companies (1996)	Quantity (1996)	Value (1996)	Number of Companies (1997)	Quantity (1997)	Value (1997)
	Electromedical and irradiation equipment	271	X	$11,219,683	259	X	$11,759,717
	Medical and diagnostic equipment:						
	X-ray equipment:						
38440 02	Digital radiography equipment	5	7,268	r/ $ 147,883	6	6,400	$159,840
38440 03	Computerized axial tomography (CT or CAT SCAN) [1]	7	r/ 2483	$524,526	6	2,660	$543,091
38440 09	Dental and conventional	12	r/ 22,713	$487,225	14	22,811	$520,549
38440 11	All other medical diagnostic x-ray equipment [1]	15	X	r/ $299,099	12	X	$367,688
38440 06	Nuclear medicine equipment (all equipment used for nuclear in vivo studies)	8	1,155	$307,263	7	1,441	$363,128
38450 18	Magnetic resonance imaging equipment (MRI)	6	D	D	7	D	D
38450 17	Ultrasound scanning devices	12	r/10,826	$935,994	14	13,863	$1,074,018
		14	46,142	$204,281	14	44,164	$202,717
38450 13	Electroencephalograph (EEG) and Electromyograph (EMG) [2]	6	D	D	6	D	D
38450 21	Audiological euipment	6	X	$11,099	6	X	$10,544
38450 23	Endoscopic equipment (bronchoscope, cystoscope, proctosigmoidoscope, colonoscope, etc.)	5	D	D	5	D	D
38450 25	Respiratory analysis equipment	6	5,295	$48,541	5	4,477	$39,291
38450 29	All other medical diagnostic equipment [2]	25	X	r/ $242,885	25	X	$229,936
	Patient monitoring equipment:						
38450 51	Intensive care/coronary care units, including component modules such as temperature, blood pressure, and pulse	23	X	$597,819	21	X	$627,084
38450 53	Prenatal monitoring [3]	6	X	D	6	X	D
38450 55	Respiratory monitoring [3]	6	X	$100,688	7	X	$88,046
38450 57	All other patient monitoring	33	X	$608,904	36	X	$588,746
	Medical therapy equipment:						
38450 43	Ultrasound therapy	7	X	$11,387	7	X	$12,203
38450 31	Pacemakers	10	204,605	r/ $ 636,018	8	220,915	$733,335
38450 33	Defibrillators	11	49,735	$528,073	10	50,024	$578,780
38450 39	Dialyzers (includes machines and equipment)	4	X	r/ $ 385,735	5	X	$458,290
38450 41	Medical laser equipment	18	X	$559,526	17	X	$600,041
38450 32	Radiation therapy (linear accelerators, x-ray, cobalt 60, brachetherapy) [4]	8	X	D	8	X	D
38450 49	All other medical therapy equipment	34	X	r/ $ 656,469	35	X	$748,190
	All other irradiation and electromedical equipment:						

11.23. Electromedical and Irradiation Equipment, Quantity and Value of Shipments, United States, 1996–97 (quantity in number of units; value in thousands of dollars) *(continued)*.

Product Code	Product Description	1996 Number of Companies	1996 Quantity	1996 Value	1997 Number of Companies	1997 Quantity	1997 Value
38440 13	Industrial and scientific x-ray equipment	18	2,182	$172,627	17	2,957	$193,030
38440 21	X-ray tubes, sold separately	11	45,870	$421,872	11	45,391	$413,196
38440 23	Other nonmedical irradiation equipment, including gamma and beta ray equipment, n.e.c.	5	X	$20,024	5	X	$20,578
38440 27	Parts and accessories for x-ray equipment and other nonmedical irradiation equipment, n.e.c. [4]	26	X	$520,332	22	X	$555,954
	Surgical systems:						
38450 35	Electrosurgical equipment	14	X	$339,205	14	X	$339,940
38450 71	Heart-lung machines, excluding iron lungs [5]	3	D	D	3	D	D
38450 73	Blood flow systems [5]	8	X	$320,307	7	X	$328,609
38450 75	All other surgical support systems	13	X	r/ $399,573	16	X	$384,314
38450 83	Other electromedical equipment (except diagnostic and therapeutic), n.e.c.	11	X	r/ $81,469	13	X	$82,067
38450 89	Electromedical parts and accessories, including diagnostic and therapeutic, n.e.c.	42	X	$784,931	42	X	$705,834

Notes:

D = Withheld to avoid disclosing data for individual companies.

n.e.c. = Not elsewhere classified.

r/ = Revised by 5 percent or more from previously published data.

X = Not applicable.

[1] For 1995, receipts for "Computerized axial tomography (CT or CAT Scan)" are combined with "All other medical diagnostic x-ray equipment" to avoid disclosing data for individual companies.

[2] Receipts for "Electroencephalograph (EEG) and Electromyograph (EMG)" are combined with "All other medical diagnostic equipment" to avoid disclosing data for individual companies.

[3] Receipts for "Prenatal monitoring" are combined with "Respiratory monitoring" to avoid disclosing data for individual companies.

[4] Receipts for "Radiation therapy (linear accelerators, x-ray, cobalt 60, brachetherapy)" are combined with "Parts and accessories for x-ray equipment and other nonmedical irradiation equipment, n.e.c." to avoid disclosing data for individual companies.

[5] Receipts for "Heart-lung machines, excluding iron lungs" are combined with "Blood flow systems" to avoid disclosing data for individual companies.

Source: U.S. Bureau of the Census. Current Industrial Reports, Series MA38R: *Electromedical Equipment and Irradiation Equipment (Including X-Ray)*. Washington, DC: U.S. Bureau of the Census, 1998. Table 2. http://www.census.gov/cir/www/ma38r.html

BIOTECHNOLOGY

11.24. Biotechnology Market Areas: Participation of Biotechnology Companies by Primary Focus, United States, 1996.

Market Area	Number of Companies	Percentage of All Companies
Therapeutics	315	29.4%
Diagnostics	187	17.4%
Reagents	84	7.8%
Plant agriculture	68	6.3%
Specialty chemicals	54	5.0%
Immunological products	36	3.4%
Environmental testing/Treatment	35	3.3%
Testing/Analytical services	32	3.0%
Animal agriculture	29	2.7%
Biotechnology equipment	26	2.4%
Veterinary	26	2.4%
Drug delivery systems	24	2.2%
Vaccines	24	2.2%

Source: Institute for Biotechnology Information. *U.S. Companies Database.* Research Triangle Park, NC: Institute for Biotechnology Information, 1996.

11.25. U.S. Biotechnology Firms, 1997.

All Biotechnology Firms	Average	Median	Minimum	Maximum
Employees	115.5	29	1	6894
Biotech Employees	73.1	24.5	1	4100
Revenues (in millions of dollars)	$23.3	$3.0	$0.0006	$1,940
R&D budget (in millions of dollars)	$9.6	$2.0	0	$451.7
Percent of R&D for biotech	84.2%	100%	0	100%

Notes:
Based on the Institute for Biotechnology Information's annual survey, annual reports, 10k's, press releases, Internet information, word of mouth, and other resources. These data are for biotechnology firms only (1,070 companies), as opposed to companies that are primarily involved in a particular industry but are using the new technologies to diversify their activities or product lines.

Source: Mark D. Dibner. *Biotechnology Guide U.S.A.: Companies, Data and Analysis,* Fourth Edition. Research Triangle Park, NC: Institute for Biotechnology Information, 1997.

11.26. 1997 U.S. Biopharmaceutical Facts and Figures.

	Number
Biotechnology companies developing drugs	379
Pharmaceutical companies developing biotechnology drugs	42
Biotechnology drugs in clinical trials (estimate)	500
Biotechnology drugs on the market	75
Biotechnology drugs with worldwide sales over $500 million (1996 data)	6
Nonbiotechnology drugs with worldwide sales over $500 million (1996 data)	51
Biotechnology companies with drugs in clinical trials	188
Biotechnology companies with more than one drug in clinical trial	70
Biotechnology companies with at least one drug on the market	20
Biotechnology companies with drugs that have $500 million or more in worldwide sales	1
Biotechnology companies with no drugs in clinical trials	191
Biotechnology companies with no drug on the market	359

Source: Institute for Biotechnology Information. U.S. Companies and Actions databases and a mid-1996 survey of all biopharmaceutical companies, Research Triangle Park, NC. Cited in Mark D. Dibner. "Biotechnology and Pharmaceuticals: 10 Years Later." *BioPharm* 10 (1997): 24+.

Chapter 12. Space

Humans have always been fascinated by the mysterious night sky, but it wasn't until Galileo first pointed a telescope upward that we were able to see phenomena not visible to the naked eye. Ever since that heretical act, which confirmed Copernicus' assertion that the earth revolved around the sun rather than vice versa, we have been learning more and more about the universe around us. Galileo discovered four moons around Jupiter. His successors found that we lived at the edge rather than the center of our galaxy; that a ninth planet, Pluto, circled the sun; that we are part of an expanding universe; that microwave background radiation permeates the universe; and that the universe may have originated with a single Big Bang.

Technology is taking us farther and farther into space. Space flight enables us not only to explore outside of Earth, but also to install satellites that enhance communications. Solar system exploration has proceeded apace in the 1990s, with increasingly smaller and less expensive missions returning data that show evidence of water on the Moon and Mars and a possible underground ocean on Jupiter's moon Europa. In microgravity, we can conduct experiments that produce usable results here on Earth—in medicine, botany, and many other disciplines—and that allow us to go farther into space. Sophisticated telescopes like the space-based Hubble afford us a view of the universe almost at its inception, because they let us see galaxies billions of miles away. Radio telescopes, which detect radio waves emitted and absorbed by celestial bodies, tell us about the nature and shape of our galaxy and others. Other telescopes, some mounted on satellites or rockets above the earth's atmosphere, perceive infrared, ultraviolet, x-rays, and gamma rays, and reveal secrets about universal structures and processes unavailable by observing visible light. Think of how much we have learned in the 1990s alone: that many extrasolar planets exist, that brown dwarf ("failed") stars are a reality, that Mars was once more earthlike

than we suspected, and that the Kuiper Belt, a swarm of comets in the outer solar system and beyond, does indeed exist.

Hot topics related to space technologies include

- Space debris and the threat it poses to satellites and vehicles. Table 12.6 shows that, as you would expect, space debris is proliferating. Both Table 12.6 and 12.7 indicate that fragmentation debris—that produced by collisions and explosions of human-made devices—is the most common type. Orbital debris is more dangerous than natural debris from meteoroids because it occurs in much higher concentrations in typical spacecraft orbits. Such debris becomes even more hazardous when it occurs in sizes too small to track, which can happen during fragmentation.

- Costs for space exploration. Tables 12.8 and 12.9 present some basic figures. Table 12.9 illustrates the dramatic drop in average spacecraft development cost, from $590 million in the early 1990s to $190 million in the late 1990s. During the same period, development time was cut almost in half.

- Commercialization of space. See Tables 12.1 through 12.3 on global positioning satellites, only one aspect of space commercialization. Table 12.1 shows the phenomenal growth in the GPS market from $541 million in 1992 to $1.5 billion in 1996, with $6.5 billion projected for 2001. Table 12.2 shows that the maritime community was the largest user of GPS in 1995, ahead of the Department of Defense, civil aviation, and civil land and mobile computing. Table 12.3 shows that car navigation is the largest worldwide market for GPS, ahead of survey and mapping, marine, military, and aviation.

- Military use of space. (Tables 12.2 and 12.3 plus 10.2, 10.7, 10.10, and 10.11 in Chapter 10.)

GLOBAL POSITIONING SYSTEMS AND SPACE COMMERCIALIZATION

12.1. U.S. Global Positioning Satellite Market, 1992 and 1996, and Forecast for 2001 (millions of dollars).

Item	1992	1996	2001	Percent Annual Growth 1992–1996	Percent Annual Growth 1996–2001
Total GPS Market	$541	$1,495	$6,500	29%	34%
Automotive	$7	$120	$4,000	103%	68%
Aerospace (including infrastructure)	$441	$515	$2,600	4%	20%
Communications	$3	$80	$4,000	127%	73%
Maritime	$41	$195	$1,250	48%	22%
Industrial	$7	$205	$1,700	133%	30%
Consumer/Recreational	$4	$95	$1,000	121%	30%
Government/Institutional	$38	$240	$1,300	59%	18%
All Other	negligible	$45	$550	—	27%

Source: The Freedonia Group, Inc., Cleveland, Ohio. Unpublished. Data courtesy of United States Global Positioning System Industry Council.

12.2. Global Positioning System Users, United States, 1995.

Civil Aviation	15,000
Civil Aviation, Handheld	80,000
Civil Land	90,000
Civil Mobile Computing	25,000
Civil Maritime	100,000
Civil Maritime, Handheld	175,000
Department of Defense	52,000
Total	537,000

Source: United States Global Positioning System Industry Council. Washington, DC. Unpublished.

12.3. Global Positioning Satellite Market Projections Worldwide, 1993–2000 (sales in millions of dollars).

	1993	1994	1995	1996	1997	1998	1999	2000
Car Navigation	$100	$180	$310	$600	$1,100	$2,000	$2,500	$3,000
Consumer/Cellular	$45	$100	$180	$324	$580	$1,000	$1,500	$2,250
Tracking	$30	$75	$112	$170	$250	$375	$560	$850
OEM*	$60	$110	$140	$180	$220	$275	$340	$425
Survey & Mapping	$100	$145	$201	$280	$364	$455	$546	$630
GIS**	$25	$35	$50	$90	$160	$270	$410	$650
Aviation	$40	$62	$93	$130	$180	$240	$300	$375
Marine	$80	$100	$110	$120	$130	$140	$150	$160
Military	$30	$60	$70	$80	$90	$100	$110	$130
Total	$510	$867	$1,266	$1,974	$3,074	$4,855	$6,416	$8,470

Notes:
*OEM = Original equipment manufacturers.
**GIS = Geographical information systems.

Source: U.S. Global Positioning System Industry Council. Washington, DC. Unpublished.

12.4. U.S. Commercial Space Revenues, 1990–99 (millions of dollars).

Industry	1990	1991	1992	1993	1994	1995	1996	1997	1998	1999
Total	3,385	4,370	4,860	5,295	6,640	-	-	-	-	-
Commercial satellites delivered	1,000	1,300	1,300	1,100	1,400	-	-	-	-	-
Satellite services	800	1,200	1,500	1,850	2,330	-	-	-	-	-
Fixed	735	1,115	1,275	1,600	1,980	-	-	-	-	-
Mobile	65	85	225	260	350	-	-	-	-	-
Satellite ground equipment	860	1,300	1,400	1,600	1,970	-	-	-	-	-
Mobile equipment	145	280	350	420	480	-	-	-	-	-
Commercial launches	570	380	450	465	580	483	553	750	870	992
Remote sensing data and services	155	190	210	250	300	340	480	590	661	713
Commercial research and development infrastructure	0	0	0	30	60	55	65	68	73	80

Notes:
Figures from 1996 on are forecasts, though 1996 and 1997 for commercial launches, remote sensing, and commercial R&D infrastructure are fairly firm.

Source: U.S. Department of Commerce, International Trade Administration. *U.S. Industry & Trade Outlook '99.* New York: McGraw-Hill, 1998. Table 29-2.

12.5. U.S. and World Satellite Revenues, 1997 (billions of dollars).

Industry Segment	U.S.	World
Satellite Manufacturing	$8.3	$13.5
Ground Equipment Manufacturing	$5.0	$11.0
Satellite Services	$6.6	$19.2

Notes:
Satellite communications services include one- and/or two-way delivery of voice, video, and data. Communications are fixed—broadcasting, data transmission, and telephony using fixed earth stations—and mobile—services that use mobile receivers, such as cellular phones.

Source: U.S. Department of Commerce, International Trade Administration. *U.S. Industry & Trade Outlook '99.* New York: McGraw-Hill, 1998. Figures 30-5 and 31-7.

SPACE DEBRIS

12.6. Number of Cataloged Space Objects in Orbit as of 27 September 1991

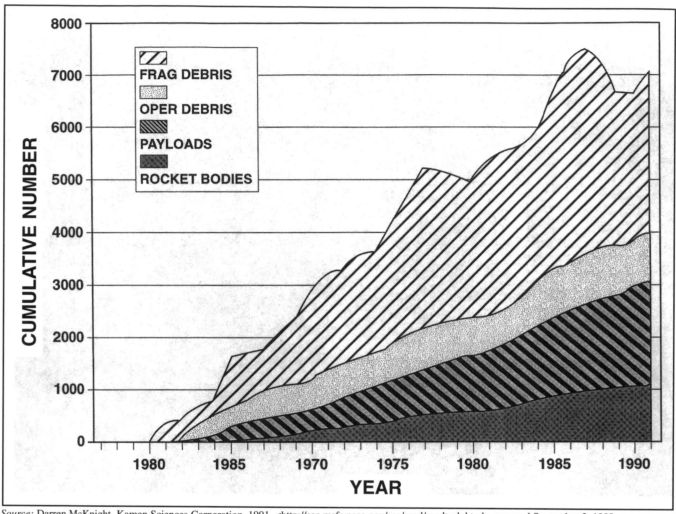

Source: Darren McKnight, Kaman Sciences Corporation, 1991. <http://see.msfc.nasa.gov/see/mod/modtech.html> accessed September 3, 1999.

12.7. Space Debris Composition.

Type of Debris	Percent of Total
Active payloads	5%
Fragmentation debris	43%
Inactive payloads	21%
Rocket body	15%
Launch debris	16%

Source: NASA Space Environments and Effects Meteroid and Orbital Debris Working Group. From the Working Group's Web site at http://see.msfc.nasa.gov/see/mod/modtech.html, April 1998.

SPACECRAFT AND SPACE PROGRAMS

12.8. U.S. Department of Defense Space Platforms Goals, Payoffs, and Time Frames: Space Vehicles, 2000 and 2005.

Space Vehicles	Fiscal Year 2000 LEO (Surv)	Geo (Comm)	Fiscal Year 2005 LEO (Surv)	Geo (Comm)
Goals				
Bus Power Load	1,050 Watts	5,560 Watts	840 Watts	5,050 Watts
Bus Mass	160 kg	1,960 kg	115 kg	1,410 kg
Control Costs	$142M	$142M	$131M	$131M
Payoffs				
Launch Mass	485 kg	6,700 kg	420 kg	5,360 kg
On-orbit Life	5 years	7 years	7.5 years	7 years
Lifecycle Cost	$835M	$1,490M	$1,010M	$1,370M
Geolocation	244 meters	260 meters	114 meters	194 meters
Launch Family Class	LLV1	Titan IV	LLV1	Titan IV

Notes:
LEO (Surv) = Low Earth Orbit Surveillance.
Geo (Comm) = Geosynchronous earth orbit (for communications satellites).
LLV1 is a small Lockheed missle and space corporation launch vehicle.
Titan IV is a large launch vehicle made by Lockheed.

Source: U.S. Department of Defense. *1997 Defense Technology Area Plan.* Ft. Belvoir, VA: Defense Technical Informaton Center, 1997.

12.9. U.S. Spacecraft Parameters, 1990–99.

Average U.S. Spacecraft Development Cost, 1990–99 (in millions of FY96 dollars)	
1990–1994	$590
1995–1999	$190
Average Spacecraft Development Time (in years)	
1990–1994	8.3
1995–1999	4.6
Expected Number of Launches, FY1999	
Space Science	7
Earth Science	4

Note:
Figures pertain to scientific craft doing space science, such as Galileo, Pathfinder, and Lunar Prospector.

Source: NASA. *FY1999 Annual GPRA (Government Performance Results Act) Performance Plan.* From telephone conversaton with NASA official.

12.10. Space Shuttle Flights, 1981–98.

Mission Number	Orbiter	Crew Size	Date	Length in Days	Orbits (number)	Main Payload
STS-88	Endeavour	6	12/4/98 to 12/15/98	11.79	185	First Space Station Flight
STS-95	Discovery	7	10/29/98 to 11/7/98	8.88	134	John Glenn's Flight, Spacelab
STS-91	Discovery	6 up/7 down	6/2 to 6/12/98	9.83	154	9th (last) Mir linkup, returned Thomas, automated measurement system, International Space Station exper'mts, 4 getaway special
STS-90	Columbia	7	4/17 to 5/3/98	15.91	255	Neurolab (16th Spacelab), bioreactor, small payloads (3 GAS)
STS-89	Endeavour	7 up/7 down	1/22 to 3/1/98	8.82	138	8th Mir linkup, crew exchange (Wolf/Thomas), cargo, SL/DM
STS-87	Columbia	6	11/19 to 12/5/97	15.69	251	4th U.S. microgravity payload USMP-4, SPARTAN-201, 2 EVAs
STS-86	Atlantis	7 up/7 down	9/25 to 10/6/97	10.80	169	Mir Docking 7, SpaceHab-DM, crew exchange (Foale/Wolf)
STS-85	Discovery	5	8/7 to 8/19/97	11.85	189	CRISTA-SPAS Satellite (2nd), Japanese Manipulator (MFD)
STS-94	Columbia	7	7/1 to 7/17/97	15.70	251	Microgravity Science lab MSL-1 reflight
STS-84	Atlantis	7 up/7 down	5/15 to 5/24/97	9.22	145	Mir Docking 6; SpaceHab-DM, crew rotation (Linenger/Foale)
STS-83	Columbia	7	4/4 to 4/8/97	3.98	63	Microgravity Science lab MSL-1—returned 12 days early (fuel cell problem); reflight as STS-94
STS-82	Discovery	7	2/11 to 2/21/97	9.98	150	Hubble service mission 2; 5 EVAs; replaced 10 instruments
STS-81	Atlantis	6	1/12 to 1/22/97	10.21	160	Mir Docking 5; SpaceHab-DM
STS-80	Columbia	5	11/19 to 12/7/96	17.66	278	ORFEUS-SPAS, WSF-3
STS-79	Atlantis	6	9/16 to 9/26/96	10.14	159	Mir Docking 4, SpaceHab-DM, crew exchange (Lucid/Blaha)
STS-78	Columbia	7	6/20 to 7/7/96	16.91	271	Life/Microgravity Spacelab (LMS), SAREX II
STS-77	Endeavour	6	5/19 to 5/29/96	10.03	161	Inflatable Antenna Experiment (IAE), SpaceHab-4, SPARTAN, TEAMS
STS-76	Atlantis	6 up/5 down	3/22 to 3/31/96	9.22	144	Mir Docking 3, SpaceHab
STS-75	Columbia	7	2/22 to 3/9/96	15.74	250	Tether Satellite TSS-1R, USMP-3, OARE
STS-72	Endeavour	6	1/11 to 1/20/96	8.92	141	SFU-Retrieval, SPARTAN/OAST Flyer, 2 EVAs
STS-74	Atlantis	5	11/12 to 11/20/95	8.19	128	2nd Mir linkup; delivered docking module, solar arrays, etc.
STS-73	Columbia	7	10/20 to 11/5/95	15.91	255	USML-2, educational experiments
STS-69	Endeavour	5	9/7 to 9/18/95	10.85	170	Wake Shield Facility (second flight), SPARTAN, 30th Shuttle EVA
STS-70	Discovery	5	7/13 to 7/22/95	8.93	142	TDRSS-G, crystal growth and biological experiments
STS-71	Atlantis	7 up/8 down	6/27 to 7/7/95	9.80	153	1st Mir docking/crew exchange, 100th U.S. human space flight, SL-M
STS-67	Endeavour	7	3/2 to 3/18/95	16.63	262	2nd UV Astronomy Spacelab (ASTRO-2), Middeck
STS-63	Discovery	6	2/3 to 2/11/95	8.27	129	"Near-Mir," SpaceHab/3, second cosmonaut, SPARTAN-4
STS-66	Atlantis	6	11/3 to 11/14/94	10.94	174	ATLAS-3, CRISTA/SPAS, ESCAPE-2, etc.
STS-68	Endeavour	6	9/30 to 10/11/94	11.24	182	Space Radar lab-2, GAS, Stamps, secondary payloads
STS-64	Discovery	6	9/9 to 9/20/94	10.95	176	Laser-Exp. LITE, SPARTAN-3, SAFER (EVA), ROMPS
STS-65	Columbia	7	7/8 to 7/23/94	14.75	235	International Microgravity Laboratory IML-2, etc.
STS-59	Endeavour	6	4/9 to 4/20/94	11.24	182	Space Radar Lab-1, CONCAP-IV, GAS, NIH experiments
STS-62	Columbia	5	3/4 to 3/18/94	13.97	223	Microgravity USMP-2, OAAST-2, SSBUV/A, DEE
STS-60	Discovery	6	2/3 to 2/11/94	8.30	130	Wake Shield facility, SpaceHab 2, cosmonaut
STS-61	Endeavour	7	12/2 to 12/13/93	10.83	162	Repair and first servicing of Hubble Telescope
STS-58	Columbia	7	10/18 to 11/1/93	14	224	2nd Spacelab for Life Sciences (SLS-2)
STS-51	Discovery	5	9/12 to 9/22/93	9.84	157	Adv. Comsat ACTS/TOS, ORFEUS-SPAS, EVA
STS-57	Endeavour	6	6/21 to 7/1/93	9.99	158	1st SpaceHab, EURECA retrieval, GAS, EVA
STS-55	Columbia	7	4/26 to 5/6/93	9.99	158	2nd German Spacelab mission (D2)
STS-56	Discovery	5	4/8 to 4/17/93	9.26	147	2nd Atmospheric Mission ATLAS-2, SPARTAN-2
STS-54	Endeavour	5	1/13 to 1/19/93	5.99	95	Sixth TDRS(F), Differential X-ray experiment DXS, EVA

12.10. Space Shuttle Flights, 1981–98 (continued).

Mission Number	Orbiter	Crew Size	Date	Length in Days	Orbits (number)	Main Payload
STS-53	Discovery	5	12/2 to 12/9/92	7.31	116	Last DOD mission, DOD satellite, laser experiment
STS-52	Columbia	6	10/22 to 11/1/92	9.87	157	LAGEOS Satellite, materials experiments, USMP, etc.
STS-47	Endeavour	7	9/12 to 9/20/92	7.94	126	First Japanese Spacelab (SL-J)
STS-46	Atlantis	7	7/31 to 8/8/92	7.97	126	EURECA platform, tether experiment TSS-1
STS-50	Columbia	7	6/25 to 7/9/92	13.81	220	US Microgravity Laboratory USML-1
STS-49	Endeavour	7	5/7 to 5/16/92	8.89	140	Intelsat-VI recovery and redeployment
STS-45	Atlantis	7	3/24 to 4/2/92	8.92	142	1st Earth Atmosphere mission ATLAS-1
STS-42	Discovery	7	1/22 to 1/30/92	8.05	128	International Microgravity Laboratory IML-1
STS-44	Atlantis	6	11/24 to 12/1/91	6.95	108	DOD satellite, contaminations research
STS-48	Discovery	5	9/12 to 9/18/91	5.35	81	Upper Atmosphere Research Satellite (UARS) launched
STS-43	Atlantis	5	8/2 to 8/11/91	8.89	140	Tracking/Data Relay Satellite TDRS-E
STS-40	Columbia	7	6/5 to 6/14/91	9.09	145	1st Spacelab for Life Sciences (SLS-1)
STS-39	Discovery	7	4/28 to 5/6/91	8.31	133	Infrared astronomy experiments
STS-37	Atlantis	5	4/5 to 4/11/91	5.98	92	Gamma Ray Observatory (GRO) launched
STS-35	Columbia	7	12/2 to 12/11/90	8.96	143	UV Astronomy Spacelab ASTRO
STS-38	Atlantis	5	11/15 to 11/20/90	4.92	79	Seventh DOD mission
STS-41	Discovery	5	10/6 to 10/10/90	4.02	65	Solar probe Ulysses launched
STS-31	Discovery	5	4/24 to 4/29/90	5.05	76	Hubble Space Telescope (HST) launched
STS-36	Atlantis	5	2/28 to 3/4/90	4.44	68	Sixth DOD mission
STS-32	Columbia	5	1/9 to 1/20/90	10.86	171	1 COMSAT, retrieval & return of LDEF facility
STS-33	Discovery	5	11/22 to 11/27/89	5	78	Fifth DOD mission
STS-34	Atlantis	5	10/18 to 10/23/89	4.98	79	Jupiter probe Galileo launched
STS-28	Columbia	5	8/8 to 8/13/89	5.04	80	Fourth DOD mission
STS-30	Atlantis	5	5/4 to 5/8/89	4.04	64	Venus probe Magellan launched
STS-29	Discovery	5	3/13 to 3/18/89	4.98	79	Data Relay Satellite TDRS-D, experiments
STS-27	Atlantis	5	12/2 to 12/6/88	4.38	68	Third DOD mission
STS-26	Discovery	5	9/29 to 10/3/88	4.04	63	Tracking/Data Relay Satellite TDRS-C
51-L	Challenger	7	1/28/86	—	—	Loss of vehicle and crew of seven
61-C	Columbia	7	1/12 to 1/18/86	6.08	97	1 COMSAT, material and astronomy experiments
61-B	Atlantis	7	11/26 to 12/3/85	6.88	108	3 COMSATs, space structures assembly test
61-A	Challenger	8	10/30 to 11/6/85	7.03	111	1st German Spacelab mission (D1)
51-J	Atlantis	5	10/3 to 10/7/85	4.07	63	Second DOD mission
51-I	Discovery	5	8/27 to 9/3/85	7.10	111	3 COMSATs, EVA to repair Syncom IV satellite
51-F	Challenger	7	7/29 to 8/6/85	7.94	126	Spacelab 2
51-G	Discovery	7	6/17 to 6/24/85	7.07	111	3 communications satellites
51-B	Challenger	7	4/29 to 5/6/85	7	110	Spacelab 3
51-D	Discovery	7	4/12 to 4/19/85	6.98	109	2 COMSATs, EVA to repair Syncom IV satellite
51-C	Discovery	5	1/24 to 1/27/85	3.90	47	First military (DOD) mission
51-A	Discovery	5	11/8 to 11/16/84	7.98	126	2 COMSATs, 2 satellite retrievals/recoveries
41-G	Challenger	7	10/5 to 10/13/84	8.23	132	Environmental satellite ERBS
41-D	Discovery	6	8/30 to 9/5/84	7	96	3 communications satellites

12.10. Space Shuttle Flights, 1981–98 (continued).

Mission Number	Orbiter	Crew Size	Date	Length in Days	Orbits (number)	Main Payload
41-C	Challenger	5	4/6 to 4/13/84	6.96	107	1st repair in space (SMM), LDEF deployment
41-B	Challenger	5	2/3 to 2/11/84	7.96	127	2 COMSATs, first EVA testing of MMU jetpack
STS-9	Columbia	6	11/28 to 12/8/83	10.33	166	First Spacelab mission (SL-1)
STS-8	Challenger	5	8/30 to 9/5/83	6.04	97	1 communications satellite
STS-7	Challenger	5	6/18 to 6/24/83	6.13	97	2 COMSATs, German platform SPAS-1
STS-6	Challenger	4	4/4 to 4/9/83	5.04	80	Tracking & Data Relay Satellite (TDRS), first EVA
STS-5	Columbia	4	11/11 to 11/16/82	5.10	81	2 communications satellites (COMSATs)
STS-4	Columbia	2	6/27 to 7/4/82	7.04	112	Last orbital test flight, beginning of operations
STS-3	Columbia	2	3/22 to 3/30/82	8	129	3rd orbital test flight, first experiments
STS-2	Columbia	2	11/12 to 11/14/81	2.25	36	Test of Canadian robot arm (RMS)
STS-1	Columbia	2	4/12 to 4/14/81	2.25	36	First orbital test flight

Source: NASA. http://www.spaceflight.nasa.gov/shuttle/index.html, April 1999. See Past missions.

12.11. Hubble Space Telescope Statistics as of April 1999.

Miles traveled: 1.4 billion.
Amount of data generated: 6 terabytes.
Number of images taken: 237,566.
Number of celestial objects observed: 12,042.
Most distant object seen: A galaxy possibly 11.4 billion light-years from Earth.
Closest object seen: The Moon.
Inherently brightest and most massive star seen: The Pistol Star, 25,000 light-years away in the direction of the constellation Sagittarius. The Pistol Star shines with the brilliance of 10 million suns and may, at one time, have had a mass 200 times that of the sun.
Biggest star seen: Betelgeuse, a red super giant in the constellation Orion.
Coolest star-like object: A brown dwarf, cool by celestial standards, at no more than 1,300 degrees Fahrenheit.
Inherently dimmest object: Several charcoal black comets that roam the Kuiper Belt, a region of the solar system beyond Neptune's orbit.

The 2.4-meter reflecting Hubble Space Telescope was deployed in low Earth orbit (600 km) by the crew of the space shuttle Discovery (STS-31) on April 25, 1990. The telescope is a joint project of the European Space Agency and NASA, with responsibility for conducting and coordinating science operations assigned to the Space Telescope Science Institute on the campus of Johns Hopkins University.

Source: Space Telescope Science Institute. "ASTROFILE: Hubble's Record Book of Celestial Objects." Baltimore, MD: Space Telescope Science Institute, 1999. Located at http://oposite.stsci.edu/pubinfo/pr/1999/13/background.html. Accessed on April 20, 1999.

12.12. Worldwide Successful Space Launches, 1957–97 (Criterion of success is attainment of Earth orbit or Earth escape).

Country	Total	1957–1964	1965–1969	1970–1974	1975–1979	1980–1984	1985–1989	1990–1994	1995	1996	1997
Total	3,892	289	586	555	607	605	550	466	75	73	86
Soviet Union/Commonwealth of Independent States	2,548	82	302	405	461	483	447	283	32	25	28
United States	1,124	207	279	139	126	93	61	122	27	33	37
Japan	52	0	0	5	10	12	11	9	2	1	2
European Space Agency*	96	0	0	0	1	8	21	33	11	10	12
China	49	0	0	2	6	6	9	15	2	3	6
France	10	0	4	3	3	0	0	0	0	0	0
India	8	0	0	0	0	3	0	3	0	1	1
Israel	3	0	0	0	0	0	1	1	1	0	0
Australia	1	0	1	0	0	0	0	0	0	0	0
United Kingdom	1	0	0	1	0	0	0	0	0	0	0

Notes:
* = Includes launches by Arianespace, a French commercial space transportation company.

Source: Library of Congress, Congressional Research Service, Science Policy Research Division. *Space Activities of the United States, CIS, and Other Launching Countries/Organizations 1957 - 1994. July 31, 1995; and Forthcoming Report.* Cited in U.S. Bureau of the Census. *Statistical Abstract of the United States, 1998.* Washington, DC: U.S. Government Printing Office, 1998. Table 1010.

Chapter 13. Transportation

Sometimes it seems as if we Americans are obsessed with our cars, lavishing ridiculous amounts of money and attention on these metal pets whose only task is to get from point A to point B.

But wait a minute. Getting from point A to point B is a big deal. Without transportation, everyone would be eternally stuck at point A, which would limit contact with other people and opportunities for trade. We'd have to be completely self-sufficient, for we couldn't acquire anything we need from elsewhere.

So maybe all that expenditure on our cars and transportation isn't so foolish after all. Without railroads, planes, ships, cars, bicycles, and even roller blades, society would have a lot bigger problems than sore feet—like empty store shelves, mail that takes months to deliver, and social isolation.

Not that transportation doesn't bring problems with it—pollution, from horse droppings to fuel exhaust, for one; congestion, risk of accident and death, and urban blight, for others. But that's what technology is for—to solve these problems and provide the perfect means of transportation.

Hot topics in transportation technology include

- Alternative-fueled vehicles. Table 13.19 shows that their number is increasing, especially for vehicles running on compressed natural gas—from 23,191 in 1992 to 92,805 in 1996—and 85% methanol—from 4,850 in 1992 to 19,636 in 1996. Do not be fooled by the huge jump in vehicles using 85% ethanol in 1996. This figure is an estimate based on a planned introduction of a line of either-or vehicles that can run on either ethanol 85 or gasoline. Table 13.22 follows the consumption of vehicle fuels in recent years, both traditional and alternative. In 1997, alternative and replacement fuels represented only 2.7% of those used, up from 1.6% in 1992.
- Safety. Tables 13.17 and 13.18 show the numbers of vehicles—cars and trucks—with airbags. In 1998, 41% of cars on the road and 33% of light trucks featured airbags. Passenger-side airbags are

gaining in popularity, up from 27,700 in 1990 to almost 9 million in 1995.

- Pollution and environmental impact. Table 13.24 illustrates the drop in U.S. transportation-related air emissions from 1970 to 1995. Lead in our air has decreased dramatically, as you would expect with the elimination of lead-based gasoline. Nitrogen oxides are the only substance that has actually increased, albeit slightly. Airborne particulates, carbon monoxide, and volatile organic compounds have all fallen somewhat. Table 13.20 illustrates the energy-intensiveness of various types of transportation. This table harbors hidden pitfalls. While cars are consuming less energy per passenger mile, our tendency to drive one person to a vehicle leads to more congestion, more pollution, and more energy consumption than if we drove together in our increasingly efficient cars. The table is also tricky in that it reports consumption as Btu per passenger mile. Btu, or British thermal unit, is a measure of heat, kind of like a calorie. It is not a measure of volume, like a gallon, or price. However, by looking at this table, you *can* compare how much energy a particular mode has used over time, and you can tell that it takes more work *per passenger* to move a transit bus than a school bus, and less to move a train (per passenger!) than a motorcycle. The "per passenger" concept is critical in evaluating the data in this table. A truck uses much more energy per passenger than a certificated air carrier, but trucks aren't designed to move lots of people, and commercial flights are.

Table 13.21 shows how much energy in volume and Btu's the U.S. transportation sector has used since 1955. Again, this table doesn't tell you anything about price, but it does show trends in the types and amounts of transportation fuels we use. It also shows the amount and percent of transportation energy consumption compared to total energy consumption. Since 1955, the transportation sector has held fairly steady with respect to the whole, moving only from 24.6% in 1955 to 27.2% in 1996. This table is also interesting in that it shows that the oil crises of the 1970s did not lead to a drop in petroleum consumption for transportation.

- Ability to handle traffic volume and flow. See Table 6.7 in Chapter 6 for the conditions of our highways, as well as Table 6.9 for highway mileage and Table 6.10 for bridge conditions.
- Noise.

Other information you will find in Chapter 13 includes

- Table 13.1, the major elements of the U.S. transportation system. This table details the number and extent of our roads, airports, and railroads, and shows numbers and capacities of vehicles that use these facilities. There are more miles of gas pipelines (1,211,500 for both transmission and distribution) than of interstate highway (46,036 miles)!
- Table 13.2 shows that there are a lot fewer general aviation aircraft (that is, not scheduled commercial or military—essentially private) being shipped now than in the years right after World War II. In 1946, there were 35,000 units shipped. In 1998, there were only 2,220. However, the value of those aircraft has climbed steadily, from $110 million in 1946 to $5.9 billion in 1998. As you might expect from those numbers, the average age of the U.S. general aviation fleet, as shown in Table 13.4, is high—an average of 28 years.

- Table 13.5 provides a fascinating look at the types of aircraft in operation by air carriers from 1984 to 1996. In 1996, there were 1,055 Boeing 737s in operation (no wonder it seems like we are always riding on those things), 457 767s, 213 777s, and 195 747s. Only five of the classic 707 model were still in operation that year.
- Table 13.7 shows how many and what type of airports exist in various states and regions. The state with the most airports is . . . Alaska.
- Tables 13.8 and 13.9 trace aerospace industry sales by product group from 1981 through 1998. Look how missile sales have dropped off since the end of the Cold War.
- Tables 13.10 through 13.12 examine the state of our urban transit system, which features more vehicles in rush hour service and more route-miles serviced from 1985 to 1995 (Table 3.12), with better physical facilities overall from 1984 to 1992 (Table 13.10).
- Tables 13.13 and 13.14 compare the materials used in and weights of passenger cars over the years. Between 1976 and 1996, iron and steel content fell 32%, contributing to an 18% overall decrease in vehicle weight. During the same period, aluminum content shot up 200%, and polymeric materials 97.5%.
- Table 13.15 presents sales, market shares, and fuel economies for light trucks between 1980 and 1996.
- Table 13.16 compares the annual growth rates of passenger cars in relation to population in the United States, Canada, Europe, and Japan. Contrary to expectation, the U.S. ranks third out of the four, with Japan in the top position.

TRANSPORTATION SYSTEM

13.1. Major Elements of the U.S. Transportation System: 1996.

Mode	Major defining elements	Components
Highways [1]	Public roads and streets; automobiles, vans, trucks, motorcycles, taxis and buses (except local transit buses) operated by transportation companies, other businesses, governments, and households; garages, truck terminals, and other facilities for motor vehicles.	*Public roads* [2] 46,036 miles of Interstate highways 112,467 miles of other National Highway System roads 3,760,947 miles of other roads *Vehicles and use* 130 million cars, driven 1.5 trillion miles 69 million light trucks, driven .8 trillion miles 7.0 million commercial trucks with 6 tires or more, driven .2 trillion miles 697,000 buses, driven 6.5 billion miles
Air	Airways and airports; airplanes, helicopters, and other flying craft for carrying passengers and cargo.	*Public use airports* 5,389 airports *Airports serving large certificated carriers* [3] 29 large hubs (72 airports), 417 million enplaned passengers 31 medium hubs (55 airports), 89 million enplaned passengers 60 small hubs (73 airports), 37 million enplaned passengers 622 nonhubs (650 airports), 15 million enplaned passengers *Aircraft* 5,961 certificated air carrier aircraft, 4.8 billion miles flown [4] *Passenger and freight companies* 96 carriers, 538 million domestic revenue passenger enplanements, 12.9 billion domestic ton-miles of freight [4] *General aviation* 187,300 aircraft, 3.5 billion miles flown [4]
Rail [5]	Freight railroads and Amtrak.	*Miles of track operated* 126,682 miles by major (Class I) [6] railroads 19,660 miles by regional railroads 27,554 miles by local railroads 24,500 miles by Amtrak *Equipment* 1.2 million freight cars 19,269 freight locomotives *Freight railroad firms* Class I: 9 systems, 182,000 employees, 1.4 trillion ton-miles of freight carried Regional: 32 companies, 10,491 employees Local: 511 companies, 13,030 employees *Passenger (Amtrak)* 23,000 employees, 1,730 passenger cars [7], 299 locomotives [7], 19.7 million passengers carried [7,8]

13.1. Major Elements of the U.S. Transportation System: 1996 *(continued).*

Mode	Major defining elements	Components
Transit [9]	Commuter trains, heavy-rail (rapid-rail) and light-rail (streetcar) transit systems, local transit buses, vans, and other demand response vehicles, and ferry boats.	*Vehicles* 43,577 buses, 17 billion passenger-miles 8,725 rapid-rail and light-rail, 11.4 billion passenger-miles 4,413 commuter trains, 8.2 billion passenger-miles 68 ferries, 243 million passenger-miles 12,825 demand response, 397 million passenger-miles
Water	Navigable rivers, canals, the Great Lakes, the St. Lawrence Seaway, Intracoastal Waterway, and ocean shipping channels; ports; commercial ships and barges, fishing vessels, and recreational boats.	*U.S.-flag domestic fleet* [10] Great Lakes: 730 vessels, 58 billion ton-miles Inland: 33,323 vessels, 297 billion ton-miles Ocean: 7,051 vessels, 408 billion ton-miles Recreational boats: 11.9 million [11] *Ports* [12] Great Lakes: 362 terminals, 507 berths Inland: 1,811 terminals Ocean: 1,578 terminals, 2,672 berths
Pipeline	Crude oil; petroleum product, and natural gas lines	*Oil* Crude lines: 114,000 miles of pipe (1995), 338 billion ton-miles transported (1996) Product lines: 86,500 miles of pipe (1995), 281 billion ton-miles transported (1996) 160 companies [13], 14,500 employees *Gas* Transmission: 259,400 miles of pipe Distribution: 952,100 miles of pipe 20.0 trillion cubic feet, 138 companies, 171,600 employees

Notes:

[1] U.S. Department of Transportation, Federal Highway Administration. *Highway Statistics.* Washington, DC, 1996.

[2] Does not include Puerto Rico.

[3] Large certificated carriers operate aircraft with a seating capacity of more than 60 seats. U.S. Department of Transportation, Bureau of Transportation Statistics, Office of Airline Information. *Airport Activity Statistics of Certificated Air Carriers, 12 Months Ending December 31, 1996.* Washington, DC, 1997.

[4] Preliminary data.

[5] Except where noted, figures are from Association of American Railroads. *Railroad Facts: 1997.* Washington, DC, 1997.

[6] Includes 891 miles of road operated by Class I railroads in Canada.

[7] Fiscal year 1996. Amtrak. *Twenty-Fifth Annual Report, 1996.* Washington, DC, 1997.

[8] Excludes commuter service.

[9] Data for 1995. U.S. Department of Transportation, Federal Transit Administration. 1997. *National Transit Summaries and Trends for the 1995 National Transit Database Section 15 Report Year.* Washington, DC,1997.

[10] Excludes fishing and excursion vessels, general ferries and dredges, derricks, and so forth used in construction work. Vessel data from U.S. Army Corps of Engineers. *Transportation Lines of the United States.* New Orleans, LA, 1998. Ton-miles data from U.S. Army Corps of Engineers. *Waterborne Commerce of the United States, 1996.* New Orleans, LA, 1997.

[11] U.S. Department of Transportation, United States Coast Guard. *Boating Statistics.* Washington, DC, 1996.

[12] Data for 1995 from U.S. Department of Transportation, Maritime Administration. *A Report to Congress on the Status of the Public Ports of the United States, 1994–1995.* Washington, DC, October 1996.

[13] Regulated by the Federal Energy Regulatory Commission.

Source: Unless otherwise noted, U.S. Department of Transportation, Bureau of Transportation Statistics. *National Transportation Statistics 1998.* Washington, DC: Bureau of Transportation Statistics, 1998. Table 1-1. http://www.bts.gov/btsprod/nts/

AIR TRANSPORTATION

13.2. Annual Shipments of New U.S. Manufactured General Aviation Aircraft by Units Shipped, Number of Companies Reporting, and Factory Net Billings, 1946–98.

Year	Units Shipped	Companies Reporting	Factory Net Billings (millions)
1946	35,000	—	$110.0
1947	15,594	15	$ 57.9
1948	7,037	12	$ 32.4
1949	3,405	11	$ 17.7
1950	3,386	13	$ 19.1
1951	2,302	12	$ 16.8
1952	3,058	8	$ 26.8
1953	3,788	7	$ 34.4
1954	3,071	7	$ 43.4
1955	4,434	7	$ 68.2
1956	6,738	8	$103.7
1957	6,118	9	$ 99.6
1958	6,414	10	$101.9
1959	7,689	9	$129.8
1960	7,588	8	$151.2
1961	6,778	8	$124.3
1962	6,697	7	$136.8
1963	7,569	7	$153.4
1964	9,336	8	$198.8
1965	11,852	8	$318.2
1966	15,768	10	$444.9
1967	13,577	14	$359.6
1968	13,698	14	$425.7
1969	12,457	14	$584.5
1970	7,292	13	$337.0
1971	7,466	11	$321.5
1972	9,774	12	$557.6
1973	13,646	12	$828.1
1974	14,166	12	$909.4
1975	14,056	12	$ 1,032.9
1976	15,451	12	$ 1,225.5
1977	16,904	12	$ 1,488.1
1978	17,811	12	$ 1,781.2
1979	17,048	12	$ 2,165.0
1980	11,877	12	$ 2,486.2
1981	9,457	12	$ 2,919.9
1982	4,266	11	$ 1,999.5
1983	2,691	10	$ 1,469.5
1984	2,431	9	$ 1,680.7
1985	2,029	9	$ 1,430.6
1986	1,495	9	$ 1,261.9
1987	1,085	9	$ 1,363.5
1988	1,143	9	$ 1,918.4
1989	1,535	11	$ 1,803.9
1990	1,144	14	$ 2,007.5
1991	1,021	14	$ 1,968.3
1992	941	16	$ 1,839.6
1993	964	16	$ 2,143.8
1994	928	13	$ 2,357.1
1995	1,077	13	$ 2,841.9
1996	1,130	13	$ 3,126.5
1997	1,569	12	$ 4,674.3
1998	2,220	NA	$5,900.0 (rounded)

Notes:
General aviation is all civil aviation except that of scheduled carriers. It includes companies using their own planes for business, transportation, charter, air taxis, personal and recreational flying, emergency medical evacuation, agricultural flight, traffic and aerial observation, and flight training.

Source: General Aviation manufacturers Association. *1997 Statistical Databook.* Washington, DC: General Aviation Manufacturers Trssociation, 1997. Page 4. 1997 data obtained via telephone.1998 data from *GAMA 1998 Statistical Review* at http://www.generalalaviation.org

13.3. General Aviation Aircraft Shipments by Type of Aircraft, U.S.-Manufactured, 1962–98.

Year	Total	Single-Engine	Multi-Engine	Total Piston	Turboprop	Jet	Total Turbine
1962	6,697	5,690	1,007	6,697	0	0	0
1963	7,569	6,248	1,321	7,569	0	0	0
1964	9,336	7,718	1,606	9,324	9	3	12
1965	11,852	9,873	1,780	11,653	87	112	199
1966	15,768	13,250	2,192	15,442	165	161	326
1967	13,577	11,557	1,773	13,330	149	98	247
1968	13,698	11,398	1,959	13,357	248	93	341
1969	12,457	10,054	2,078	12,132	214	111	325
1970	7,292	5,942	1,159	7,101	135	56	191
1971	7,466	6,287	1,043	7,330	89	47	136
1972	9,774	7,913	1,548	9,446	179	134	313
1973	13,646	10,788	2,413	13,193	247	198	445
1974	14,166	11,579	2,135	13,697	250	202	452
1975	14,056	11,441	2,116	13,555	305	194	499
1976	15,451	12,785	2,120	14,905	359	187	546
1977	16,904	14,054	2,195	16,249	428	227	655
1978	17,811	14,398	2,634	17,032	548	231	779
1979	17,048	13,286	2,843	16,129	639	282	921
1980	11,877	8,640	2,116	10,756	778	326	1,104
1981	9,457	6,608	1,542	8,150	918	389	1,307
1982	4,266	2,871	678	3,549	458	259	717
1983	2,691	1,811	417	2,228	321	142	463
1984	2,431	1,620	371	1,991	271	169	440
1985	2,029	1,370	193	1,563	321	145	466
1986	1,495	985	138	1,123	250	122	372
1987	1,085	613	87	700	263	122	385
1988	1,143	628	67	695	291	157	448
1989	1,535	1,023	87	1,110	268	157	425
1990	1,144	608	87	695	281	168	449
1991	1,021	564	49	613	222	186	408
1992	941	552	41	593	177	171	348
1993	964	516	39	555	211	198	409
1994	928	444	55	499	207	222	429
1995	1,077	515	61	576	255	246	501
1996	1,130	530	70	600	289	241	530
1997	1,569	905	80	985	236	348	584
1998	2,220	1,436	98	1,534	271	415	686

Notes:

General aviation is all civil aviation except that of scheduled carriers. It includes companies using their own planes for business, transportation, charter, air taxis, personal and recreational flying, emergency medical evacuation, agricultural flight, traffic and aerial observation, and flight training.

Source: General Aviation Manufacturers Association. *1997 Statistical Databook.* Washington, DC: General Aviation Manufacturers Association, 1997. Page 6. 1997 data obtained via telephone. 1998 data from *GAMA 1998 Statistical Review* at http://www.generalaviation.org.

13.4. Average Age of U.S. General Aviation Fleet, 1996 and 1997.

Aircraft Type	Engine Type	Seats	Average Age in Years 1996	Average Age in Years 1997
Single-Engine	Piston	1 to 3	30	31
		4	29	30
		5 to 7	24	25
		8+	38	39
		All	29	30
	Turboprop	All	8	9
	Jet	All	32	29
Multi-Engine	Piston	1 to 3	32	34
		4	29	30
		5 to 7	27	28
		8+	27	28
		All	27	28
	Turboprop	All	17	17
	Jet	All	16	16
All Aircraft			28	28

Notes:
General aviation is all civil aviation except that of scheduled carriers. It includes companies using their own planes for business, transportation, charter, air taxis, personal and recreational flying, emergency medical evacuation, agricultural flight, traffic and aerial observation, and flight training.

Source: General Aviation Manufacturers Association. *1997 Statistical Databook.* Washington, DC: General Aviation Manufacturers Association, 1997. Page 11. 1997 data obtained via telephone.

13.5. Aircraft Reported in Operation by Air Carriers, Sorted by Manufacturer, and Model, 1984–96.

Aircraft Make and Model	1984	1985	1986	1987	1988	1989	1990	1991	1992	1993	1994	1995	1996
Total	4370	4678	4909	5250	5660	5778	6083	6054	7320	7297	7370	7411	7478
Turbojet, 4-engine Total	349	322	322	382	427	428	432	410	389	410	420	435	440
Boeing B707	22	27	35	31	31	27	25	27	20	13	16	6	5
Boeing B747	156	151	150	156	171	180	190	184	178	183	186	189	195
British Aerospace Aircraft													
Group BAE146	14	29	25	57	57	53	44	17	23	20	15	18	21
Douglas DC8	157	115	112	138	168	168	173	182	168	194	203	219	219
Turbojet, 3-engine Total	1438	1488	1466	1469	1542	1459	1438	1376	1381	1292	1236	1210	1212
Boeing B727	1161	1195	1172	1168	1246	1167	1152	1073	1029	953	906	877	856
Douglas DC10/MD-11	174	179	180	185	184	185	185	203	239	239	244	236	254
Lockheed L1011	103	114	114	116	112	107	101	100	113	100	86	97	102
Turbojet, 2-engine Total	1172	1354	1495	1724	1946	2055	2278	2381	2676	2882	2980	3189	3270
Airbus A300	38	46	52	52	57	63	67	63	58	58	63	53	62
Airbus A310	—	4	7	13	19	19	21	42	21	27	17	23	27
Airbus A320	—	—	—	—	—	11	10	35	54	75	86	104	113
Boeing B737	391	476	555	633	706	756	812	835	915	1013	1012	1055	1055
Boeing B757	19	48	73	95	122	146	199	234	328	375	395	440	457
Boeing B767	53	59	69	83	126	111	120	136	170	187	194	210	213
Boeing B777	—	—	—	—	—	—	—	—	—	—	—	7	15
British Aircraft BAC111	33	32	45	39	30	—	—	—	—	—	—	—	—
Canadair CL-600	1	2	—	—	—	—	3	1	—	5	—	35	53
Cessna C500/C501	—	—	—	—	—	—	0	—	2	3	—	—	—
Cessna C550	—	—	—	—	—	5	7	—	—	—	—	—	—
Cessna C650	—	—	—	—	—	—	—	—	1	—	—	—	—
Dassault Falcon	—	—	—	—	—	—	—	—	—	—	—	—	—
Dassault MD10	2	—	—	—	—	2	1	2	1	—	2	3	—
Dassault MD20	9	2	1	2	1	1	2	—	3	1	—	—	4
Douglas DC9/MD-80	594	641	643	760	837	888	967	953	1002	1009	1061	1102	1114
Fokker F28	23	41	50	47	47	53	68	75	117	129	148	155	155
Grumman G1159	1	—	—	—	—	—	1	3	1	—	—	—	—
Gulfstream G111	—	—	—	—	—	—	—	—	—	—	2	2	—
Israel Aircraft 1121	—	—	—	—	—	—	—	—	1	—	—	—	—
Learjet LR25	—	—	—	—	1	2	1	2	3	—	2	3	—
Learjet LR35	8	3	1	2	—	1	2	—	3	1	—	—	4
Turboprop, 4-engine Total	109	108	96	102	95	96	88	75	107	102	87	81	56
Canadair CL44	5	6	2	6	6	5	5	—	5	1	1	1	—
DeHavilland DHC7	46	42	40	41	39	41	40	33	40	38	27	16	12
Lockheed L188	34	38	33	34	30	30	24	24	44	45	41	43	23
Lockheed L382	22	22	21	21	20	20	19	18	18	18	18	21	21
Vickers V745	2	—	—	—	—	—	—	—	—	—	—	—	—
Turboprop, 2-engine Total	847	965	1108	1139	1280	1380	1507	1523	1787	1751	1695	1634	1639
Beech BE65	2	—	1	4	1	—	—	—	16	—	1	—	—
Beech BE90	2	3	1	4	1	—	—	—	1	3	1	1	3
Beech BE99	85	103	95	52	84	53	54	32	39	29	41	36	27
Beech BE100	2	1	1	—	—	1	2	1	4	1	1	—	2
Beech BE200	6	1	2	5	7	10	16	8	11	9	7	4	11
Beech BE1900	17	42	60	48	80	109	147	167	231	251	281	289	254
Beech STC18	1	—	—	—	—	—	—	—	—	—	—	—	—

13.5. Aircraft Reported in Operation by Air Carriers, Sorted by Manufacturer, and Model, 1984–96 (continued).

Aircraft Make and Model	1984	1985	1986	1987	1988	1989	1990	1991	1992	1993	1994	1995	1996
British Aerospace Aircraft													
Group Jetstream	10	46	69	113	135	165	222	214	240	247	237	174	223
British Aerospace BA ATP	—	—	—	—	—	—	4	10	10	9	9	10	10
Cessna C425	—	—	—	—	—	—	—	—	1	2	2	2	2
Cessna C441	3	1	3	2	3	4	2	2	2	1	1	1	2
Construcciones Aeronautics CA212	27	24	19	16	18	16	16	13	—	1	1	1	—
Convair CV580/CV640/CV600	107	100	91	77	72	58	33	37	19	16	29	34	23
DeHavilland DHC6	107	86	68	71	63	69	67	69	74	67	53	44	38
DeHavilland DHC8	—	10	26	34	44	64	74	81	115	120	142	137	151
Dornier DO228	—	6	12	18	33	34	32	31	13	13	7	—	—
Douglas DC3	—	1	—	—	—	—	—	—	—	—	—	—	—
Embraer EM110	81	79	91	97	77	59	48	23	16	14	15	14	3
Embraer EM120	—	—	16	36	62	105	156	167	195	217	223	217	235
Fairchild FH27	23	28	20	13	7	7	9	7	2	1	2	1	1
Fairchild FH227	9	8	7	8	11	4	3	3	—	—	—	—	—
Fokker F27	14	27	36	26	33	42	46	40	51	49	35	34	35
Grumman G73	—	—	—	—	7	5	7	4	5	—	5	5	5
Grumman G159	21	23	15	14	5	6	7	2	1	—	—	—	—
Grumman G500	—	—	—	—	1	—	—	—	—	—	—	—	—
Hawker-Siddeley HS748	2	—	—	—	—	—	—	—	—	2	2	2	4
McKinnon G-21	—	—	—	—	—	—	—	—	—	—	—	—	3
Mitsubishi MU2	1	3	6	1	—	—	1	1	10	—	2	2	—
Nihon YS11	30	42	36	36	22	21	21	22	31	25	25	11	11
Nord ND262	14	14	15	12	9	2	1	—	1	—	—	—	—
Piper 31T	8	4	5	6	9	12	8	8	99	79	1	5	9
Piper 42	4	4	—	—	—	—	—	1	1	—	1	1	2
Rockwell AC690	3	4	4	1	1	—	—	—	—	—	—	—	—
Saab-Fairchild SF340	1	17	34	51	68	85	109	153	195	209	202	219	226
Short SC7	—	—	1	—	—	—	2	2	6	6	5	3	3
Short SD3	78	77	110	110	110	118	103	93	88	74	63	38	39
S.N.I.A.S. ATR42	—	—	8	20	35	62	77	101	108	108	111	110	99
S.N.I.A.S. ATR72	—	—	—	—	—	—	—	31	14	27	44	51	51
Swearingen SA226	121	113	122	101	90	57	22	31	14	14	11	13	9
Swearingen SA227	70	101	135	163	191	212	218	200	174	158	138	144	116
Turboprop, 1-engine Total	NA	NA	NA	NA	NA	NA	NA	NA	NA	15	0	0	5
Piston, 4-engine Total	50	38	32	38	36	35	31	26	20	22	19	15	18
DeHavilland DH114	6	3	1	—	—	—	—	—	—	—	—	—	—
Douglas DC4	3	3	1	3	3	1	1	5	5	—	5	1	—
Douglas DC6	41	34	30	37	35	34	30	25	19	21	18	15	18
Douglas DC7	—	—	—	1	3	1	1	1	1	1	1	—	—
Piston, 3-engine Total	4	4	3	3	3	5	6	5	5	—	5	1	7
Britten Norman MK3	4	4	3	3	3	5	6	5	5	—	5	1	7
Piston, 2-engine Total	389	394	385	380	323	313	292	252	415	293	335	329	313
Aero Commander 500	—	—	—	—	—	—	—	—	—	—	2	1	—
Beech BE18	15	7	9	5	6	5	3	5	18	16	16	18	17
Beech BE36	—	—	—	—	3	1	—	—	5	—	—	—	—

13.5. Aircraft Reported in Operation by Air Carriers, Sorted by Manufacturer, and Model, 1984–96 (continued).

Aircraft Make and Model	1984	1985	1986	1987	1988	1989	1990	1991	1992	1993	1994	1995	1996
Beech BE55	—	—	1	2	—	—	—	—	1	—	—	—	—
Beech BE58	9	9	4	7	15	6	4	4	14	6	5	6	5
Beech BE65	—	—	3	2	2	2	2	2	—	19	18	19	16
Beech BE76	3	3	2	—	—	—	—	—	—	—	—	—	—
Beech BE80	8	4	—	—	—	—	—	—	—	—	—	—	—
Beech BE95	—	—	—	—	3	1	1	1	3	1	—	—	—
Britten Norman BN2A	27	7	29	29	30	16	15	14	18	25	21	23	21
Cessna C210	—	—	1	1	1	—	—	—	6	6	—	—	—
Cessna C303T	—	—	1	1	—	—	—	—	—	1	—	—	—
Cessna C310	2	1	1	—	—	2	2	2	5	5	1	2	2
Cessna C320	1	—	—	—	—	—	—	—	1	1	1	—	—
Cessna C340	—	—	—	—	—	—	—	—	—	—	—	—	—
Cessna C401	—	—	—	—	4	1	1	1	1	—	1	—	1
Cessna C402	112	155	147	143	101	98	110	91	126	117	107	104	97
Cessna C404	4	5	6	4	4	1	1	1	3	3	—	—	—
Cessna C411	1	—	—	—	—	—	—	—	1	1	—	—	—
Cessna C414	1	1	2	—	1	—	1	1	2	1	—	—	—
Cessna C421	1	—	—	—	—	—	—	—	2	2	—	—	13
Convair CV240	15	12	9	10	9	9	11	13	19	23	20	16	13
Convair CV340/CV440	14	18	17	23	21	26	25	24	30	29	27	15	19
Curtiss-Wright C46	2	3	—	—	—	—	—	—	—	—	—	—	—
Douglas DC3	30	39	43	38	20	19	15	12	21	21	6	11	6
Grumman G21	4	3	—	—	—	—	—	—	7	5	4	4	—
Grumman G44	1	1	1	1	1	—	—	—	4	4	6	4	2
Grumman G73	5	3	11	12	4	3	2	2	—	6	—	—	—
Grumman G111	—	6	3	2	—	—	—	—	—	2	—	—	2
Martin M404	1	—	—	1	2	2	—	—	—	—	—	—	—
Partenivia PT68	—	—	—	2	—	2	—	—	—	—	—	—	—
Piper P23	10	3	9	11	9	9	9	8	16	11	12	12	9
Piper P28	—	—	—	—	—	—	—	—	18	18	—	—	—
Piper P30	1	—	—	—	—	—	—	—	—	—	—	—	—
Piper P31	110	100	73	77	71	100	81	66	78	78	79	83	86
Piper P32	—	—	9	2	2	2	2	4	—	—	—	—	—
Piper P34	11	12	1	4	12	9	7	3	16	8	9	9	10
Piper P44	1	1	2	1	1	—	—	—	—	—	—	—	—
Piper PA600	—	—	2	2	1	1	—	—	1	1	1	2	1
Piper PA1020T	—	—	—	—	—	—	—	—	1	—	—	—	—
Piston, 1-engine Total	NA	NA	NA	NA	NA	NA	NA	NA	407	406	465	399	397
Helicopter, Total	12	5	2	13	8	7	11	6	133	124	128	118	121
Aerospatiale AS332	NA	NA	NA	NA	NA	NA	NA	NA	NA	NA	NA	2	2
Aerospatiale AS350	NA	NA	NA	NA	NA	NA	NA	NA	NA	NA	NA	16	16
Bell B206	NA	NA	NA	NA	NA	NA	NA	NA	NA	NA	NA	1	4
Bell B212	NA	NA	NA	NA	NA	NA	NA	NA	NA	NA	NA	35	35
Bell B208	NA	NA	NA	NA	NA	NA	NA	NA	NA	NA	NA	14	14
Bell B412	NA	NA	NA	NA	NA	NA	NA	NA	NA	NA	NA	16	—

13.5. Aircraft Reported in Operation by Air Carriers, Sorted by Manufacturer, and Model, 1984–96 *(continued).*

Aircraft Make and Model	1984	1985	1986	1987	1988	1989	1990	1991	1992	1993	1994	1995	1996
MBB BO105	NA	NA	NA	NA	NA	NA	NA	NA	NA	NA	NA	—	16
Enstrom F28	NA	NA	NA	NA	NA	NA	NA	NA	NA	NA	NA	28	28
Sikorsky S58	NA	NA	NA	NA	NA	NA	NA	NA	NA	NA	NA	2	2

Notes:

Air carrier aircraft are those carrying passengers or cargo for hire under Code of Federal Regulations Title 14, Part 121 (large aircraft—more than 30 seats) and Part 135 (small aircraft—30 seats or fewer). This definition is more encompassing than that used in the FAA Aviation Forecast (jet aircraft 60 seats or more, carrying passengers or cargo for hire). Beginning in 1987, the number of aircraft is the monthly average of the number of aircraft reported in use for the last three months of the year. Before 1987, the number of aircraft was the number reported in use during December of the year.

Does not include the aircraft listed below that are operated by on-demand air taxis:

1993:	
Piston multiengine	2,669
Piston single engine	3,043
Turboprop single engine	321
Turboprop multiengine	1,662
Helicopter	1,977
Total	10,692

Source: 1983–1991 Air Carrier Aircraft Utilization and Propulsion Reliability Report; Aviation Standards National Field Office, Federal Aviation Administration. Beginning in 1992, the source is the Vital Information System. All numbers cited in Federal Aviation Administration. *Statistical Handbook of Aviation.* Various years. Washington, DC.

13.6. U.S. FAA Air Route Facilities and Services, 1972–97.

Year	VOR VORTAC *	Non-directional Beacons	Air Route Traffic Control Centers	Airport Traffic Control Towers	Flight Service Stations	International Flight Service Stations	Instrument Landing Systems	Airport Surveillance Radar
1972	991	706	27	355	324	7	403	125
1973	995	739	27	403	315	7	467	142
1974	1,000	793	26	417	320	7	490	156
1975	1,011	848	25	487	321	7	580	177
1976	1,020	920	25	488	321	7	640	175
1977	1,021	959	25	495	319	7	678	182
1978	1,020	988	25	494	319	6	698	185
1979	1,028	1,015	25	499	318	6	753	192
1980	1,037	1,055	25	502	317	6	796	192
1981	1,033	1,123	25	501	316	6	840	199
1982	1,029	1,143	25	492	316	6	884	197
1983	1,032	1,183	25	494	316	5	934	197
1984	1,035	1,211	25	497	310	5	955	197
1985	1,039	1,222	25	500	302	4	968	198
1986	1,043	1,239	25	686	293	3	977	312
1987	1,039	1,212	25	500	302	4	968	198
1988	1,043	1,239	25	686	293	3	977	312
1989	1,046	1,263	25	686	255	3	1,100	312
1990	1,045	1,271	25	686	235	3	1,120	311
1991	1,045	1,295	24	694	192	3	1,114	318
1992	1,044	1,314	24	691	179	3	1,177	312
1993	1,041	1,344	24	684	165	3	1,231	310
1994	1,041	1,342	24	685	156	3	1,245	303
1995	1,039	1,333	24	683	91	3	1,280	298
1996	1027**	1,165**	21	664**	94***	0	1,197	228
1997	1,041	1,344	24	684	135	3	1,231	310

Notes:
* = VOR = a ground station that transmits a signal that allows an aircraft to determine its bearing relative to the station; VORTAC = a VOR co-located with a military station (TACAN, which stands for Tactical Air Navigation).
** = Includes nonfederal and military.
*** = Includes Automated Flight Service Stations (AFSS).

Source: U.S. Federal Aviation Administration, Washington, DC. Cited in General Aviation Manufacturers Association. *1997 Statistical Databook.* Washington, DC: General Aviation Manufacturers Association, 1997. Page 22. 1997 data obtained via telephone.

13.7. Airports by Geographic Area: U.S. Civil and Joint Use Airports, Heliports, STOLports, and Seaplane Bases on Record By Type of Ownership, 1997.

FAA Region and State	Total Facilities	Total Facilities By Ownership		Airports Open to the Public				Total Airports
		Public	Private	Paved Airports[1]		Unpaved Airports[1]		
				Lighted	Unlighted	Lighted	Unlighted	
Grand Total	18,345	5,134	13,211	3,597	297	392	789	5,075
United States—Total*	18,268	5,083	13,185	3,573	294	392	776	5,035
Alaskan—Total	553	385	168	46	5	100	152	303
Alaska	553	385	168	46	5	100	152	303
Central—Total	1,493	496	997	382	19	38	53	492
Iowa	306	135	171	97	1	10	15	123
Kansas	387	132	255	100	10	17	19	146
Missouri	505	135	370	109	6	5	11	131
Nebraska	295	94	201	76	2	6	8	92
Eastern—Total	2,383	349	2,034	330	28	49	83	490
Delaware	35	4	31	6	0	3	1	10
District of Columbia	17	8	9	2	0	0	0	2
Maryland	200	21	179	27	1	3	3	34
New Jersey	362	48	314	36	2	5	7	50
New York	537	91	446	83	13	22	32	150
Pennsylvania	760	74	686	88	6	14	28	136
Virginia	368	73	295	62	2	1	3	68
West Virginia	104	30	74	26	4	1	9	40
Great Lakes—Total	4,226	906	3,320	740	22	140	154	1,056
Illinois	872	122	750	92	0	18	5	115
Indiana	599	87	512	77	4	6	22	109
Michigan	469	134	335	122	6	43	55	226
Minnesota	476	151	325	108	0	21	17	146
North Dakota	432	96	336	67	5	11	10	93
Ohio	737	136	601	121	4	14	24	163
South Dakota	157	77	80	54	1	14	5	74
Wisconsin	484	103	381	99	2	13	16	130
New England—Total	703	142	561	115	16	4	26	161
Connecticut	136	16	120	17	2	0	3	22
Maine	150	46	104	30	7	2	7	46
Massachusetts	224	35	189	35	3	1	5	44
New Hampshire	94	16	78	16	3	1	6	26
Rhode Island	26	9	17	7	0	0	0	7
Vermont	73	20	53	10	1	0	5	16
Northwest, Mountain—Total	1,930	685	1,245	408	50	20	145	623
Colorado	382	92	290	64	4	3	8	79
Idaho	226	131	95	44	9	1	63	117
Montana	243	123	120	73	7	7	32	119
Oregon	411	103	308	59	17	4	19	96
Utah	126	59	67	41	6	0	0	47
Washington	438	128	310	93	9	4	18	124
Wyoming	104	49	55	34	1	1	5	41
Southern—Total	2,641	839	1,802	634	45	26	49	754
Alabama	237	102	135	86	3	5	4	98
Florida	781	163	618	100	5	7	15	127

13.7. Airports by Geographic Area: U.S. Civil and Joint Use Airports, Heliports, STOLports, and Seaplane Bases on Record By Type of Ownership, 1997 *(continued)*.

FAA Region and State	Total Facilities	Total Facilities By Ownership		Airports Open to the Public				Total Airports
		Public	Private	Paved Airports[1]		Unpaved Airports[1]		
				Lighted	Unlighted	Lighted	Unlighted	
Georgia	400	136	264	97	9	2	2	110
Kentucky	174	72	102	53	8	0	3	64
Mississippi	226	91	135	71	8	2	2	83
North Carolina	355	95	260	87	4	6	15	112
Puerto Rico	32	17	15	10	1	0	0	11
South Carolina	166	69	97	57	1	4	6	68
Tennessee	261	88	173	71	6	0	2	79
Virgin Islands	9	6	3	2	0	0	0	2
Southwest—Total	3,011	819	2,192	631	52	11	74	768
Arkansas	275	116	159	90	6	0	4	100
Louisiana	437	112	325	70	3	1	7	81
New Mexico	164	73	91	44	8	0	10	62
Oklahoma	418	159	259	111	13	5	17	146
Texas	1,717	359	1,358	316	22	5	36	379
Western—Total	1,405	513	892	311	60	4	53	428
Arizona	272	88	184	55	7	0	11	73
California	931	319	612	207	42	1	11	261
Hawaii	46	18	28	11	2	0	0	13
Nevada	120	60	60	26	7	3	18	54
South Pacific**	36	28	8	12	2	0	13	27

Notes:
*Excludes Puerto Rico, Virgin Islands, Northern Mariana Islands, and South Pacific
**American Samoa, Guam, and Trust Territories
/1 Includes all airports open to the public, both publicly and privately owned

Source: U.S. Federal Aviation Administration, Washington, DC. Cited in General Aviation Manufacturers Association. *1998 Statistical Databook.* Washington, DC: General Aviation Manufacturers Association, 1998. Page 21.

13.8. U.S. Aerospace Industry Sales by Product Group in Current Dollars, 1981–98.

	Total Sales	Aircraft, Total	Aircraft, Civil	Aircraft, Military[a]	Missiles[a]	Space[a]	Related Products and Services
1981	$63,974	$36,062	$16,427	$19,635	$7,640	$9,388	$10,884
1982	$67,756	$35,484	$10,982	$24,502	$10,368	$10,514	$11,390
1983	$79,975	$42,431	$12,373	$30,058	$10,269	$13,946	$13,329
1984	$83,486	$ 41,905	$10,690	$31,215	$11,335	$16,332	$13,914
1985	$96,571	$50,482	$13,730	$36,752	$11,438	$18,556	$16,095
1986	$106,183	$56,405	$15,718	$40,687	$11,964	$20,117	$17,697
1987	$110,008	$ 59,188	$15,465	$43,723	$10,219	$22,266	$18,335
1988	$114,562	$60,886	$19,019	$41,867	$10,270	$24,312	$19,094
1989	$120,534	$61,550	$21,903	$39,646	$13,622	$25,274	$20,089
1990	$134,375	$71,353	$31,262	$40,091	$14,180	$26,446	$22,396
1991	$39,248	$75,918	$37,443	$38,475	$10,970	$29,152	$23,208
1992	$138,591	$73,905	$39,897	$34,008	$11,757	$29,831	$23,099
1993[r]	$123,183	$65,829	$33,116	$32,713	$8,451	$28,372	$20,531
1994[r]	$110,558	$57,648	$25,596	$32,052	$7,563	$26,921	$18,426
1995[r]	$107,782	$55,048	$23,965	$31,082	$7,386	$27,385	$17,964
1996	$116,489	$59,908	$26,869	$33,039	$8,044	$29,122	$19,415
1997[p]	$129,622	$69,121	$38,643	$30,478	$8,284	$30,613	$21,604
1998[e]	$144,450	$79,545	$49,104	$30,441	$8,020	$32,811	$24,075

Notes:
a = Includes funding for research, development, test, and evaluation.
r = Revised.
p = Preliminary.
e = Estimate.
Based on company reports; the budget of the United States government; data from the National Aeronautics and Space Administration and the Departments of Commerce and Defense; and AIA estimates.

Source: Aerospace Industries Association. Washington, DC. Unpublished.

13.9. U.S. Aerospace Industry Sales by Product Group in Constant Dollars 1981–98[b] (millions of dollars).

	Total Sales	Aircraft, Total	Aircraft, Civil	Aircraft, Military[a]	Missiles[a]	Space[a]	Related Products and Services
1981	$80,470	$45,361	$20,663	$24,698	$9,610	$11,809	$13,691
1982	$ 77,083	$40,369	$12,494	$27,875	$11,795	$11,961	$12,958
1983	$ 86,741	$46,021	$13,420	$32,601	$11,138	$15,126	$14,457
1984	$ 83,653	$41,989	$10,711	$31,278	$11,358	$16,365	$13,942
1985	$ 97,843	$51,147	$13,911	$37,236	$11,589	$18,800	$16,307
1986	$106,396	$56,518	$15,749	$40,769	$11,988	$20,157	$17,732
1987	$110,008	$59,188	$15,465	$43,723	$10,219	$22,266	$18,335
1988	$112,426	$59,751	$18,664	$41,086	$10,079	$23,859	$18,738
1989	$113,604	$58,011	$20,644	$37,367	$12,839	$23,821	$18,934
1990	$121,606	$64,573	$28,291	$36,281	$12,833	$23,933	$20,268
1991	$121,508	$66,246	$32,673	$33,573	$9,572	$25,438	$20,251
1992[r]	$117,251	$62,525	$33,754	$28,772	$9,947	$25,238	$19,542
1993[r]	$101,636	$54,314	$27,323	$26,991	$6,973	$23,409	$16,940
1994[r]	$89,160	$46,490	$20,642	$25,848	$6,099	$21,710	$14,860
1995[r]	$85,473	$43,654	$19,005	$24,649	$5,857	$21,717	$14,246
1996	$91,364	$46,987	$21,074	$25,913	$6,309	$22,841	$15,227
1997[p]	$99,480	$53,048	$29,657	$23,391	$6,358	$23,494	$16,580
1998[e]	$108,121	$59,540	$36,754	$22,785	$6,003	$24,559	$18,020

Notes:
a = Includes funding for research, development, test, and evaluation.
b = Based on AIA's aerospace composite price deflator (1987 = 100)
e = Estimate.
p = Preliminary.
r = Revised.
Based on company reports; the budget of the United States government; data from the National Aeronautics and Space Administration and the Departments of Commerce and Defense; and AIA estimates.

Source: Aerospace Industries Association. Washington, DC. Unpublished.

RAIL AND URBAN TRANSPORT

13.10. Physical Condition of U.S. Transit Rail Systems, 1984 and 1992.

Type of System	Bad 1984	Bad 1992	Poor 1984	Poor 1992	Fair 1984	Fair 1992	Good 1984	Good 1992	Excellent 1984	Excellent 1992
Stations	0%	0%	15%	5%	56%	29%	23%	63%	6%	3%
Track	0%	0%	7%	5%	49%	32%	31%	49%	12%	14%
Power systems										
Substations	6%	2%	23%	19%	5%	17%	43%	56%	23%	6%
Overhead	20%	0%	12%	33%	27%	10%	36%	52%	5%	5%
Third rail	13%	0%	26%	21%	19%	20%	36%	53%	6%	6%
Structures										
Bridges	1%	0%	16%	11%	51%	28%	28%	54%	4%	7%
Elevated	0%	0%	1%	1%	80%	72%	3%	15%	16%	12%
Tunnels	0%	0%	5%	5%	49%	34%	35%	51%	11%	10%
Maintenance										
Facilities	4%	2%	54%	34%	14%	12%	24%	35%	4%	17%
Yards	4%	2%	53%	7%	26%	26%	16%	55%	1%	9%

Source: U.S. Department of Transportation, Federal Transit Administration, 1992. *Modernization of the Nation's Rail Transit Systems: A Status Report.* Washington, DC. Cited in U.S. Department of Transportation, Bureau of Transportation Statistics. *Transportation Statistics Annual Report, 1997.* Washington, DC: Bureau of Transportation Statistics, 1997. Table 1–9.

13.11. Average Age of Urban Transit Vehicles, United States, 1985 and 1995.

Type of Vehicle	1985 Years	1995 Years
Commuter train locomotives	16.3	15.9
Unpowered commuter railcars	19.1	21.4
Powered commuter railcars (self-propelled)	12.3	19.8
Heavy rail	17.1	19.3
Light rail (streetcars)	20.6	16.8
Articulated buses	3.4	10.9
Full-size buses	8.1	8.7
Mid-size buses	5.6	6.9
Small buses	4.8	4.1
Vans	3.8	3.1
Trolleys	NA	13.1
Ferry boats	NA	23.4

Notes:
NA = Not available.

Sources: (1) U.S. Department of Transportation, Federal Transit Administration. *National Urban Mass Transportation Statistics, 1985.* Washington, DC, 1986. (2) U.S. Department of Transportation, Federal Transit Administration. *1995 National Transit Database.* Washington, DC, 1997. (3) Bureau of Transportation Statistics. *National Transportation Statistics 1998.* Washington, DC: Bureau of Transportation Statistics, 1998. Based on Table 1-40. http://www.bts.gov/btsprod/nts/

13.12. Growth in U.S. Urban Transit Service, 1985 to 1995.

Transit Mode	Vehicles in Rush-hour Service			Vehicle-miles Operated (millions)			Route-miles serviced		
	1985	1995	Change	1985	1995	Change	1985	1995	Change
All modes	54,437	57,183	5%	2,101	2,377	13%	143,606	163,941	14%
All rail modes	11,832	13,120	11%	626	773	23%	5,251	6,185	18%
Heavy rail	7,673	7,973	4%	445	522	17%	1,322	1,458	10%
Light rail	534	734	37%	16	34	113%	384	568	48%
Commuter rail	3,625	4,413	22%	165	217	32%	3,545	4,159	17%
Bus	42,605	44,063	3%	1,475	1,604	9%	138,355	157,756	14%

Sources: (1) U.S. Department of Transportation, Federal Transit Administration. *National Urban Mass Transportation Statistics, 1985.* Washington, DC, 1986. (2) U.S. Department of Transportation, Federal Transit Administration. *1995 National Transit Database.* Washington, DC. Cited in U.S. Department of Transportation Bureau of Transportation Statistics. *Transportation Statistics Annual Report 1997.* Washington, DC: Bureau of Transportation Statistics, 1997. Table 1-5.

CARS AND TRUCKS

13.13. A Typical U.S. Family Vehicle, Material Content and Total Weight, 1976, 1986, and 1996.

Material/year	1976		1986		1996		1976–1996
	Pounds	Percent of total	Pounds	Percent of total	Pounds	Percent of total	Percent change
Iron and steel	2785	74.1%	2,190	69.0%	1,890	61.0%	-32.0%
Polymeric materials	325	8.6%	433	13.7%	642	21.0%	97.5%
Aluminum	85.5	2.3%	139.5	4.4%	257	8.3%	200.0%
All other	564.5	15.0%	407.5	12.9%	301	9.7%	-47.0%
Total vehicle weight	3,760	100%	3,170	100%	3,090	100%	18.0%

Source: Recent Trends in Automobile Recycling: An Energy and Economic Assessment. ORNL/TM/12628. March 1996. (ORNL = Oak Ridge National Laboratory.) Cited in Vincent DeSapio et al. *Commercialization of New Manufacturing Processes for Materials.* Publication 3100. U.S. International Trade Commission Office of Industries: Washington, DC, April 1998. ftp://ftp.usitc.gov/pub/reports/studies/PUB3100.pdf

13.14. Material Usage per Passenger Car Built in the United States, 1980 and 1993.

1980	
Conventional steel	53.3%
Other materials	2.9%
Aluminum	3.9%
High strength/stainless steel, powder metals	6.5%
Iron	14.2%
Other metals/nonmetals	19.1%
1993	
Conventional steel	45.2%
Other materials	2.9%
Aluminum	5.6%
High strength/stainless steel, powder metals	10.4%
Iron	14.2%
Other metals/nonmetals	19.1%

Note:
Average weight of passenger vehicles was 3,363 pounds in 1980; 3,150 pounds in 1993.

Source: Compiled from data presented in *AAMA Motor Vehicle Facts and Figures, 1993.* Presented in *Recent Trends in Automobile Recycling: An Energy and Economic Assessment.* ORNL/TM/12628, March 1996. (ORNL = Oak Ridge National Laboratory.) Cited in Vincent DeSapio et al. *Commercialization of New Manufacturing Processes for Materials.* Publication 3100. U.S. International Trade Commission Office of Industries: Washington, DC, April 1998. ftp://ftp.usitc.gov/pub/reports/studies/PUB3100.pdf

13.15. Sales, Market Shares, and Fuel Economies of New Domestic (U.S.) and Import Light Trucks, 1980–96.

	1980	1985	1990	1991 r	1992 r	1993 r	1994 r	1995	1996 p
Small Pickup									
Total sales, units	516,412	863,584	678,488	628,098	586,752	332,470	365,322	356,856	391,000
Market share, percent	23.3%	20.4%	15.0%	15.5%	13.4%	6.6%	6.4%	6.0%	6.3%
Fuel economy, mpg	25.5	26.8	25.2	25.7	25	24.9	25.3	25.6	26.3
Large Pickup									
Total sales, units	1,115,248	1,690,931	1,573,729	1,309,283	1,452,192	1,877,806	2,199,224	2,183,793	2,202,000
Market share, percent	50.3%	39.9%	34.9%	32.3%	33.1%	37.1%	38.4%	36.8%	35.4%
Fuel economy, mpg	17	19	18.9	18.8	18.9	19.6	20.1	19.4	19
Small Van									
Total sales, units	13,649	437,660	932,693	888,165	968,361	1,129,459	1,263,933	1,257,116	1,230,000
Market share, percent	0.6%	10.3%	20.7%	21.9%	22.0%	22.3%	22.1%	21.2%	19.8%
Fuel economy, mpg	19.6	23.9	23.1	22.6	22.5	22.9	22.1	22.8	22.7
Large Van									
Total sales, units	328,065	536,242	398,877	308,317	350,013	388,435	407,737	401,056	370,000
Market share, percent	14.8%	12.7%	8.8%	7.6%	8.0%	7.7%	7.1%	6.8%	6.0%
Fuel economy, mpg	16.3	16.4	16.9	17.4	16.9	17.3	17.4	17.1	17.2
Small Utility									
Total sales, units	75,875	477,706	738,294	782,588	867,934	948,797	1,042,584	1,225,131	1,379,000
Market share, percent	3.4%	11.3%	16.4%	19.3%	19.8%	18.8%	18.2%	20.6%	22.2%
Fuel economy, mpg	16.9	22.1	21.9	21.1	20.9	21.3	20.7	20.8	21.3
Large Utility									
Total sales, units	167,288 r	229,242	192,544	131,740	167,199	378,710	445,601	509,914	641,000
Market share, percent	7.5%	5.4%	4.3%	3.3%	3.8%	7.5%	7.8%	8.6%	10.3%
Fuel economy, mpg	14.5 r	16.6	16.1	16.4	16.9	17.5	17.8	17.4	18.1
Fleet									
Total sales, units	2,216,537	4,235,365	4,514,625	4,043,191	4,392,451	5,055,677	5,724,401	5,933,866	6,213,000
Fuel economy, mpg	18.1	20.4	20.5	20.6	20.4	20.5	20.4	20.2	20.3

Notes:
r = revised
p = preliminary

Source: 1980 through 1995: Oak Ridge National Laboratory. Light Duty Vehicle MPG and Market Shares Systems.1996. Cited in United States Department of Transportation, Bureau of Transportation Statistics. *National Transportation Statistics 1997.* Washington, DC: Bureau of Transportation Statistics, 1996. Table 1-29. 1996: United States Department of Transportation, Bureau of Transportation Statistics. *National Transportation Statistics 1998.* Washington, DC: Bureau of Transportation Statistics, 1998. Table 1-33.

13.16. Average Annual Growth Rate of Passenger Cars and Their Use, In Relation to Population, United States, Canada, European Union, and Japan, 1970–90.

	Number of cars	Vehicle-kilometers traveled	Passenger-kilometers traveled	Population
United States	3.6%	3.3%	2.0%	1.0%
Canada	3.3%	2.3%	NA	1.3%
European Union*	4.3%	3.9%	3.5%	0.6%
Japan	7.1%	5.7%	5.7%	0.9%

Notes:
* = Member states of the European Union are Austria, Belgium, Denmark, Finland, France, Germany, Greece, Ireland, Italy, Luxembourg, the Netherlands, Portugal, Spain, Sweden, and the United Kingdom.
NA = Not available.

Sources: United States: U.S. Department of Transportation, Federal Highway Administration. *Highway Statistics.* Washington, DC, various years. U.S. Department of Transportation, Bureau of Transportation Statistics. *National Transportation Statistics 1997.* Washington, DC. Organization for Economic Cooperation and Development. *Environmental Data 1995.* OECD Publication Services. Paris, 1995. For other countries: Organization for Economic Cooperation and Development. *Environmental Data 1992.* OECD Publication Services. Paris, 1992. Organization for Economic Cooperation and Development. *Environmental Data 1995.* OECD Publication Services. Paris, 1995. Japan Ministry of Transport. *National Transportation Statistics Handbook 1995.* Tokyo, 1996. European Commission. *The Trans-European Transportation Network: Transforming a Patchwork into a Network.* Brussels, 1995. Cited in U.S. Department of Transportation, Bureau of Transportation Statistics, *Transportation Statistics Annual Report 1997.* Washington, DC.

13.17. U.S. Cars and Light Trucks with Airbags, 1990–95.

	Driver-side	Passenger-side
1990	2,314.5	27.7
1991	3,197.3	NA
1992	5,083.8	435.0
1993	7,597.1	1,351.7
1994	9886.0	5,547.1
1995	14,064.8	8,848.8

Note:
Based on vehicles built in North America or imported into the U.S.

Source: Reprinted with permission from Automotive News, 1997, Copyright Crain Communications, Inc. All rights reserved. Cited in "Air Bag Industry Explodes with Growth." *Rubber & Plastics News* 26 (February 24, 1997): 25+.

13.18. Number of Vehicles with Airbags as of September 1, 1998.

Cars	53 million; 42% of cars on the road
Light trucks	26 million; 35% of light trucks
Total	79 million

Source: National Highway Transportation Safety Administration. Washington, DC. http://www.nhtsa.dot.gov, July 1999.

ALTERNATIVE FUELS, FUEL EFFICIENCY, AND VEHICLE EMISSIONS

13.19. Alternative Fuel Vehicles by Fuel Type, United States, 1992–99.

Fuel Used	1992	1993	1994	Number of Vehicles 1995	1996	1997	1998	1999
Total	251,352	314,848	324,472	333,049	352,421	369,807	395,625	418,128
Liquefied petroleum gases /1	221,000	269,000	264,000	259,000	263,000	263,000	269,000	274,000
Non-LPG subtotal	30,352	45,848	60,472	74,049	89,421	106,807	126,625	144,128
Compressed natural gas	23,191	32,714	41,227	50,218	60,144	70,852	85,730	96,017
Liquefied natural gas	90	299	484	603	663	813	1,358	1,517
Methanol, 85% /2	4,850	10,263	15,484	18,319	20,265	21,040	21,578	21,829
Methanol, 100%	404	414	415	386	172	172	378	378
Ethanol, 85% /2 /3	172	441	605	1,527	4,536	9,130	11,743	17,892
Ethanol, 95%	38	27	33	136	361	347	14	14
Electricity	1,607	1,690	2,224	2,860	3,280	4,453	5,824	6,481

Notes:
/1 The Energy Information Administration (EIA) rounds its liquefied petroleum gases estimates
to the nearest thousand.
/2 The remaining part of 85 percent methanol and ethanol fuels is gasoline.
/3 Does not include recently-announced plans of some major automakers to make available large numbers of vehicles capable of operating on E85 fuel in
the near future.
Estimates for 1997 are revised. Estimates for 1998 are preliminary and estimates for 1999 are based on plans or projections.

Sources: 1992-1995: Science Applications International Corporation, "Alternative Transportation Fuels and Vehicles Data Development," unpublished final report prepared for the Energy Information Administration (McLean, VA, July 1996) and U.S. Department of Energy, Office of Energy Efficiency and Renewable Energy.
1996-1999: Energy Information Administration, Office of Coal, Nuclear, Electric, and Alternate Fuels and U.S. Department of Energy Office of Energy Efficiency and Renewable Energy.
Cited in U.S. Department of Energy, Energy Information Administration. *Alternatives to Traditional Transportation Fuels 1997—Advance Information.* Washington, DC: U.S. Department of Energy, Energy Information Administration, 1998. Table 1.

13.20. Energy Intensiveness of Passenger Modes, United States, 1960–96 (Btu per passenger mile).

Year	Air: Certificated Air Carrier — Single-unit Domestic Operations	Air: International Operations	Highway: Passenger Car[1]	Highway: Motorcycle	Truck: Other 2-axle 4-tire Vehicle	Truck: 2-axle 6-tire or more Truck	Truck: Combination Truck	Bus: Transit Motor Bus	Bus: School Bus	Amtrak
1960	8,483	8,529	3,986	**	—	—	—	—	—	—
1965	9,863	8,646	4,094	**	—	—	28,417	—	—	—
1970	9,781	7,627	4,624	2,030	8,003	20,323	29,008	—	—	—
1975	7,502	7,049	4,717	1,976	7,717	19,298	28,658	—	—	2,383
1980	6,009	4,129	4,372	1,920	6,771	19,359	25,655	2,742	1,200	2,148
1985	4,918	4,566	4,243	1,883	6,730	19,884	26,625	3,389	800	2,089
1990	4,855	4,249	4,085	1,952	4,966	18,930	25,143	3,718	800	2,071
1991	4,647	4,253	3,394	2,278	4,872	18,396	24,547	3,767	800	1,977
1992	4,515	3,963	3,419	2,271	4,793	18,546	24,757	4,045	800	2,023
1993	4,522	3,861	3,495	2,271	4,757	20,218	23,832	3,946	800	1,995
1994	4,345	3,917	3,277	2,275	5,448	20,462	23,777	4,161	1,000	NA
1995	4,283	3,934	3,721	2,273	4,538	20,301	23,621	4,161	NA	NA
1996p	4,098	3,902	3,690	2,271	4,545	NA	NA	4,029	NA	NA

Notes:

[1]This table is based on official U.S. Department of Transportation/Federal Highway Administration data; however, over time, the Nationwide Personal Transportation Survey (NPTS) consistently shows declining vehicle occupancy rates.

** = Data included with passenger car information.

NA = Data not available at press time.

p = preliminary.

In 1995, the Federal Highway Administration revised its vehicle type categories from 1993 and later. These new categories include Passenger Cars, Other 2-axle 4-tire Vehicles, Single-unit 2-axle 6-tire or more Trucks, and Combination Trucks. Other 2-axle 4-tire Vehicles include vans, pickup trucks, and sport utility vehicles. In previous years, some minivans and sport utility vehicles were included in the Passenger Car category. Single-unit 2-axle 6-tire or more Trucks are on a single frame with at least 2 axles and 6 tires. Pre-1993 data have been reassigned to the closest available category.

Sources: Certificated Air Carrier: 1960-1975: Civil Aviation Board. *Handbook of Airline Statistics,* 1969, 1973.
1975-1980: Civil Aviation Board. *Air Carrier Traffic Statistics,* 1976, 1981.
1980: Civil Aviation Board. *Fuel Cost and Consumption, Twelve Months Ended Dec. 31, 1984.*
1985-1995: U.S. Dept. of Transportation, Office of Airline Information. *Air Carrier Traffic Statistics,* annual issues, and personal communication.
Passenger Car: 1960-1980: U.S. Dept. of Transportation/Federal Highway Administration. *Highway Statistics, Summary to 1985.*
1985-1995: U.S. Dept. of Transportation/Federal Highway Administration. *Highway Statistics,* annual issues.
Motorcycle: 1970-1980: U.S. Dept. of Transportation/Federal Highway Administration. *Highway Statistics, Summary to 1985.*
1985-1995: U.S. Dept. of Transportation/Federal Highway Administration. *Highway Statistics,* annual issues.
Other 2-axle 4-tire Vehicles: 1965-1980: U.S. Dept. of Transportation/Federal Highway Administration. *Highway Statistics, Summary to 1985.*
1985-1995: U.S. Dept. of Transportation/Federal Highway Administration. *Highway Statistics,* annual issues.
Single-unit 2-axle 6-tires or more Trucks: 1970-1980: U.S. Dept. of Transportation/Federal Highway Administration. *Highway Statistics, Summary to 1985.*
1985-1995: U.S. Dept. of Transportation/Federal Highway Administration. *Highway Statistics,* annual issues.
Combination Trucks: 1965-1980: U.S. Dept. of Transportation/Federal Highway Administration. *Highway Statistics, Summary to 1985.*
1985-1995: U.S. Dept. of Transportation/Federal Highway Administration. *Highway Statistics,* annual issues.
Transit Motor Buses: 1960-1994: American Public Transit Association (APTA). *Transit Fact Book.* 1996. Eno Transportation Foundation, Inc. *Transportation in America.* 1996.
School Buses: 1980-1994: Eno Transportation Foundation, Inc. Transportation in America. 1996 and earlier editions. National Safety Council. *Accident Facts.* 1996 and earlier editions. Amtrak: 1975-1993:
Amtrak, State and Local Affairs Department.
All cited in Department of Transportation, Bureau of Transportation Statistics. *National Transportation Statistics 1997 and National Transportation Statistics 1998.* Washington, DC. Department of Transportation, Bureau of Transportation Statistics, 1996 and 1998. Tables 4-16 and 4-23, respectively.

13.21. U.S. Energy Consumption by the Transportation Sector, 1955–96.

Year	Coal — Million Short Tons	Coal — Quadrillion Btu	Natural Gas[a] — Trillion Cubic Feet	Natural Gas[a] — Quadrillion Btu	Petroleum — Million Barrels	Petroleum — Quadrillion Btu	Total Fossil Fuels[b] — Quadrillion Btu	Electricity — Quadrillion Btu	Renewables — Quadrillion Btu	Net Transportation Consumption — Quadrillion Btu[c]	Electrical System Energy Losses — Quadrillion Btu	Gross Transportation Consumption — Quadrillion Btu	Gross Transportation Consumption — Percent of Total Energy Consumption[r]	Total Energy Consumption — Quadrillion Btu
1955	17.0	0.422	0.25	0.259	1,627.9	8.804	9.484 /r	—	—	9.484	—	9.550	24.6%	38.82
1960	3.0	0.076	0.35	0.359	1,881.2	10.126 /r	10.561 /r	0.011	—	10.572	0.027	10.60	24.1%	43.80
1965	0.7	0.016	0.50	0.518	2,204.6	11.867 /r	12.401 /r	0.010	—	12.411	0.023	12.43	23.6%	52.68
1970	0.3	0.007	0.72	0.740	2,839.7	15.31 /r	16.057	0.009	—	16.066	0.021	16.09	24.2%	66.43
1975	—	<0.001	0.58	0.595	3,266.8 /r	17.614 /r	18.209 /r	0.010	—	18.219	0.025	18.24	25.9%	70.55
1980	—	—	0.63	0.650 /r	3,495.3 /r	19.008	19.658 /r	0.011	—	19.669	0.026	19.70	25.9%	75.96
1985	—	—	0.50	0.519	3,595.3 /r	19.504	20.024 /r	0.013	—	20.037	0.030	20.07	27.1%	73.98
1990	—	—	0.66	0.680	4,004.1 /r	21.810	22.490	0.014	0.082	22.586	0.031	22.54	27.7%	81.27
1991	—	—	0.60	0.620	3,942.0 /r	21.456	22.076	0.014	0.065	22.155	0.030	22.12	27.3%	81.12
1992	—	—	0.59	0.606	4,007.7 /r	21.812	22.418	0.014	0.079	22.511	0.029	22.46	27.3%	82.14
1993	—	—	0.63	0.642	4,080.7	22.201	22.842	0.013	0.088	22.943	0.028	22.88	27.3%	83.86
1994 /r	—	—	0.69	0.705	4,193.9	22.824	23.529	0.014	0.098	23.641	0.028	23.57	27.5%	85.59
1995	—	—	0.70	0.735	4,263.2	23.182	23.917	0.013	0.105	24.035	0.027	23.96	27.5%	87.19
1996 /p	—	—	0.71	—	4,348.0	—	—	0.014	—	—	0.029	24.51	27.2%	89.97

Notes:

/p = Preliminary.

/r = Revised.

/a = Pipeline fuel, includes supplemental natural gas.

/b = Sum of coal, petroleum, and natural gas.

/c = Sum of total fossil fuels, electricity, and renewables (from 1990 onward).

Source: Petroleum (barrels), natural gas (cubic feet), renewables, gross transportation consumption, and total energy consumption:
1955–95: U.S. Department of Energy, Energy Information Administration. *Annual Energy Review 1995.*
Coal (Btus), petroleum (Btus), natural gas (Btus): 1955: U.S. Department of Energy, Energy Information Administration. *Annual Energy Review 1995.*
Coal (Btus), petroleum (Btus), natural gas (Btus), electricity, electrical system energy losses:
1960–70: U.S. Department of Energy, Energy Information Administration. *State Energy Data Report 1993.*
Coal (Btus), petroleum (Btus), natural gas (Btus), total fossil fuels: 1975–95: U.S. Department of Energy, Energy Information Administration. *Monthly Energy Review.* April 1996.
Cited in United States Department of Transportation, Bureau of Transportation Statistics. *National Transportation Statistics 1997.* Washington, DC: U.S. DOT/Bureau of Transportation Statistics, 1996. Table 4-2;
and *National Transportation Statistics 1998.* Washington, DC: U.S. DOT/Bureau of Transportation Statistics, 1998. Table 4-8.

13.22. Estimated Consumption of Vehicle Fuels in the United States: 1992–97 (thousand gasoline-equivalent gallons).

Fuel	1992	1993	1994	1995	1996	1997 (estimated)
Total fuel consumption [1]	134,230,631	135,912,964	139,847,642	143,019,659	145,634,659	148,289,767
Total alternative and replacement fuels	2,105,631	3,122,534	3,145,852	3,879,407	3,707,131	4,032,889
Alternative fuels						
Subtotal	229,631	293,334	281,152	277,507	297,231	321,389
Liquefied petroleum gas (LPG)	208,142	264,655	248,467	232,701	239,158	244,612
Compressed natural gas (CNG)	16,823	21,603	24,160	35,182	46,923	63,258
Liquefied natural gas (LNG)	585	1,901	2,345	2,759	3,247	4,567
Methanol, 85% [2] (M85)	1,069	1,593	2,340	2,887	3,390	3,625
Methanol, 100% (M100)	2,547	3,166	3,190	2,150	347	347
Ethanol, 85% [2] (E85)	21	48	80	190	694	1,416
Ethanol, 95% [2] (E95)	85	80	140	995	2,699	2,628
Electricity	359	288	430	663	773	936
Replacement fuels (oxygenates)						
Subtotal	1,876,000	2,829,200	2,864,700	3,601,900	3,409,900	3,711,500
MTBE (methyl-tertiary-butyl-ether) [3]	1,175,000	2,069,200	2,108,800	2,691,200	2,749,700	2,923,700
Ethanol in gasohol	701,000	760,000	845,900	910,700	660,200	787,800
Traditional fuels						
Subtotal	134,001,000	135,619,630	139,566,490	142,741,750	145,334,920	147,950,950
Gasoline [4]	110,135,000	111,323,000	113,144,000	115,943,000	117,768,000	120,125,000
Diesel	23,866,000	24,296,630	26,422,490	26,798,750	27,566,920	27,825,950

Notes:
[1] Total fuel consumption is the sum of alternative and traditional fuels. Replacement fuel consumption is included in gasoline consumption.
[2] The remaining portion of 85% methanol and both ethanol fuels is gasoline.
[3] Includes a very small amount of other ethers, primarily tertiary-amyl-methyl-ether (TAME) and ethyl-butyl-ether.
[4] Gasoline consumption includes ethanol in gasohol and MTBE (methyl-tertiary-butyl-ether).
Fuel quantities are expressed in a common base unit of gallons of gasoline-equivalent (GGE) to allow comparisons of different fuel types. Totals may not equal sum of components due to rounding.

Source: U.S. Department of Energy, Energy Information Administration. *Alternatives to Traditional Transportation Fuels 1996.* Washington, DC: 1997. Table 10. Cited in U.S. Department of Transportation, Bureau of Transportation Statistics. *Transportation Statistics Annual Report 1998.* Washington, DC: Bureau of Transportation Statistics, 1998. Table 4-1.

13.23. Sources of Nonpetroleum Transportation Energy, United States, 1981–95.

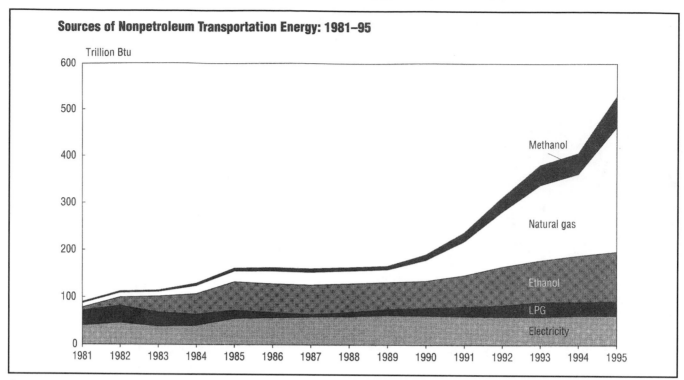

Sources of Nonpetroleum Transportation Energy: 1981–95

Sources: Methods of estimation developed by D.L. Greene. Oak Ridge National Laboratory.

Data from: 1) U.S. Department of Energy, Energy Information Administration. 1982–85 (annual volumes). *Petroleum Supply Annual,* DOE/EIA-0340(95)/1. Washington, DC. Vol. 1, tables 16 and 17.

2) U.S. Department of Energy, Energy Information Administraton. 1996. *Alternatives to Traditional Fuls 1994.* DOE/EIA-0585(94)/1. Washington, DC. February. Vol 1, table 14.

3) S.C. Davis and D.N. McFarlin. 1996. *Transportation Energy Book, Edition 16,* ORNL-6898. Oak Ridge, TN: Oak Ridge National Laboratory. July. Tables 2.10 and 5.14.

13.24. U.S. Transportation-related Air Emissions: 1970–95.

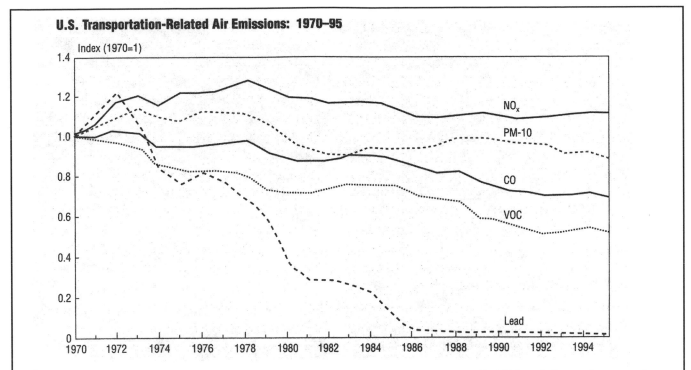

Key: NO$_x$ = oxides of nitrogen; PM-10 = airborne particulates of less than 10 microns; CO = carbon monoxide; VOC = volatile organic compounds.
Note: Transportation emissions include all onroad mobile sources and the following nonroad mobile sources: recreational vehicles, recreational marine vessels, airport service equipment, aircraft, marine vessels, and railroads. Lead estimates include onroad mobile sources only.

Source: U.S. Environmental Protection Agency, Office of Air Quality Planning and Standards. 1996. *National Air Quality and Emissions Trends, 1900 to 1995.* Research Triangle Park, NC. October

Glossary

Application software. Programs that allow users to perform specialized functions not related to the running of the computer itself. Some types of applications are word processors, database management systems, and spreadsheets.

Artificial intelligence. Programming that allows a computer to assume capabilities resembling those of human intelligence, like reasoning, learning, and adaptation.

Bitumens. Certain mixtures of hydrocarbons and other materials, either naturally occurring or derived through distillation of coal or petroleum. Used for waterproofing and road surfacing.

Btu. British Thermal Unit. The amount of heat required to raise the temperature of a pound of water one degree Fahrenheit.

Center pivot sprinkler systems (agriculture). The primary irrigation technology involving pressurization. Center pivot sprinklers are self-propelled systems involving a single pipeline supported by a row of mobile towers suspended above the ground. They may be a quarter of a mile long and can irrigate up to 132-acre circular fields.

Central processor. The brains of the computer. The central processor executes instructions provided by the software.

Computer chip. An integrated circuit consisting of interconnected semiconductor devices.

Computer telephony. A wedding of computers and telephone technology that allows voice communication over the Internet.

Conservation tillage (agriculture). A tillage or planting system that covers 30% or more of the soil with crop residue to reduce erosion by water, or that maintains at least 1,000 pounds of residue per acre to reduce wind erosion.

Constant dollars. A number that simulates how prices would look if the dollar maintained unvariable purchasing power. Changes in constant dollar figures are supposed to reflect only real changes in volume of output, income, expenditures, and so on. Constant dollars are often contrasted with current dollars, which reflect actual prices.

Conventional tillage (agriculture). Tillage that leaves less than 30% residue cover.

Crop residue management. A method of soil and water preservation that relies on reduced or special tillage practices.

Current dollars. Numbers that reflect actual prices, unadjusted for changes in the purchasing power of the dollar. Current dollar figures vary with inflation and may or may not reflect real changes in volumes of output, income, expenditures, and other measures. Current dollars are often contrasted with constant dollars to illustrate the effects of inflation on the numbers.

Data mining. The use of artificial intelligence techniques to look for hidden or unexpected patterns in a database. Data mining can yield vital strategic business information.

Database management system. A software program that allows a user to build and manage (change, add to, sort, format, and otherwise manipulate) a set of related data.

Device manager. A software program that controls a piece of hardware such as a printer, scanner, or other add-on.

Dial equipment minutes (communications). A measure of telephone call volumes. Dial equipment minutes are measured as calls enter and leave telephone switches. Two dial equipment minutes are counted for each conversation minute.

Equal access lines (communications). A class of service under which all long distance service providers receive equivalent connections to the local exchange carrier's network. Customers must select a long distance carrier, after which calls are automatically routed through the selected carrier. Where equal access lines are not available, long distance calls are automatically routed through AT & T unless the customer invokes a special dialing arrangement.

Extranet. A private Internet-based network accessible to a select group of people, both within and outside an organization.

Fiber miles (communications). A measure of capacity. The number of fiber cables times the number of fibers times the length of the route.

Grafted plant (biotechnology). A plant created by joining pieces of two plants so that the tissues unite and grow together.

Gravity-flow systems (agriculture). Irrigation in which the water is propelled by gravity. Types of gravity-flow systems include the release of water through ditches or pipes, runoff control and return, and management of land and furrow shape and length.

Hand move sprinkler systems (agriculture). Portable sprinkler systems used for small, irregular fields with low crops. Hand move systems are labor-intensive.

Interlata (communications). Calls directed to and carried by interexchange carriers.

Internet. An interconnected global collection of networks and computers accessible through service providers and other gateways.

Intralata (communications). Calls carried by the local operating company within a given local access transport area (lata).

Intranet. A private network within the Internet used for organizational internal communication.

Isotonics (medicine, nutraceuticals). A type of beverage used by athletes, which replenishes nutrients depleted during exercise and provides nutrients to optimize performance. Isotonics contain sodium and potassium for electrolyte replacement and carbohydrates in the form of fructose, glucose, and sucrose.

Local area network. A hard-wired network (as opposed to one that operates over phone lines) used to connect computers within a building.

Mechanical move sprinkler systems (agriculture). Sprinklers mounted on carts, trailers, or wheels.

Micronutrients (agriculture). Nutrients that plants need only in tiny amounts.

Mulch till (agriculture). A method of tillage in which the soil is disturbed prior to planting (as opposed to no-till and ridge-till, in which the soil is left undisturbed). Herbicides and/or cultivation are used to control weeds.

Mutant plant (biotechnology). A type of plant created by inducing a change in the genes or chromosomes of another.

No-till (agriculture). An agricultural technique in which the soil is left undisturbed between harvest and planting, except for the injection of nutrients. Seeds are applied in narrow beds or slots created by a variety of tools. Cultivation is used only for emergency weed control.

Nonequal access lines (communications). A class of long distance telephone service under which all lines are presubscribed to AT & T. Customers in areas with nonequal access lines may redirect calls through other carriers by dialing special codes.

Nutraceuticals. Foods and beverages that purport to replenish nutrients, optimize bodily performance, and rectify or mitigate the body's susceptibility to a variety of pathogens.

Operating system. The software that controls the operation of a computer's hardware (printers, disks, memory, and so on); provides file management functions like copy and delete; and forms a base upon which specialized programs, such as word processing and spreadsheets, can run.

Peripheral. A device connected to a computer and controlled by its central processing unit. Common peripherals are printers and scanners.

Pick and place robots (manufacturing). Programmable devices on an assembly line that extract or pick up components from bins or hoppers and place them on a moving track.

Primary nutrients (agriculture). The major nutrients that plants need: nitrogen, phosphate, and potash.

Product class. Product classes used by the U.S. Census Bureau correspond to four-, six-, and eight-digit SIC (Standard Industrial Classification) codes. The tables in this book do not reflect the new NAICS (North American Industry Classification System) codes, which will be phased in over the period 1999 to 2004. The new system adds 358 new industries, revises 390, and keeps 422 essentially the same.

Programming language. An artificial language for constructing sets of computer instructions. COBOL, C++, Java, and BASIC are common programming languages.

Reagents. Substances used in chemical reactions to make or assess other substances.

Recombinant plant (biotechnology). A plant created from the DNA of more than one organism.

Reduced tillage (agriculture). A type of tillage that leaves 15 to 30% residue on the soil after planting (to protect against water erosion), or 500 to 1,000 pounds of residue per acre (to guard against wind erosion).

Ridge till (agriculture). An agricultural technique in which the soil is left undisturbed between harvest and planting except for the injection of nutrients. Seeds are deposited in beds on ridges, which are built at planting time and reconstructed during weeding. Residue is left on the surface between ridges.

Route miles (communications). A measure of length rather than capacity. The length of the cable from point to point, without regard to the number of fibers or of parallel cables.

Scouting (agriculture). A method of pest monitoring that helps farmers determine when to apply pesticides.

Search engine. A software program that lets you query a system by keyword or other criteria such as date, author, publication, and so on.

Secondary nutrients (agriculture). Nutrients necessary to plants, but other than primary nutrients—nitrogen, phosphate, and potash.

Security. Techniques for assuring that only authorized users can get to particular networks, computers, and data and that the data is whole and undamaged.

Server. A computer or software package that provides a specific type of service over a network.

Solid set and permanent sprinkler systems (agriculture). Stationary sprinkler systems in which water supply pipelines are fixed—usually below the soil—with nozzles above the surface. Solid set systems are used in orchards and vineyards, and for turf production and landscaping.

Somatic cell fusion-derived plant (biotechnology). A plant created by combining nuclei of cells from the bodies, rather than the germ cells (reproductive cells), of other plants.

Specialty chemicals (materials). A segment of the chemical industry. As opposed to the industrial chemicals (alkalis and chlorine, inorganic pigments, industrial gases, plastics, rubber, petrochemicals, etc.) and pharmaceuticals segments of the chemical industry, specialty chemicals include soaps and detergents; surfactants; specialty cleaning, polishing, and sanitary preparations; perfumes, cosmetics, and other toilet preparations; paints, varnishes, enamels, and other allied products; fertilizers, pesticides, and other agricultural chemicals; and adhesives and sealants, explosives, printing ink, and other specialty chemicals and chemical preparations.

Storage device. Hardware that holds digital data. Common storage devices are floppy disks, hard disks, zip disks, and tapes.

Subirrigation (agriculture). A method of irrigation involving regulating water tables through the use of drainage systems.

Switched access lines (communications). A telephone service in which a loop is used to provide access to both interexchange carriers and incumbent local exchange carriers (LEC).

Utility. A software program that performs some kind of management task. Common utilities include tape backup programs, image viewers, and virus checkers.

Virtual reality. A computer simulation using three-dimensional graphics and devices like a data glove and goggles. The user feels as though he or she is immersed in the environment depicted by the computer.

Wide area network. Two or more local area networks connected together.

Index

by Linda Webster

Academic libraries. *See* College and university libraries

Acupuncturists, 211

Adhesives, 64–65

Advertising, computer animation for, 145

Aerodynamics, 9. *See also* Aircraft

Aerospace. *See* Space technologies

AFIS. *See* Automated Fingerprint Identification System (AFIS)

Agriculture and food technologies
biotechnology in, 34, 43–44, 204, 229
communication equipment used by farmers, 34, 39
crop residue management practices, 34, 39
farm equipment, 34, 39–42
fertilizer use, 33, 35
field test permits for genetically modified plants, 44
genetics and, 34, 43–44
highest degree received by agricultural scientists, 21
hot topics, 33
introduction to, 33–34
irrigation systems, 34, 36
methods in, 33–39
patents, 33, 43
pest management practices, 33–34, 37–38
pesticides, 33–34, 38
sustainable agriculture production, 7
U.S. national critical technologies, 7

Aided/Automatic Target Recognition (ATR), 182, 187

Air conditioning and heating systems
in construction industry, 119, 124
heating by various types of energy, 166
heating with solar collectors, 171

Air emissions from transportation, 265

Air Force. *See* Military

Air transportation. *See also* Aircraft
air passenger and freight companies, 243
airports, 242, 243, 253–54
energy consumption, 261
FAA route facilities and services, 252

Airbags, 241, 259

Aircraft
age of general aviation fleet, 242, 247
avionics, 9
composites shipments for, 58
control systems, 194

cost per R&D scientist or engineer, 21
Defense Department budget for, 184
factory net billings for, 242, 245
integration, 10
miles flown by, 243
military aircraft, 184, 186, 255
number of, 243
R&D scientists and engineers, 17
research and development funding, 12, 14
sales, 242, 245, 255
shipments by type of, 246
shipments of new aircraft, 242, 245–46
structures, 9
turbines, 9
types of aircraft in operation by manufacturer and model, 242, 248–51

Airports, 242, 243, 253–54

Alarm systems
for buildings, 75, 124
vehicle security systems, 136

Alloys, 8, 46. *See also* Materials

Alpha amylase, 204, 212, 213, 216, 218

Alpha interferon, 204

Alternative fuel vehicles, 241, 260, 263

Alternative medicine, 203, 211, 225

Amtrak, 243, 261

Angiocardiography, 209

Animation. *See* Computer animation industry

Anthropologists, 21

Antibiotics, 203, 213–15, 223, 224

Antidepressants, 203, 213, 215, 219

Antihistamines, 203, 216, 220

Antivirals, 203, 223

Apparel industry
cost per R&D scientist or engineer, 21
online sales of apparel and footwear, 114
R&D scientists and engineers, 17
research and development funding, 12, 14

Appliances. *See* Household appliances

Aquaculture, 7

Architectural uses of computer animation, 145

Architecture, highest degree received by architectural engineers, 21

Area codes, 66, 82–86

Army. *See* Military

Arteriography, 209

Arthroscopy of knee, 209

Artificial intelligence, 4

Artificial structuring methods, 8

Assistive technology devices, 204, 225–26

Asthma medications, 214

Astronomers, 21

ATMs, 45, 47

ATR. *See* Aided/Automatic Target Recognition (ATR)

Audio systems, 137–38

Audiovisual materials in libraries, 150, 151

Automated Fingerprint Identification System (AFIS), 199

Automated teller machines (ATMs), 45, 47

Automation. *See also* Computers and computing; Manufacturing
high-tech companies formed in U.S. from 1980–94, 32
in manufacturing, 45, 48–50
in school library media centers, 149

Automobiles. *See* Motor vehicles

Aviation aircraft. *See* Aircraft

Avionics and controls, 9

Bacterial/viral detection and screening, 6

Bar codes, 50, 106

Batteries, 2

Beauty products, online sales of, 114

Biocompatible materials, 6, 9

Biological/chemical warfare, 186, 188–91, 194

Biological scientists, highest degree received, 21

Biomass energy and biofuels
consumption, 160–65
electricity generation from, 165, 167
prices, 170

Biomedicine. *See* Medicine

Bioprocessing, 6

Bioremediation, 3

Biotechnology. *See also* Medicine
in agriculture, 34, 43–44, 204, 229
employees in biotechnology firms, 229
equipment, 229
funds for research and development, 229
high-tech companies formed in U.S. from 1980–94, 32
hot topics, 204
introduction to, 204
market areas, 229

Biotechnology *(continued)*
    in pharmaceuticals industry, 204, 229
    revenues from biotechnology firms, 229
    trade in advanced technology products, 31
    U.S. national critical technologies, 6
Blacks, home computer use by, 155–57
Boats. *See* Ships and boats
Bone marrow transplants, 210
Books
    in college and university libraries, 151
    online sales of, 114
    in school library media centers, 150
Bridges, 128, 256
Broadcasting. *See also* Entertainment products; Radio; Television
    computer animation for, 145
    FCC auctions summary, 94–95
    licensing of broadcast stations, 93
    value of shipments of radio, television, and studio equipment, 96–97
Building technologies. *See* Construction
Buses, 243, 244, 256, 261
Business technology
    automated teller machines (ATMs), 45, 47
    composites shipments for, 58
    computer animation, 145
    cost per R&D scientist or engineer, 21, 22
    hot topics, 45
    introduction to, 45
    manufacturers' shipments of business appliances, 45, 47
    R&D scientists and engineers, 17, 18
    research and development funding, 12, 13, 14, 15
    smart cards, 45, 47

CAD. *See* Computer-aided design (CAD)
Caesarean section, 208
Calculators, 47, 135
CAM. *See* Computer-aided manufacturing (CAM)
Camcorders, 130, 138
Cameras, digital, 47
Cancer and cancer-related treatments, 215
Capacitors, 2
Carbon, 46
Carbon monoxide detectors, 124
Cardiac catheterization, 207, 208
Cars. *See* Motor vehicles
Carson, Rachel, 158
CAT scan, 207, 209, 227
Catalysts, 7
CD players, 130, 139
CDs in libraries, 149, 150
Cellular telephones
    average monthly bill, 92
    farmers' use of, 34, 39
    number of systems and subscribers, 91
Cement production, 51
Census Bureau, xxi–xxii
Center for Defense Information, 183
Central cleaning systems, 130, 135

Ceramics, 8, 46
CERCLA (Comprehensive Environmental Response, Compensation, and Liability Act), 177–78
Chemical/biological warfare, 186, 188–91
Chemical industry
    cost per R&D scientist or engineer, 21
    highest degree received by chemists and chemical engineers, 21
    R&D scientists and engineers, 17
    research and development funding, 12, 14
    sales of chemical specialties, 53
Childbirth, 209
Chiropractors, 211
Cholesterol reducers, 203, 213, 215
CIM. *See* Computer-integrated manufacturing (CIM)
Civil engineering
    Defense Technology Area Plan, 186
    highest degree received by civil engineers, 21
Clay products industry. *See* Stone, clay, and glass products industry
Clocks, 135
Cloning, 1, 204
CNC. *See* Computer numerical control (CNC)
Coal energy
    consumption, 160–61, 162, 262
    housing units heated by, 166
    prices, 170
    for transportation, 262
Coconut oil consumption, 59
Cold medications. *See* Cough and cold medications
College and university libraries, 146, 151
Color TV receivers, 135, 140
Combines, 40–42
Combustion propulsion systems, 2
Communication equipment. *See also* Communications
    cost per R&D scientist or engineer, 21
    farmers' use of, 34, 39
    fax machines, 39, 78
    R&D scientists and engineers, 17
    research and development funding, 12, 14
Communications. *See also* Communication equipment
    broadcasting, 93–97
    cellular telephones, 91–92
    computer animation for, 145
    cost per R&D scientist or engineer, 21
    data communications market, 111–12
    Defense Technology Area Plan, 186
    fiber cable systems, 66, 67–72
    high-tech companies formed in U.S. from 1980–94, 32
    hot topics, 66
    international telephone service, 89–90
    introduction to, 66
    in manufacturing, 49, 50
    mobile telephone industry, 93

paging industry, 92
    quantity of shipments of communication equipment, 76
    R&D scientists and engineers, 18
    research and development funding, 12, 14
    telephone area codes, 66, 82–86
    telephone calls and telephone carriers, 73–74
    telephone dial equipment minutes, 88
    telephone lines by state, 66, 72
    telephone penetration, 86–87
    telephone sales, 77–78
    telephone service expenditures, 66, 79–82
    trade in advanced technology products, 31
    transatlantic cable systems, 66, 71
    U.S. national critical technologies, 3–6
    value of shipments of communication equipment, 75–76
Commuters. *See* Urban transit systems
Compact discs (CDs) in libraries, 149, 150
Compactors, 135
Companies
    fiber cable systems of communications companies, 67–71
    high-tech companies formed in U.S. from 1980–94, 2, 32
    patents of specific corporations, 25–30
Composites, 8, 46, 58
Computer-aided design (CAD), 45, 49, 50, 120
Computer-aided manufacturing (CAM), 45, 50, 120
Computer and data processing services
    cost per R&D scientist or engineer, 22
    R&D scientists and engineers, 18
    research and development funding, 13, 15
Computer and information scientists, 19–20
Computer animation industry, 144–145
Computer-integrated manufacturing (CIM), 50
Computer numerical control (CNC), 45, 50
Computer printers
    quantity and value of shipments, 106
    sales of, 47, 110
Computerized axial tomography (CAT), 207, 209, 227
Computers and computing. *See also* Automation; Software and toolkits
    animation using, 144–145
    in college and university libraries, 151
    crime mapping, 202
    data communications market, 111–12
    data mining market forecast, 118
    databases, 115–17
    Defense Technology Area Plan, 186
    electronic components for, 110
    electronic mail, 34, 39, 99, 113–14
    equipment, 100–12
    farmers' use of, 34, 39
    high-tech companies formed in U.S. from 1980–94, 32

history of microprocessors, 100
home computer use by students, 155–57
Intel chip sizes, 102
Internet, 34, 39, 99, 114, 146–49, 153, 157, 182
introduction to, 98–99
Java software, 99, 118
in law enforcement, 199–202
manufacturers' shipments by type of computer, 47
for manufacturing, 7, 45, 49–50
milestones, 101–02
modems/fax modems, 109
networks, 112–14
online sales of PC hardware and software, 114
printers, 47, 106, 110
quantity of shipments, 104–08
sales of computer printers, 47, 110
sales of modems/fax modems, 109
sales of personal computers, 109, 114
in school library media centers, 149
school use of computers, 154–55
security and, 118
trade in advanced technology products, 31
U.S. national critical technologies, 4
value of shipments, 98, 103–08
Construction
air conditioning and heating systems, 119, 124
alarm systems, 124
CAD/CAM used in, 120
carbon monoxide detectors, 124
composites shipments for, 58
energy-efficient design and materials, 119, 120
fire detection and prevention equipment, 119, 124
hot topics, 119
introduction to, 119
modernization projects, 120–21
new construction, 120–23
number of stories, 123
products used in, 120–21
smart buildings, 119
smoke detectors, 124
technologies in new construction, 121
types of buildings, 120, 122
U.S. national critical technologies, 2
Consumer products. *See also* Entertainment products
composites shipments for, 58
electronics products, 114, 135–43
hot topics, 130
household appliances, 130, 131–35
Internet revenues, 99, 114
introduction to, 130
photovoltaic modules and cells for, 172
telephone answering devices, 136
vehicle security systems, 136
Continuous materials, 7–8
Contraceptives and contraceptive agents, 214, 215, 222

Cooking ovens and ranges, 131
Copernicus, Nicolaus, 230
Copiers, 47
Cordless telephones, 78, 130
Cornea transplants, 210
Corporations. *See* Companies
Corrosion-resistant equipment, 58
Cough and cold medications, 220–21
Crime mapping, 202
Crop residue management practices, 34, 39
Cystoscopy, 209

Data communications market, 111–12
Data compression, 3
Data fusion, 4
Data mining, 118
Data processing services. *See* Computer and data processing services
Data storage, high-density, 3
Databases
by form of representation of data, 116
home computer use by students, 157
media for distribution/access, 117
number of, 115
number of entries, 115
in school library media centers, 149
vendors and producers, 115
Defense Department, U.S., 182, 184–85, 235
Defense technologies. *See* Military
Defense Technology Area Plan, 182, 186–87
Defibrillators, 227
Dehumidifiers, 124
Desktop publishing, 157
Diagnostic imaging, 6, 203, 227
Dialyzers, 227
Dictation equipment, 47
Diesel fuel, 263
Digestive system medications, 212, 221, 224
Digital cameras, 47
Direct to home (DTH) satellite systems, 39, 130, 139
Discrete product manufacturing, 7
Dishwashers, 134
Disk drives, 104–05
Disks, 107
Display terminals, 103, 105
Displays, high-definition, 3
Distance education, 147, 149, 151–52
Distillate fuel prices, 170
DNA samples at forensic laboratories, 201
DNA technologies. *See* Recombinant DNA technologies
Drugs and medicines. *See also* Medicine
biotechnology for development of, 204, 229
cancer and cancer-related treatments, 215
cost per R&D scientist or engineer, 21
dispensed prescription volume, 215–18
drug delivery systems, 203, 229
drug mentions by therapeutic classification, according to ambulatory care setting, 212

generic substances most frequently used at ambulatory care visits, 211–12
leading prescription drugs by name of drug, 213–18
leading therapeutic classes of prescription drugs, 215–16
online sales of, 114
pain management, 203, 212, 215, 219, 225
R&D scientists and engineers, 17
research and development funding, 12, 14
sales, 203, 213–17
value of product class shipments of, 224
value of shipments of pharmaceutical preparations, 218–24
Dryers, 130, 133
DTH satellite systems, 39, 130, 139

Earth scientists, 21
Economists, 21
Education/training
college and university libraries, 146, 151
computer animation for, 145
computer use by students, 154–55
distance education, 147, 149, 151–52
hot topics, 146–47
introduction to, 146–47
public schools with Internet access, 146–47, 153
school library media centers, 146, 149–50
of scientists and engineers, 2, 19–20
software for, 5
800 telephone service, 85–86
Elastomers, 46, 65
Electric, gas, and sanitary services
cost per R&D scientist or engineer, 21
electricity generation by renewable energy source, 167
R&D scientists and engineers, 18
research and development funding, 12, 15
value of construction, 129
Electrical appliances, 130, 131–35
Electrical engineers, 21
Electrical equipment
composites shipments for, 58
cost per R&D scientist or engineer, 21
R&D scientists and engineers, 17
research and development funding, 12, 14
Electrically powered vehicles, 9, 260, 262, 263, 264
Electricity
consumption, 160–62, 164–65, 262, 263
generation from photovoltaic modules and cells, 172
generation from renewable energy sources, 158, 165, 167
housing units heated by, 166
light bulbs by type of room, 173
from nuclear power plants, 168
prices, 170
solar thermal collectors for, 171, 172–73
for vehicles, 9, 260, 262, 263, 264

Electronic mail (E-Mail), 34, 39, 99, 113–14
Electronics industry
    composites shipments for, 58
    consumer electronics, 114, 135–43
    cost per R&D scientist or engineer, 21
    defense technology and, 187, 194–95
    high-tech companies formed in U.S. from
        1980–94, 32
    R&D scientists and engineers, 17
    research and development funding, 12, 14
    trade in advanced technology products, 31
    U.S. national critical technologies, 8
    value of shipments of electronic compo-
        nents of computers, 110
Elementary schools. *See* Education/training;
    School library media centers
E-Mail, 34, 39, 99, 113–14
Emerson, Ralph Waldo, xxi
Endoscopy, 207–09, 227
Endurance unmanned aerial vehicles (UAVs),
    182, 185
Energy
    alternative fuel vehicles, 241, 260, 263
    biomass energy, 160–65, 167, 170
    coal energy, 160–61, 162, 166, 170, 262
    conservation of, in construction, 119, 120
    consumption by type, 160–62
    fossil fuels, 160–61, 162, 262
    gasoline consumption, 169–70
    gasoline prices, 158, 169, 170
    geothermal energy, 160–62, 164–65, 167
    housing units heated by types of energy,
        166
    hydroelectric power, 160–62, 164–65, 167
    introduction to, 158
    light bulbs by type of room, 173
    nuclear energy, 158, 168, 170
    petroleum, 160–63, 170, 262
    photovoltaic modules and cells, 172–73
    prices by type of fuel, 158, 169, 170
    renewable energy, 3, 158, 160–62, 164–
        65, 167, 262
    solar energy, 158, 160–62, 164–67, 171–
        73
    solar thermal collectors, 171, 172–73
    transportation, 260–64
    U.S. national critical technologies, 2–3
    wind energy, 160–62, 164–65, 167
Energy Department, U.S., 158
Engineering
    computer animation for, 145
    cost per R&D scientist or engineer by in-
        dustry and size of company, 21–22
    employment status of engineers by occu-
        pation and highest degree received, 2,
        19–20
    number of R&D scientists and engineers,
        2, 17–18
    research and development funding, 13, 15
Entertainment products. *See also* Broadcast-
    ing; Consumer products
    audio systems, 137–38
    camcorders, 130, 138

CD players, 130, 139
color TV receivers, 135, 140
computer animation, 144–45
direct to home (DTH) satellite systems,
    130, 139
hot topics, 130
introduction to, 130
online sales of, 114
portable CD equipment, 136
projection television, 130, 141
radios in homes, 142–43
VCR decks, 141
Environment
    air emissions from transportation, 265
    biotechnology for, 204
    greenhouse gases, 181
    hazardous waste generators by state, 159,
        175–76
    hazardous waste treatment and defense
        technology, 183, 193
    introduction to, 158–59
    number and types of National Priorities
        List (NPL) cleanup sites, 159, 174
    pollution control equipment, 179–80
    priority list of hazardous substances, 159,
        177–78
    recycling of wood and wood-fiber mate-
        rials, 181
    U.S. national critical technologies, 3, 8
Environmental life scientists, 21
Ethanol, 260, 263, 264
Ethnicity. *See* Race and ethnicity
Exports. *See* Trade

FAA route facilities and services, 252
Facsimile equipment, 47, 76
Family income level. *See* Socioeconomic sta-
    tus
Fans, 135
Farm equipment, 34, 39–42
Farming. *See* Agriculture
Fatty acids, 59–63
Fax machine
    farmers' use of, 39
    sales of, 78
    in school library media centers, 149
Fax modems, sales of, 109
FCC
    auctions summary, 94–95
    licensing of broadcast stations, 93
Federal Aviation Administration (FAA), 252
Federal Communications Commission, 93–
    95
Females. *See* Gender
Fertilizers, 33, 35
Fetal monitoring, 209
Fiber cable systems for communications, 66,
    67–72
Film/video production, computer animation
    for, 145
Finance, insurance, and real estate
    computer animation for insurance, 145

cost per R&D scientist or engineer, 22
R&D scientists and engineers, 18
research and development funding, 13, 15
Fire detection and prevention equipment, 119,
    124
Fisheries, 7
Flexible manufacturing systems (FMS), 50
Flowers, online sales of, 114
FMS. *See* Flexible manufacturing systems
    (FMS)
Food industry
    cost per R&D scientist or engineer, 21
    online sales of food and beverages, 114
    R&D scientists and engineers, 17
    research and development funding, 12, 14
    safety assurance, 7
Food technologies. *See* Agriculture
Foreign countries. *See* International statistics
Forensic laboratories, DNA samples at, 201
Fossil fuels, 160–61, 162, 262
Freezers, 132
Fuel cells, 2
Fuels. *See* Energy; Gasoline
Fungicides, 38
Furniture industry. *See* Lumber and wood
    products industry

Galileo, 230
Games
    computer animation for, 145
    home computer use by students, 157
Gardening items, online sales of, 114
Gas appliances, 130, 131, 133, 134
Gas furnaces, 124
Gas pipelines, 242, 244
Gas services. *See* Electric, gas, and sanitary
    services
Gas turbines, 2
Gasoline
    consumption, 169–70, 263
    prices, 158, 169, 170
Gasoline power motors, 135
Gender, home computer use by, 157
Genetics and agriculture, 34, 43–44
Geothermal energy, 160–62, 164–65, 167
Gifts, online sales of, 114
Glass products industry. *See* Stone, clay, and
    glass products industry
Global positioning systems (GPS), 230, 231–
    32
Gold production, 51
GPS. *See* Global positioning systems (GPS)
Graphics, home computer use of, 157
Greenhouse gases, 181

Hair dryers, 135
Hazardous wastes
    defense technology and, 183, 193
    number and types of National Priorities
        List (NPL) cleanup sites, 159, 174
    priority list of, 159, 177–78
    by state, 175–76

Health information systems and services, 6
Health products, online sales of, 114
Health services. *See also* Drugs and medi-
cines; Medicine
cost per R&D scientist or engineer, 22
photovoltaic modules and cells for, 172
R&D scientists and engineers, 18
research and development funding, 13, 15
as U.S. national critical technology, 4
Hearing devices, 226
Heart transplants, 210
Heating systems. *See* Air conditioning and
heating systems
Hepatitis A vaccine, 204
Herbicides, 34, 38
High energy-density materials, 9
Higher education
academic libraries, 146, 151
distance education, 147, 151–52
Highways
funding, 126
mileage, 127, 242, 243
pavement conditions, 125
as U.S. national critical technology, 9
value of construction, 129
Hispanics, home computer use by, 155–57
Hobbies. *See* Toys and hobbies, online sales
of
Home diagnostic tests, 203, 206
Home furnishings, online sales of, 114
Home page
farmers' use of, 39
public libraries' use of, 146, 148
Hormones, 204, 212, 213, 216, 218
Hot water heaters, 130, 134
Household appliances, 130, 131–35
Hubble space telescope, 230, 239
Human factors engineering, 10
Human-machine interface, as U.S. national
critical technology, 7, 10
Human systems, 7
Hydroelectric power, 160–62, 164–65, 167
Hypertension drugs, 203, 213–14, 220

Imaging equipment, 6, 203, 227
Imports. *See* Trade
Income level. *See* Socioeconomic status
Industrial engineers, 21
Industrial research and development. *See*
Research and development
Information scientists. *See* Computer and in-
formation scientists
Information systems and technology
defense technology and, 186, 195
U.S. national critical technologies, 3–6
Infrastructure
bridges, 128
highway funding, 126
highway mileage, 127, 242, 243
highway pavement conditions, 125
introduction to, 119
as U.S. national critical technology, 9

value of construction, 129
Insecticides, 34, 38
Insurance. *See* Finance, insurance, and real
estate
Integrated signal processing, 5
Intel chip sizes, 102
International statistics
computer animation industry, 144–45
global positioning systems (GPS), 232
intranet market, 112–13
passenger cars, 242, 259
patents, 2, 23–24
petroleum consumption, 162–63
plastics consumption, 57
prescription drug sales, 203, 213–15
satellite revenues, 233
smart cards, 45, 47
speech and voice recognition, 111
telephone service, 89–90
Internet
farmers' use of, 34, 39
home computer use by students, 157
origin of, 182
public library access to, 146, 147–48
public school access to, 146–47, 153
retail revenues, 99, 114
school library media center access to, 149
Intranet market, 99, 112–13
Iron production, 51
Irrigation systems, 34, 36

Java software, 99, 118
Jet fuel prices, 170
Jewelry, online sales of, 114
JIT manufacturing. *See* Just-in-time (JIT)
manufacturing
Joint Stand-Off Weapon (JSOW), 182–83,
184
JSOW. *See* Joint Stand-Off Weapon (JSOW)
Jupiter, 230
Just-in-time (JIT) manufacturing, 45, 50

Kerosene, 166
Kidney transplants, 210
Kuiper Belt, 230

LANs. *See* Local area networks (LANs)
Lard consumption, 60
Laser-based weapons, 195
Laser eye protection, 183, 193
Lasers
for manufacturing, 49
medical equipment, 227
Law enforcement
Automated Fingerprint Identification System (AFIS), 199
body armor worn by officers, 198
cash allowance for officers, 198
computers used by, 199–202
county and municipal officers, 197–200
crime mapping by, 202
DNA samples at forensic laboratories, 201

hot topics, 183
introduction to, 182–83
911 emergency system, 199
state police, 197–200
vehicles other than cars, 199
weapons used by, 197–98
wireless voice and/or data security used
by, 201
Lead production, 51
Legal uses of computer animation, 145
Libraries
college and university libraries, 146, 151
hot topics, 146–47
Internet access in public libraries, 146,
147–48
introduction to, 146–47
public libraries hosting own Web sites,
146, 148
school library media centers, 146, 149–
50
virtual libraries, 147–48
Life sciences
employment status and highest degree received by life scientists, 19–20
trade in advanced technology products, 31
U.S. national critical technologies, 6–7
Light bulbs, 173
Linseed oil consumption, 60
Liquid petroleum gas. *See* LPG (liquid petroleum gas)
Liver transplants, 210
Local area networks (LANs), 49, 50
LPG (liquid petroleum gas)
housing units heated by, 166
prices, 170
for transportation, 260, 263, 264
Lubricants, 59–63
Lumber and wood products industry
R&D scientists and engineers, 17
recycling of wood and wood-fiber mate-
rials, 181
research and development funding, 12, 14
Lung transplants, 210

Machinery. *See also* Communication equipment; Computers and computing; Electrical equipment; Farm equipment;
Medical equipment; and specific machines and equipment
cost per R&D scientist or engineer, 21
R&D scientists and engineers, 17
research and development funding, 12, 14
Magnetic and optical recording media, 107
Magnetic resonance imaging (MRI), 227
Males. *See* Gender
Management services
R&D scientists and engineers, 18
research and development funding, 13, 15
Manufacturing
computer-aided design (CAD), 45, 49, 50
computer-aided manufacturing (CAM),
45, 50

Manufacturing *(continued)*
    computer numerical control (CNC), 45, 50
    computer-integrated manufacturing (CIM), 50
    cost per R&D scientist or engineer, 21
    Defense Technology Area Plan, 186
    flexible manufacturing systems (FMS), 50
    hard technologies, 45, 48–50
    hot topics, 45
    introduction to, 45
    just-in-time (JIT) manufacturing, 45, 50
    manufacturing cells, 45, 50
    pollution control equipment, 179–80
    R&D scientists and engineers, 17
    research and development funding by industry and size of company, 12, 14
    robots, 7, 48–50
    soft technologies, 49–50
    software for, 7, 45, 50
    trade in advanced technology products, 31
    U.S. national critical technologies, 7–8
Manufacturing cells, 45, 50
Marine industry, composites shipments for, 58
Mars, 230
Material requirements planning (MRP), 50
Materials
    adhesives, 64–65
    ceramics, 46
    chemicals, 53
    composites, 46, 58
    defense technology and, 183, 186, 193
    high-tech companies formed in U.S. from 1980–94, 32
    hot topics, 46–47
    introduction to, 45–47
    metals, 46, 51
    minerals, 51–52
    oils and fats for industrial use, 59–63
    phase-change materials, 47
    plastics, 46, 54, 56, 57, 59–63
    polymers, 46, 54–57
    refractories, 46, 53
    resins, 54, 55, 59–63
    smart materials, 47
    superconducting materials, 47
    thermoplastics, 46, 54, 56, 65
    thermosets, 46, 54, 56, 65
    trade in advanced technology products, 31
    types of, 46
    U.S. national critical technologies, 8–9
Mathematicians, 19–20
Mechanical engineers, 21
Medical equipment, 225–28
Medicine. *See also* Biotechnology; Drugs and medicines; Health services
    alternative medicine, 203, 211, 225
    ambulatory and inpatient procedures, 207–09
    assistive technology devices, 204, 225–26
    cancer and cancer-related treatments, 215

    computer animation in, 145
    defense technology and, 183, 186–87, 191–92
    dispensed prescription volume, 215–18
    drug mentions by therapeutic classification, according to ambulatory care setting, 212
    electromedical and irradiation equipment, 227–28
    generic substances most frequently used at ambulatory care visits, 211–12
    home diagnostic tests, 203, 206
    hot topics, 203–04
    introduction to, 203–04
    medical testing chemicals, 206
    organ and tissue transplants, 210
    pain management, 203, 212, 215, 219, 225
    prescription drug sales, 203, 213–17
    surgical procedures, 208
    telemedicine, 203, 205
    U.S. national critical technologies, 6–7
    value of product class shipments of pharmaceutical preparations, 224
    value of shipments of pharmaceutical preparations, 218–24
Metals industry
    cost per R&D scientist or engineer, 21
    production of major metals, 46, 51
    R&D scientists and engineers, 17
    research and development funding, 12, 14
Methanol, 260, 263, 264
Micro/nanofabrication and machining, 8
Microdevice manufacturing technologies, 8
Microforms in libraries, 151
Microprocessors. *See* Computers and computing
Microwave ranges and ovens, 131
Military
    aircraft, 184, 186, 255
    aircraft control systems, 194
    chemical/biological weapons, 186, 188–91, 194
    Defense Department budget, 184–85
    Defense Department strategic forces highlights, 196
    Defense Technology Area Plan, 182, 186–87
    funding for Defense Technology Area Plan, 186–87
    ground and sea vehicles, 193
    hot topics, 182–83
    introduction to, 182–83
    joint warfighting capability objectives funding, 196
    laser-based weapons, 195
    missiles, 184, 196, 255
    nuclear weapons, 183, 186, 188, 194
    stealth technology, 9, 182
    types of military technologies, 182
Milking machines, 41
Missiles, 184, 196, 255
Mobile telephone industry, 93

Modeling software
    Defense Technology Area Plan, 186
    for manufacturing, 50
    as U.S. national critical technology, 5
Modems/fax modems, sales of, 109
Monitors, 106–07
Monoclonal antibody production, 6, 204
Motor vehicles
    airbags in, 241, 259
    alternative fuel vehicles, 241, 260, 263
    buses, 243, 244, 256, 261
    cost per R&D scientist or engineer, 21
    energy consumption, 261
    fuel consumption, 169–70
    fuel prices, 158, 169, 170
    international comparisons on passenger cars, 242, 259
    materials used in passenger cars, 242, 257
    miles driven by, 243, 259
    motorcycles, 261
    number of, 243, 259
    passenger car energy consumption, 261
    R&D scientists and engineers, 17
    research and development funding, 12, 14
    trucks, 243, 258, 259, 261
    weight of passenger cars, 242, 257
Motorcycles, 261
MRI, 227
MRP. *See* Material requirements planning (MRP)
Music, online sales of, 114
Mustache/beard trimmers, 135

NAICS (North American Industry Classification System), xxii
Nanotechnology, 46–47
Natural gas
    consumption, 162, 262
    housing units heated by, 166
    prices, 170
    for transportation, 262, 264
Naturopaths, 211
Navigation systems, 4
Navy. *See* Military
Net shape processing, 7
Networks. *See also* Internet
    data communications market, 111–12
    LANs in manufacturing, 49–50
    as U.S. national critical technology, 5
    worldwide intranet market, 99, 112–13
911 emergency system, 199
North American Industry Classification System (NAICS), xxii
Nuclear energy
    military use of, 183, 186, 188, 194
    nuclear power plants, 158, 168
    nuclear reactors as U.S. national critical technology, 2
    prices, 158, 170
    trade in advanced technology products, 2, 31
Nuclear reactors, 2

Nuclear waste storage/disposal, 3
Nuclear weapons, 183, 186, 188, 194

OCR equipment, 106
Oil pipelines, 244
Oils and fats for industrial use, 59–63
OKT3, 204
Online services, farmers' use of, 34, 39
Optical scanning devices, 106
Optics companies, 32
Opto-electronics trade, 31
Outdoor cooking equipment, 131, 135
Ovens, 131

Pacemakers, 227
Pager, farmers' use of, 39
Paging industry, 92
Pain management, 203, 212, 215, 219, 225
Paint or varnish, 59–63
Pancreas transplants, 210
Paper and allied products industry
    cost per R&D scientist or engineer, 21
    R&D scientists and engineers, 17
    research and development funding, 12, 14
Parallel processing, as U.S. national critical
    technology, 4
Pasta makers, 135
Patents
    in agriculture, 33, 43
    by date and by country of origin, 23–24
    by organization and government agency,
        25–30
Pattern recognition, 5
Penicillin, 203, 223
Peripherals for computers, 103, 106
Personal computers. *See* Computers and com-
    puting
Pest management practices in agriculture, 33–
    34, 37–38
Pesticides, 33–34, 38
Petroleum industry
    consumption of petroleum, 160–63, 262
    cost per R&D scientist or engineer, 21
    price of petroleum, 170
    R&D scientists and engineers, 17
    research and development funding, 12, 14
    transportation use of petroleum, 262
    value of petroleum pipelines, 129
Pharmaceutical industry. *See* Drugs and medi-
    cines
Phase-change materials, 47
Photonics, 8, 32
Photovoltaic modules and cells, 172–73
Physical scientists, 19–20
Physicists, 21
Pipelines, 242, 244
Plastics, 46, 54, 56, 57, 59–63
Plotters, 106
Pluto, 230
Police. *See* Law enforcement
Political scientists, 21
Pollution avoidance, 8

Pollution control. *See also* Environment
    defense technology and, 183, 193
    equipment, 179–80
    hazardous wastes, 159, 174–78
    U.S. national critical technologies, 3, 8
Pollution from transportation, 265
Polymers, 46, 53, 54, 56
Portable CD equipment, 136
Ports, 244
Postsecondary teachers, 21
Power electronics, 2
Power supplies, 2
Predictive process control, 8
Prescription drugs. *See* Drugs and medicines
Prilosec/Losec, 203, 213
Printed circuit boards, 110
Printers. *See* Computer printers
Professional instruments. *See* Scientific and
    professional instruments
Programming languages. *See* Java software
Projection television, 130, 141
Propulsion and power, 9
Protein engineering, 6
Prozac, 213, 216, 217
Psychologists, 21
Public libraries
    Internet access in, 146, 147–48
    with own Web sites, 146, 148
Public schools. *See* Education/training

R&D. *See* Research and development
Race and ethnicity
    home computer use and, 155–57
    Internet access and, 153
Radiation therapy, 227
Radio
    cost per R&D scientist or engineer, 21
    FCC auctions, 94–95
    licensing of broadcast stations, 93
    number of home radios, 135, 142–43
    ownership statistics on radios, 135
    R&D scientists and engineers, 17
    research and development funding, 12, 14
    value of shipments of, 96–97
Railroads, 129, 243, 261
Ranges for cooking, 131
Rapeseed oil consumption, 61
Rapid solidification processing, 7
Real estate. *See* Finance, insurance, and real
    estate
Recombinant DNA technologies, 6
Refining methods, 8
Refractories, 46, 53
Refrigerators, 130, 132, 135
Renewable energy
    consumptipn, 158, 160–62, 164–65, 262
    electricity generation from, 167
    for transportation, 262
    as U.S. national critical technology, 3
Research and development
    biotechnology, 229
    cost per scientist or engineer by industry

and size of company, 21–22
    employment status of scientists and engi-
        neers by occupation and highest de-
        gree earned, 2, 19–20
    funding by industry and size of company,
        1, 12–15
    funding for basic research, applied re-
        search, and development, 16
    funding from federal government and pri-
        vate sources, 1, 11–16
    high-tech companies formed in U.S. from
        1980–94, 2, 32
    patents, 2, 23–30
    scientists and engineers by industry and
        size of company, 2, 17–18
Resins, 49–65, 54, 55
Respiratory therapy, 207, 209
Retail revenues from Internet, 99, 114
Retail sales. *See* Consumer products
Roads. *See* Highways
Robotic devices, 4, 7, 48–50
Rubber materials, 46, 64
Rubber products industry
    cost per R&D scientist or engineer, 21
    R&D scientists and engineers, 17
    research and development funding, 12, 14

Sanitary services. *See* Electric, gas, and sani-
    tary services
Satellite systems
    farmers' use of, 39
    sales of, 139
    in school library media centers, 149
Scanners, 3, 47, 106
School library media centers, 146, 149–50
Schools. *See* Education/training
Scientific and professional instruments
    cost per R&D scientist or engineer, 21
    R&D scientists and engineers, 17–18
    research and development funding, 12, 14
Scientific uses of computer animation, 145
Scientists and engineers
    cost per R&D scientist or engineer by in-
        dustry and size of company, 21–22
    employment status by occupation and
        highest degree received, 2, 19–20
    number of R&D scientists and engineers,
        2, 17–18
Secondary schools. *See* Education/training;
    School library media centers
Security
    computers and, 118
    law enforcement use of wireless voice
        and/or data security, 201
    vehicle security systems, 136
Sedatives, 219
Semiconductors
    integration technologies, 8
    manufacturing, 8
    as U.S. national critical technology, 8
    value of shipments, 110

Sensors
  Defense Technology Area Plan and, 187, 194–95
  U.S. national critical technologies, 4–5
Serial subscriptions in libraries, 150, 151
Sewer systems. *See* Electric, gas, and sanitary services
Sheriffs. *See* Law enforcement
Ships and boats, 244
SIC (Standard Industrial Classification), xxii
Signal conditioning and validation, 3
Silver production, 51
Simulation software
  Defense Technology Area Plan, 186
  for manufacturing, 50
  as U.S. national critical technology, 5
Smart cards, 45, 47
Smart materials, 47
Smoke detectors, 124
Soap, 59–63
Social scientists, 19–20
Socioeconomic status
  and home computer use by students, 155–56
  and Internet access, 153
Sociologists, 21
Software and toolkits. *See also* Computers and computing
  in college and university libraries, 151
  high-tech companies formed in U.S. from 1980–94, 32
  home computer use by students, 157
  Java software, 99, 118
  for manufacturing support software, 7, 45, 49–50
  network and system software, 5
  online sales of PC hardware and software, 114
  quantity and value of shipments, 107
  in school library media centers, 150
  security software, 118
  trade in advanced technology products, 31
  U.S. national critical technologies, 5–6, 7
Solar energy
  consumption, 158, 160–62, 164–65
  electricity generation from, 167
  housing units heated by, 166
  solar thermal collector shipments, 171, 172–73
Solar thermal collectors, 171, 172–73
Solidification. *See* Rapid solidification processing
Soybean oil consumption, 61
Space debris, 230, 234
Space shuttle flights, 236–38
Space technologies
  commercialization of space, 230, 233
  costs, 230, 235
  Defense Department budget for, 185
  global positioning systems, 230, 231–32
  highest degree received by aerospace engineers, 21
  hot topics, 230

Hubble space telescope, 230, 239
  introduction to, 230
  revenues from satellites, 233
  space debris, 230, 234
  space shuttle flights from 1981–98, 236–38
  space vehicle reentry, 183, 193
  spacecraft, 9–10, 235, 255
  successful space launches from 1957–97, 240
  trade in advanced technology products, 31
  U.S. national critical technologies, 9, 10
Space vehicle reentry, 183, 193
Spacecraft. *See also* Space technologies
  avionics, 9
  cost, 235
  development time, 235
  integration, 10
  life support, 10
  number of launches in 1999, 235
  power systems, 9
  reentry, 183, 193
  sales, 255
Speech recognition equipment, 111
Sporting goods, online sales of, 114
Spreadsheets, 157
SQC/SPC. *See* Statistical quality control (SQC/SPC)
Standard Industrial Classification (SIC), xxii
State statistics
  airports, 253–54
  area codes, 66, 82–84
  hazardous waste generators, 159, 175–76
  highway mileage, 127
  telephone lines, 66, 72
Statistical quality control (SQC/SPC), 50
Stealth technology, 9, 182
Stomach acid drugs, 203
Stone, clay, and glass products industry
  cost per R&D scientist or engineer, 21
  R&D scientists and engineers, 17
  research and development funding, 12, 14
Structures, 9
Studio equipment, 96–97
Submarines, 185, 193, 196
Superconductors, 9, 47
Surface cooking units, 131
Surface transport controls, 9
Surface treatments, 8
Surface vehicle aerodynamics, 9
Surgery
  equipment, 228
  procedures, 208
Sustainable agriculture production, 7

Tall oil consumption, 62
Tallow consumption, 59
Teachers. *See* Postsecondary teachers
Technology. *See also* specific areas of technology
  adoption factors for, xxv
  cost per R&D scientist or engineer by industry and size of company, 21–22

definition of technological advancement, 1
employment status of scientists and engineers by occupation and highest degree received, 2, 19–20
funding for basic research, applied research, and development, 16
high-tech companies formed in U.S. from 1980–94, 2, 32
patents, 23–30
R&D scientists and engineers, 2, 17–18
research and development funding by industry and size of company, 1, 12–15
research and development funding from federal government and private sources, 1, 11–16
trade in advanced technology products, 31
U.S. national critical technologies, 1, 2–10
Telecom/data routing, 3–4
Telecommunications. *See* Communications
Teleconferencing equipment, 76
Telegraph, 90
Telemedicine, 203, 205
Telephone answering devices, 136
Telephone communications
  answering devices for telephone, 136
  area codes, 66, 82–84
  business rates, 80
  cellular telephones, 91–92
  corded telephones, 77
  cordless telephones, 78, 130
  cost per R&D scientist or engineer, 21
  dial equipment minutes, 88
  800 service, 85–86
  expenditures for phone service, 66, 79–82
  households with telephones, 86–87
  international service, 89–90
  long distance rates, 81
  mobile telephone industry, 93
  number of calls, 66, 73–74
  number of carriers, 73–74
  paging industry, 92
  quantity of shipments of communication equipment, 76
  R&D scientists and engineers, 18
  research and development funding, 12, 14
  residential rates, 79
  sales of telephones, 77–78
  in school library media centers, 149
  special access lines, 73
  switched access lines, 73
  telehone lines by state, 66, 72
  value of shipments of communication equipment, 75–76
  wireless telephones, 77
Telescopes, 230, 239
Television
  color television ownership, 135, 140
  cost per R&D scientist or engineer, 21
  licensing of broadcast stations, 93
  projection television, 130, 141

R&D scientists and engineers, 17
research and development funding, 12, 14
in school library media centers, 149
value of shipments of, 96–97
Telex, 90
Textile industry
cost per R&D scientist or engineer, 21
R&D scientists and engineers, 17
research and development funding, 12, 14
Thermoplastics, 46, 54, 56, 65
Thermosets, 46, 54, 56, 65
Tobacco industry
cost per R&D scientist or engineer, 21
R&D scientists and engineers, 17
research and development funding, 12, 14
Tomahawk Cruise Missile, 182
Tools, online sales of, 114
Total Quality Management (TQM), 50
Toys and hobbies, online sales of, 114
TQM. *See* Total Quality Management (TQM)
Tractors, 40, 42
Trade
in advanced technology products, 31
cost per R&D scientist or engineer, 21
electricity generation from renewable energy, 167
R&D scientists and engineers, 18
research and development funding, 13, 15
Training. *See* Education/training
Tranquilizers, 219
Transatlantic cable systems, 66, 71
Transit systems, 244, 256–57
Transporation equipment
cost per R&D scientist or engineer, 21
R&D scientists and engineers, 17
research and development funding, 12, 14
Transportation. *See also* Space technologies
air emissions, 265
air transportation, 9–10, 12, 14, 243–55, 261

alternative fuel vehicles, 241, 260, 263
composites shipments for, 58
cost per R&D scientist or engineer, 21
energy consumption, 261–63
FAA route facilities and services, 252
highways, 9, 125–27, 129, 243
hot topics, 241–42
intelligent transportation systems, 9
introduction to, 241–42
major elements of transportation system, 243–44
photovoltaic modules and cells for, 172
pipelines, 242, 244
R&D scientists and engineers, 18
railroads, 129, 243, 261
research and development funding, 12–15
U.S. national critical technologies, 9–10
urban transit systems, 244, 256–57
water transportation, 244
Trash compactors, 130, 135
Travel, online sales of, 114
Trucks, 243, 258, 259, 261
Tung oil consumption, 62
Typewriters, electronic, 47

UAVs. *See* Unmanned aerial vehicles (UAVs)
Ulcer medications, 203, 213, 215, 216, 217
Ultrasonic equipment, 75
Ultrasound, 207, 209, 227
U.S. Department of Defense, 182, 184–85, 235
U.S. Department of Energy, 158
University and college libraries, 146, 151
Unmanned aerial vehicles (UAVs), 182, 185
Urban transit systems, 244, 256–57
Utlities. *See* Electric, gas, and sanitary services

Vaccines, 6, 204, 229

Vacuum cleaners, 135
VCR decks, 141
Vehicle security systems, 136
Veterinary medicine, 224, 229
Viagra, 203
Video disc players, 135
Video laser discs libraries, 149, 150
Video materials in libraries, 150
Video production. *See* Film/video production
Video teleconferencing equipment, 76
Viral detection and screening. *See* Bacterial/viral detection and screening
Virtual libraries, 147–48
Vision devices, 226
Vitamins, 203, 222–23, 224
Voice mail equipment, 76
Voice recognition equipment, 106, 111

Washers and dryers, 130, 133
Water heaters, 130, 134
Water supply facilities, 129
Water transportation, 244
Weapons. *See also* Military
chemical/biological weapons, 186, 188–91, 194
laser-based weapons, 195
law enforcement use of, 197–98
missiles, 184, 196, 255
nuclear weapons, 183, 186, 188, 194
trade in, 31
Wind energy, 160–62, 164–65, 167
Wood products industry. *See* Lumber and wood products industry
Word processing, 157

X-ray equipment, 227, 228

Zantac, 203, 213, 216, 217
Zovirax, 203